S0-CAH-118

Americium and Curium Chemistry and Technology

TOPICS IN *f*-ELEMENT CHEMISTRY

Americium and Curium Chemistry and Technology

Papers from a Symposium given at the 1984 International Chemical Congress of Pacific Basin Societies, Honolulu, HI, December 16-21, 1984

Edited by

NORMAN M. EDELSTEIN
Lawrence Berkeley Laboratory, Berkeley, California, U.S.A.

JAMES D. NAVRATIL
Rockwell International, Golden, Colorado, U.S.A.

and

WALLACE W. SCHULZ
Rockwell Hanford, Richland, Washington, U.S.A.

D. Reidel Publishing Company

A MEMBER OF THE KLUWER ACADEMIC PUBLISHERS GROUP 

Dordrecht / Boston / Lancaster

Library of Congress Cataloging in Publication Data

Main entry under title:
Americium and curium chemistry and technology.

(Topics in f-element chemistry)
Includex index.
1. Americium–Congresses. 2. Curium–Congresses. I. Edelstein, Norman M., 1936- . II. Navratil, James, D., 1941- . III. Schulz, Wallace W. IV. International Chemical Congress of Pacific Basin Societies (1984: Honolulu, Hawaii) V. Series.
QD181.A5A44 1985 546'.441 85-14238
ISBN 90-277-2097-5

Published by D. Reidel Publishing Company
P.O. Box 17, 3300 AA Dordrecht, Holland

Sold and distributed in the U.S.A. and Canada
by Kluwer Academic Publishers,
190 Old Derby Street, Hingham, MA 02043, U.S.A.

In all other countries, sold and distributed
by Kluwer Academic Publishers Group,
P.O. Box 322, 3300 AH Dordrecht, Holland

Printed in The Netherlands

TABLE OF CONTENTS

PART IV
Nuclear Studies, Environmental Studies, Separations, and Other Technological Aspects

PREFACE

The papers included in this volume were presented at the symposium on "Americium and Curium Chemistry and Technology" at the International Chemical Congress of Pacific Basin Societies in Honolulu, Hawaii, December 16-21, 1984. This symposium commemorated forty years of research on americium and curium. Accordingly, the papers included in this volume begin with historical perspectives on the discovery of americium and curium and the early characterization of their chemical properties, and then cover a wide range of subjects, such as thermodynamic properties, electronic structure, nuclear reactions, analytic chemistry, high pressure phase transitions, and technological aspects. Thus, this volume is a review of the chemistry of americium and curium, and provides a perspective on the current research on these elements forty years after their discovery.

The editors would like to thank the participants in this symposium for their contributions. It is a pleasure to acknowledge the assistance of Ms. Barbara Moriguchi in handling the administrative aspects of the symposium and of the production of this volume.

April 2, 1985

Norman M. Edelstein
Materials and Molecular Research Division
Lawrence Berkeley Laboratory
University of California
Berkeley, California 94720, U.S.A.

James D. Navratil
Rockwell International
Rocky Flats Plant
P.O. Box 464
Golden, Colorado 80402-0464, U.S.A.

Wallace W. Schulz
Rockwell Hanford
P.O. Box 800
Richland, Washington 99352, U.S.A.

Part I

Historical Aspects and General Papers

THE 40TH ANNIVERSARY OF THE DISCOVERY OF AMERICIUM AND CURIUM

Dr. Glenn T. Seaborg
Lawrence Berkeley Laboratory
University of California
Berkeley, California 94720

ABSTRACT: The recognition, in July 1944, that the unknown elements with atomic numbers 95 and 96 should be members of an actinide transition series was the key to their chemical identification following nuclear synthesis, i.e., the breakthrough that led to their discovery. That same month ^{239}Pu was bombarded with 32 MeV helium ions in the Berkeley 60-Inch Cyclotron and sent to the wartime Metallurgical Laboratory in Chicago where chemical separations led to the identification of $^{242}96$, produced in the reaction $^{239}Pu\ (\alpha,n)\ ^{242}96$. Neutron bombardments of ^{239}Pu in the reactors at Clinton Laboratories, Tennessee, and the Hanford Engineer Works, Washington, followed by chemical separations in Chicago, led to the identification, completed in January 1945, of $^{241}95$ and $^{242}96$ (again), produced by the reactions $^{239}Pu\ (n,\gamma)\ ^{240}Pu\ (n,\gamma)$

$$^{241}Pu \xrightarrow{\beta^-} {}^{241}95\ (n,\gamma)\ {}^{242}95 \xrightarrow{\beta^-} {}^{242}96.$$ The names

americium (95) and curium (96) were suggested at an ACS annual Spring Meeting in April, 1946.

The discovery of two very interesting synthetic elements--the transuranium elements with the atomic numbers 95 and 96--took place 40 years ago. The discovery experiments for these elements, which were given the names americium and curium, were performed at the University of Chicago in the wartime Metallurgical Laboratory in the so-called New Chemistry Building.

The discovery experiments took place during 1944 and early 1945. In the course of my remarks, I shall attempt to trace for you a somewhat detailed account, including the dates when the critical observations were made. The preparation of this talk was a moving experience, bringing back as it did the many emotional reactions, including heartaches and triumphs, of that time. The understanding and interpretation of the results took place over a longer period of time and had more of the ingredients of a detective story than was the case for the other elements in whose discovery I had the privilege to participate.

N. M. Edelstein et al. (eds.), Americium and Curium Chemistry and Technology, 3–17.

A number of factors contributed to the special status that this research holds in my memory. The experimental tools available at that time were crude in relation to the deductions that had to be made. And a real breakthrough in thinking had to be made in order to devise the chemical procedures that became necessary in order to identify the new elements. I believe that I derived more personal satisfaction from these experiments than from those concerned with the discovery of the other transuranium elements.

New Chemistry, the building where this work was done, stood on the east side of Ingleside Avenue extending south from 56th Street. It was hastily constructed during the fall of 1942, when it became apparent that the quarters on the fourth (top) floor of George Herbert Jones Laboratory were becoming grossly inadequate for our group of chemists, working as part of the Metallurgical Laboratory. I was responsible for the chemists working on the chemical processes to be used in the extraction of plutonium from neutron-irradiated uranium and on the basic chemistry and purification of plutonium. I, of course, was devoting my full time and energy to this task.

By December of 1943, however, so much progress had been made on these problems that I felt that I could devote part of my efforts to the synthesis and identification of the transplutonium elements with the atomic numbers 95 and 96. I asked Ralph James, a young chemist from Berkeley who had been especially proficient in the investigations of the radiochemistry of plutonium, to devote himself to this problem.

We began by attempting to produce isotopes of element 95 through bombardment of ^{239}Pu with deuterons. The target material ^{239}Pu was then just becoming available in milligram amounts as a result of its production in the uranium-graphite reactor at the Clinton Laboratories in Tennessee.

Albert Ghiorso was the Group Leader of our Instruments and Physical Measurements Group and during this period he began to give more and more of his time to this problem. As the work progressed, he devised instrumentation and techniques of increasing complexity as required. Other members of his group, particularly Arthur H. Jaffey, also gave us very valuable help.

Ralph and I went to St. Louis at the end of January, 1944, and with the help of Harry Fullbright, we bombarded 0.1 mg and 1 mg samples of ^{239}Pu with the deuterons furnished by the cyclotron at Washington University. Measurements were made on the bombarded ^{239}Pu, immediately at St. Louis and later in Chicago, without any attempt to chemically separate the products. Alpha particle range measurements were made in order to try to detect the presence of alpha particle emitters with higher or lower energy than that of ^{239}Pu. These crude experiments, and later crude experiments performed by Ralph and Al also utilizing the deuterons from the St. Louis cyclotron, gave negative results.

The 1 mg of irradiated ^{239}Pu and two other larger samples of ^{239}Pu that were bombarded with deuterons, one in the cyclotron at Washington University and the other in the 60-inch cyclotron in the Radiation Laboratory of the University of California, Berkeley, during the following months, were subjected to a number of chemical procedures. Chemical fractions were isolated, and their radiations examined, on the basis of several hypotheses concerning the chemical properties of element 95. These included various assumptions concerning the solubility properties of the compounds of element 95 and the potential for its oxidation from a lower state, in which its fluoride is insoluble, to an upper one in which its fluoride is soluble, and included the assumption that it could not be so oxidized at all. (The elements immediately preceding element 95--uranium, neptunium, and plutonium--each has a III and a IV oxidation state that have insoluble fluorides and a VI state that has a soluble fluoride. Increasingly strong oxidizing agents--as we go from uranium to neptunium to plutonium--are required to attain the VI state.) Unique alpha particles, with energies different than those of ^{239}Pu, were sought and in no case were they found. In retrospect, this is not surprising because we now know that the istopes that would have been produced, such as $^{240}95$ and $^{239}95$, decay overwhelmingly by electron capture and to such a small extent by alpha particle emission that this could not have been detected with the available techniques. These techniques did not permit detection of electron capture decay in the presence of the tremendous amount of rare earth fission product (beta and gamma) activity.

Also beginning in January, 1944, in parallel experiments, we used the reactor at the Clinton Laboratories in Tennessee to irradiate samples of ^{239}Pu with neutrons. Although it was thought at that time to be unlikely, we wanted to look for the production of ^{240}Pu. If such an n,γ reaction could occur to an appreciable extent in competition with fission, then successive n,γ reactions might occur leading to a beta-particle-emitting plutonium isotope and hence to an element 95 daughter. Again, in retrospect, we now know that the near identity of the alpha particle decay properties of ^{240}Pu and ^{239}Pu made impossible the detection of the small concentration of ^{240}Pu present in those neutron irradiated samples.

An important factor in the measurement of the radiation from the isotopes of elements 95 and 96 that were being sought was the determination of the energy of the alpha particles by absorption or range measurements. The early absorption measurements that I shall report here were performed by Ghiorso with thin mica absorbers and were translated into equivalent range in terms of centimeters of air at standard conditions.

During the spring of 1944, our search also had some of the human elements dramatized by James D. Watson in his book, The Double Helix. Visitors from Berkeley brought the news that John W. Gofman and his group had good evidence for the identification of element 95 among the products of the neutron irradiation of ^{239}Pu. I had good reason to feel that Jack Gofman might actually have beaten us to the discovery of element 95. I knew him as my brilliant graduate student whose thesis problem was the discovery and proof of the fissionability of ^{233}U, a discovery that opened the use of the world's supply of thorium as a nuclear fuel. On a trip to Berkeley at the end of March, I learned the details of Jack's work. He and his co-workers had identified a 40-60 day beta-particle-emitter that had chemical properties similar to the rare earth elements (which was reasonable for element 95) and which was at least partially chemically separable from the rare earths. Two weeks later, after I had returned to Chicago, I received a copy of Gofman's first written description of his work. It immediately occurred to me that his activity was actually the fission product, 57-day ^{91}Y, and when I communicated this thought to him, Gofman accepted this interpretation immediately. Also, at this period of time, Joe Kennedy, during his visits to Chicago, hinted, or at least seemed to me to hint, that he, Art Wahl, and co-workers at Los Alamos had uncovered evidence for transplutonium elements. Again, knowing the extraordinary competence of my co-workers in the discovery of plutonium, I had more than adequate reason to believe that this might have occurred, but nothing happened to confirm this possibility.

Following these experiments performed in the first half of 1944, which gave negative results but provided much valuable experience, the first breakthrough came in July. The first bombardment of ^{239}Pu with helium ions (32 MeV) took place in the Berkeley 60-inch cyclotron on July 8 to 10. By this time, the actinide concept had crystallized in my mind to the extent that we decided to proceed solely on the assumption that elements 95 and 96 could not be oxidized in aqueous acidic solution to soluble fluoride states (i.e., they should exhibit only the III or IV oxidation states that have insoluble fluorides). The sample was sent to Chicago by air and the chemical procedures started on July 12. A 2.2 mg portion of the target ^{239}Pu was oxidized in aqueous solution to the plutonium (VI) oxidation state (whose fluoride is soluble) and lanthanum fluoride was precipitated. This precipitate, presumably containing the insoluble fluoride of an element 95 or 96 isotope and a small fraction of the plutonium (nearly all of which remained in solution in its soluble oxidized form), was then dissolved. Most of the small amount of remaining plutonium was oxidized and lanthanum fluoride was again precipitated to carry the element 95 or 96. The cycle was then repeated a third time. This final sample (labeled 49αA - #9) contained only about 0.004% of the original 2.2 mg of ^{239}Pu, that is, about 0.09 microgram or about 12,000 alpha particle disintegrations per minute.

On July 14, 15 and 16, careful absorption measurements in mica sheets were made on this sample. A plot of these data revealed distinctly the presence of alpha particles of longer range than those of ^{239}Pu. The original graph, a photographic reproduction taken from our notebook, is shown in Figure 1.
This uses the wartime code designation "49" for the target isotope ^{239}Pu. The significance of this observation is perhaps best summarized by quoting here the entry we made in our notebook (No. 221B) on July 14, 1944:

> "Plotting and comparing this data shows that a 94^{239} sample 20% larger than 49αA - #9 falls to zero c/m much faster than does 49αA - #9 itself. This definitely shows the presence of a long-range alpha emitting isotope in sample 49αA - #9. This is undoubtedly due to a product of the nuclear reaction of α's on 94^{239} and is probably one of the following: 95^{242} (by α,p reaction) or 96^{242} (by an α,n reaction) or 96^{241} (by α,2n

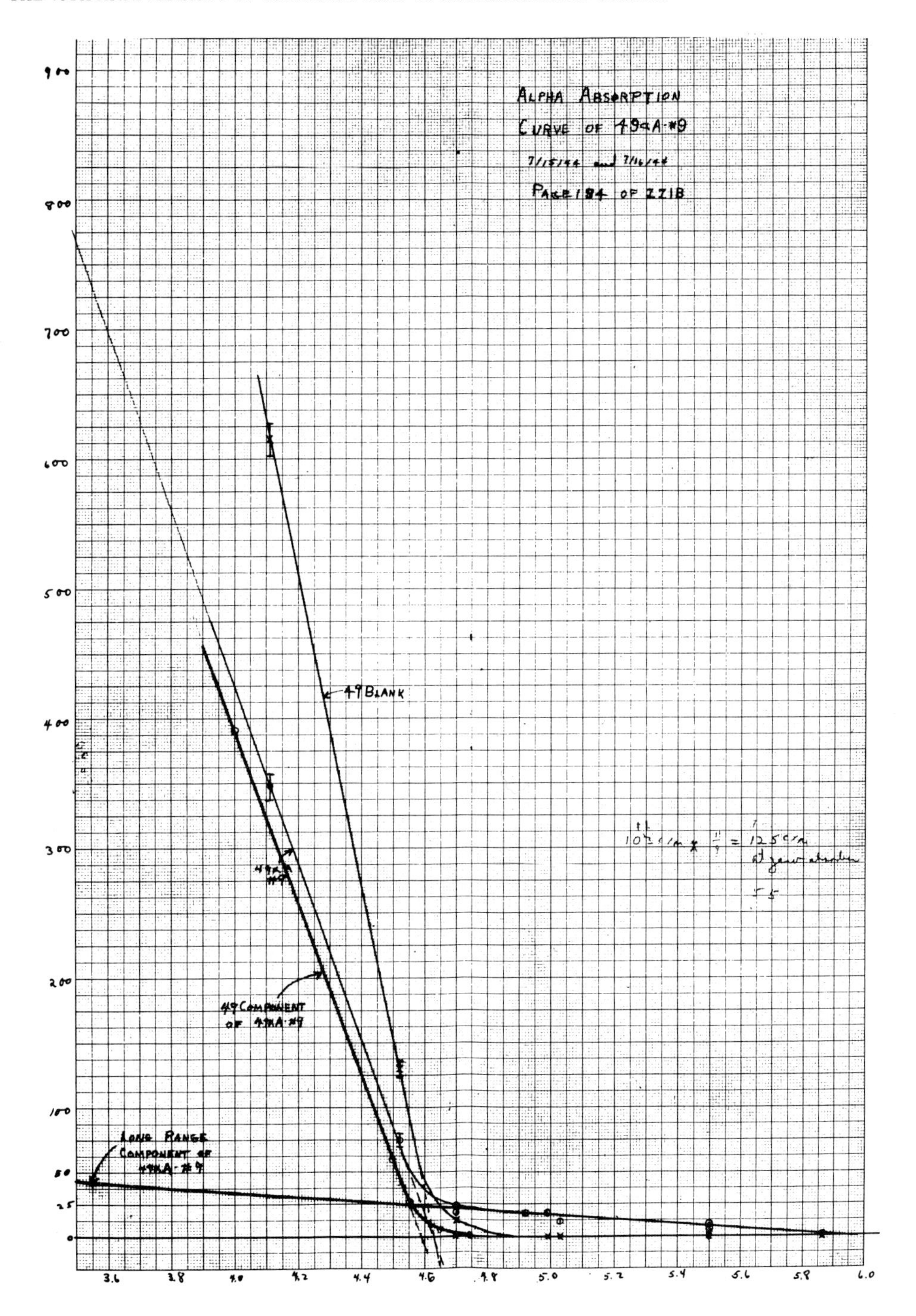

Figure 1. Original mica absorption data showing presence of $^{242}96$ alpha particles in helium-irradiated ^{239}Pu. Ordinate--counts per minute, Abscissa--Mg/Cm^2 of mica. July 15-16, 1944

reaction). Other isotopes are possible, but these seem the most likely. The isotope 95^{241} from α;n,p is fairly probable.

It is difficult to estimate the range of the α-particles from this new isotope very accurately, but it would seem to be about 4.65 ± 0.15 cm of air."

Incidentally, the range of 4.65 cm corresponds very well with the 6.1 MeV energy of the alpha particles from $^{242}96$ as known today.

The sample 49αA - #9, a mixture of a few hundred disintegrations per minute of the new long-range alpha activity and 12,000 disintegrations per minute of ^{239}Pu alpha activity, was then subjected to several additional oxidation cycles to further remove the ^{239}Pu. These were successful and by August 10, 1944, the sample was essentially free of ^{239}Pu. The sample was used to study the tracer chemical properties of the new isotope. Following is a verbatim extract from the Metallurgical Laboratory progress report (CS-2135) covering the period of August, 1944:

"About 10 mg of Pu^{239} were bombarded to the extent of 40 μahr with 32 MeV helium ions in the Berkeley sixty-inch cyclotron. There was found in this material a new radioactivity, emitting alpha-particles of range about 4.7 cm, which seems to be due to an isotope of element 96 or 95. This new radioactivity is carried quantitatively by lanthanum fluoride, even in the presence of oxidizing agents such as dichromate or silver persulfate, indicating that this isotope cannot be oxidized to a +6 oxidation state in aqueous solution. The activity is carried by lanthanum oxalate in the presence of excess alkaline oxalate and not carried from acid solution by zirconium phenylarsonate, lead sulfate, or bismuth phosphate."

Let me interrupt the narrative momentarily to note that as I was studying the notebooks which recorded all of these technical events, I came across the following entry, dated June 15, 1944, in Ralph James' notebook: "Time out to get married!" Apparently, even discovering new elements wasn't allowed to interfere with romance, although Ralph was back at work on June 19.

In the following months, it became increasingly apparent that this activity must be due to $^{242}96$. The definitive evidence for this isotope assignment came later through the observation of this same alpha activity as a product of the neutron bombardment of ^{239}Pu.

As the volume of the work and the complexity of the problem increased, we felt the need for more help. Early in September I asked Tom Morgan, a young chemist who had come to us from the University of Texas and who, like James, had distinguished himself in the investigations of the radiochemistry of plutonium, to join James, Ghiorso, and me in our search. He immediately joined Ralph in performing the chemical separations on a large (200 mg) sample of plutonium which in August had received a very intensive deuteron bombardment in the St. Louis cyclotron. (Plutonium was now available in such quantities as the result of its production in the Clinton reactor.) The chemical procedure consisted of isolating a fraction presumed to contain element 95 by separating it from plutonium through repeated oxidation of plutonium to its soluble fluoride form and carrying the non-oxidizable element 95 as its insoluble fluoride on lanthanum fluoride.

There were some indications of an alpha particle of range longer than that of ^{239}Pu in this element 95 fraction. On the basis of Ghiorso's absorption measurements, this alpha particle seemed to have a range of about 4.0 cm in air (compared to 3.7 cm for ^{239}Pu) corresponding to an energy of 5.5 MeV. Tracer experiments were carried out during September and October using a total of about one hundred disintegrations per minute of this 4.0 cm alpha emitter. These showed that it could be chemically separated from all the natural radioactive elements (lead and above) except actinum, thorium and possibly protactinium. The range of the alpha particles seemed to be inconsistent with their being due to isotopes of any of these elements.

In retrospect, it seems we were catching a glimpse of $^{241}95$ (whose alpha particles have a range of 4.0 cm) produced by a d,n reaction on the tiny amount of ^{240}Pu (approximately 10^{-2}%) present in the plutonium. The observed intensity of 4.0 cm alpha particles (about 100 disintegrations per minute) was consistent with the now known yield for this reaction. Another source can now also account for the observed 4.0 cm alpha particles, suggesting that both sources made appreciable contributions to the observed intensity. Originally present in the plutonium (before the deuteron bombardment) was a very, very small concentration of ^{241}Pu (approximately 10^{-5}%). As we now know, the ^{241}Pu is long-lived and thus its beta particle decay would continually produce the daughter isotope $^{241}95$ before, during, and after the deuteron bombardment. The amount of ^{241}Pu and time of its decay are consistent with the observed intensity of 4.0 cm alpha particles.

As shown in Figure 2, which is a photographic reproduction of the original plot in our notebook, the presence of an alpha particle emitter other than ^{239}Pu could be clearly distinguished by alpha particle absorption experiments with mica absorbers. In the graph, the mica absorption range for the new isotope was again converted to its equivalent air thickness--4.05 cm. Quoting from our notebook (No. 727B) entry for October 17, 1944, the following statement is made:

> "Among the heavy isotopes (Z=80-94) Ac, Th, and Pa might follow the observed chemistry. This activity would have to be a new isotope of these elements, however, since none of the known ones would have this range, or behave in this way (growth, decay). Element 95 would also be expected to have this chem[istry]."

However, this activity had somewhat of a "will o' the wisp" character, and the tiny supply was finally frittered away in the course of the chemical manipulations.

It was the neutron irradiation of ^{239}Pu to a relatively large total exposure, first at Clinton Laboratories and then at the Hanford Engineer Works in the state of Washington, that led to the definite observation of an isotope of element 95 and the definite identification of the above described product of the helium ion bombardment of ^{239}Pu as an isotope of element 96.

Two samples of ^{239}Pu, one of 4.4 mg and the other 8.2 mg, were placed in the reactor at Clinton Laboratories on June 5, 1944. The

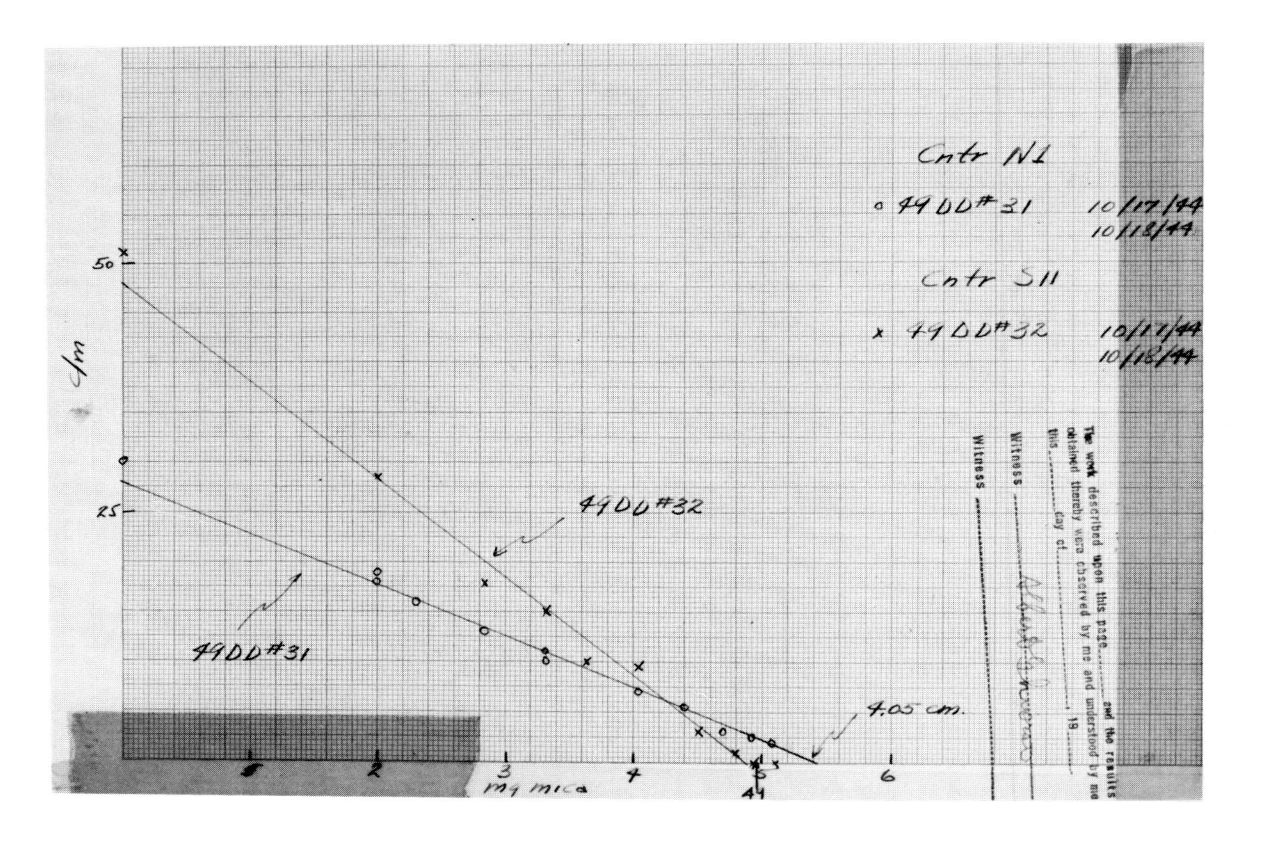

Figure 2. Original mica absorption data showing presence of $^{241}95$ alpha particles in deuteron-irradiated plutonium. Ordinate--counts per minute, abscissa--Mg/Cm^2 of mica. October 17-18, 1944.

first of these was removed on November 10 and returned to Chicago on November 19. This sample was put through the same type of chemical procedure described above, in which the plutonium was oxidized to the VI oxidation state whose fluoride is soluble, and insoluble lanthanum fluoride was precipitated with the intention of carrying any non-oxidizable isotopes of element 95 and 96 that might be present. After several such cycles had been completed, alpha particle absorption measurements were made on the fraction containing any non-oxidizable components to determine the ranges of the emitted alpha particles. A plot of these data on December 6, 1944, showed that, in addition to the alpha particles from the ^{239}Pu remaining in the sample, a component with alpha particles of longer range than those of ^{239}Pu was present at an intensity of a couple hundred disintegrations per minute.

This was very exciting. Here again, as in the case of the long-range alpha particles that had been observed the previous July in the non-oxidizable fraction isolated from ^{239}Pu bombarded with helium ions, and during September and October in plutonium bombarded with deutrons, was evidence that radiation from an isotope (or isotopes) of element 95 or 96 had been observed. This was confirmed during the following week when additional chemical cycles of the same type continued to remove ^{239}Pu while the intensity of the long-range alpha particles remained constant.

By the end of December, continuing very careful absorption measurements on the long-range alpha particles indicated the possible presence of two components.

The 8.2 mg ^{239}Pu sample was removed from the Clinton reactor on December 3 and returned to Chicago. This was also subjected to the now well developed oxidation and lanthanum fluoride precipitation procedure to remove the ^{239}Pu and concentrate the long-range alpha particle emitter (or emitters). Again, long-range alpha particles were found and in greater quantity, as would be expected on the basis of the longer neutron irradiation.

Now we were sure that we had observed transplutonium isotopes, but there remained the task of putting together all of the pieces of the puzzle. These began to fit together at about the turn of the year, early in January, 1945. It was becoming increasingly clear that the long-range alpha particles from the neutron-bombarded ^{239}Pu consisted of two components. But it was exceedingly difficult to define the two ranges with any accuracy.

Gradually, we were able to conclude that the ranges of these two alpha particles in air were about 4.0 cm and 4.7 cm (compared to 3.7 cm for ^{239}Pu). It became clear to us, and to our delight, that the 4.7 cm alpha particle was the same as that found as a product of helium ion bombardment of ^{239}Pu. Thus it appeared that the 4.0 cm alpha particle emitter was ^{241}95 and the 4.7 cm alpha particle emitter was ^{242}96, produced in neutron-irradiated ^{239}Pu by the following reactions:

$$^{239}Pu\ (n,\gamma)\ ^{240}Pu\ (n,\gamma)\ ^{241}Pu$$

$$^{241}Pu \xrightarrow{\beta^-} {}^{241}95$$

$$^{241}95\ (n,\gamma)\ ^{242}95$$

$$^{242}95 \xrightarrow{\beta^-} {}^{242}96$$

The same isotope $^{242}96$ had been produced in helium-ion-bombarded ^{239}Pu by the reaction:

$$^{239}Pu\ (\alpha,n)\ ^{242}96$$

Aiding us in coming to this conclusion was the information we obtained by use of a 25 mg sample of ^{239}Pu which had received a relatively short neutron irradiation in November in the much higher neutron flux of one of the large plutonium production reactors at the Hanford Engineer Works in the state of Washington. This sample, which was received on January 25, 1945, was subjected to our by now standard chemical separation procedure. The element 95 and 96 fraction showed the same two component long-range alpha particles as had been observed in the ^{239}Pu samples irradiated in the Clinton reactor. Consistent with the higher neutron flux, the yields were much larger than had been obtained in the Clinton irradiations. In fact, it was an enormous quantity by our standards, of the order of fifty thousand alpha disintegrations per minute. However, there were some aspects of the yield of these alpha emitters that puzzled us at first. Because $^{242}96$ is produced in a higher order neutron absorption sequence than is the case for $^{241}95$, it was expected that the ratio of the yield of $^{242}96$ to that of $^{241}95$ would be substantially higher in the high flux Hanford irradiation than in the lower flux Clinton irradiation. (The total integrated exposure was much greater at Hanford than at Clinton.) Actually this ratio was nearly the same in Hanford and Clinton irradiations. However, we soon understood the reason for this. The short duration of the Hanford irradiation and the relatively long interval from the end of this irradiation until we received the irradiated plutonium favored the relative buildup of $^{241}95$ whereas the long duration of the Clinton irradiation allowed the $^{241}95$ to have a relatively longer exposure to neutrons which favored the production of $^{242}96$.

These conclusions were presented at a meeting of the Basic Chemistry, Recovery, and Instrument groups of my Chemical Section at the Metallurgical Laboratory on January 31, 1945. Figure 3 is a photographic reproduction of a graph which was shown at that meeting and which was printed in the minutes of the meeting (MUC-GTS-1346, Feb. 2, 1945). The graph summarizes the alpha particle absorption measurements made on samples chemically separated from the neutron bombardments of plutonium. This represents a clear recognition of the formation of the two isotopes $^{241}95$ and $^{242}96$, with ranges in air for the alpha particle, of 4.0 and 4.7 - 4.8 cm., respectively.

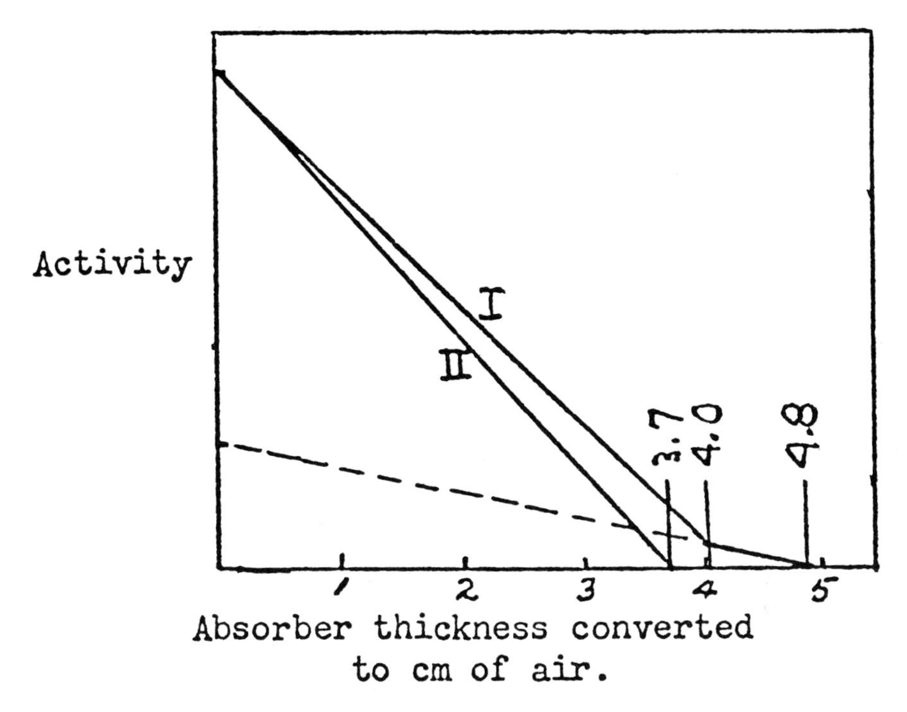

I. Range Curve for $^{241}95$ and $^{242}96$

II. Comparison Range Curve for ^{239}Pu

Figure 3. Summary alpha particle absorption curves presented at Metallurgical Laboratory meeting, Chicago, January 31, 1945.

A good summary of our understanding at that time can also be seen in the following verbatum quotation from the Metallurgical Laboratory progress report (CS-2741) covering the period of February, 1945:

> "Some very interesting new alpha-radioactivity has been found, both in plutonium irradiated with neutrons in the Clinton pile and plutonium irradiated with neutrons in the Hanford pile. This alpha-activity exhibits just the sort of chemical behavior which has been predicted for the transplutonium elements. For example, it is carried by rare earth fluorides and it has not yet been possible to oxidize it to a state or states where its fluoride is soluble. It seems to be chemically separable from all the rest of the 94 elements, and the best present interpretation is that it is due to elements 95 and/or 96. The alpha-activity is composed of two components, one of range 4.0 cm and the other of range 4.7 cm. A very attractive possibility is that the 4.0 cm alpha-activity corresponds to 95^{241} formed from the beta-decay of 94^{241} which comes from the reaction 94^{240} (n,γ) 94^{241}, and the 4.7 cm alpha-activity corresponds to 96^{242} (cf. reports CS-2124 and CS-2135) from beta-decaying 95^{242} following the reaction 95^{241}

(n,γ) 95^{242}. The ratio of the yields in the Hanford as compared to the Clinton bombardment seems to be proportional to the second power of the total neutron irradiation, as would be expected on the basis of these particular isotopic assignments, although the accuracy of the estimation of the neutron fluxes is not sufficient to make this at all certain."

Soon after our results had been communicated to scientists at Los Alamos, they made a mass spectrographic identification in purified plutonium of a relatively volatile isotope with the mass number 241, which they identified with our $^{241}95$. Since this could be extracted by volatilization from pure plutonium (from which element 95 which was originally present had been removed in the purification process), we knew that the parent isotope, ^{241}Pu, must have a relatively long half-life (of the order of months or longer). Our own work had not given us any information concerning the half-life of ^{241}Pu. In order to test for the presence of ^{241}Pu, we subjected 250 mg of purified plutonium, which had been produced in a Hanford reactor, to our chemical procedure for the isolation of element 95. By March 20, we had a purified fraction containing the extracted $^{241}95$. The intensity of the 4.0 cm alpha particles corresponded to about 30,000 disintegrations per minute. This established beyond doubt the identity of our 4.0 cm alpha particle emitter. Also, we now had a large sample with which to study the chemical properties of element 95 and, beyond that, we knew that we had an inexhaustible supply in our large stock of plutonium. Subsequent "milkings" led to continued extractions of the 4.0 cm alpha emitter from plutonium samples and we were able to establish the half-life of ^{241}Pu (now known to be 14 years).

Also, in March, we had another bombardment of ^{239}Pu with helium ions in the Berkeley 60-inch cyclotron. Again, we chemically separated an element 96 fraction using our well-established procedures. Imagine, then, our surprise and consternation when the alpha particles emitted in this fraction had a range of 5.0 cm rather than the 4.7 cm of the expected $^{242}96$. However, we were able soon to solve this additional puzzle. It happened that the Berkeley cyclotron had been modified since our July, 1944, bombardment so that it now delivered 40 MeV helium ions rather than the previous 32 MeV energy. This meant that isotopes such as $^{241}96$ and $^{240}96$ would now be produced in predominating yields, due to α,2n and α,3n reactions, in addition to the previously produced $^{242}96$. It was possible later to assign the 5.0 cm alpha emitter to the new isotope $^{240}96$ and to detect the lower intensity alpha particles due to the $^{242}96$.

A few months later, in July, 1945, the isotope ^{238}Pu was identified as the alpha particle decay product of $^{242}96$, thus lending further confirmation to this isotopic assignment. Decay measurements showed that $^{242}96$ had a half-life of about five months (and $^{240}96$ a half-life of about one month) while $^{241}95$ exhibited no measurable decay (now known to have a half-life of 432 years). As soon as a sufficient quantity of separated $^{241}95$ was available as a decay product of ^{241}Pu, it was irradiated with reactor neutrons to confirm the production of $^{242}96$ by the reactions $^{241}95\ (n,\gamma)$

$$^{242}95 \xrightarrow{\beta^-} {}^{242}96.$$ Soon thereafter W. M. Manning and L. B.

Asprey measured the half-life of $^{242}95$ as about 16 hours.

Of special interest is the fact that we were able to produce and definitely identify the isotope $^{241}95$ also by charged particle bombardments. Using helium ions from the Berkeley cyclotron to bombard uranium, we were able to identify ^{241}Pu and its beta-decay daughter $^{241}95$ produced by the reactions $^{238}U\ (\alpha,n)$

$$^{241}Pu \xrightarrow{\beta^-} {}^{241}95.$$ The significance of this lies in the fact

that we were able to obtain declassification (permission to publish) of this method of production much before the original method for the production of $^{241}95$ (via the neutron irradiation of ^{239}Pu) could be declassified. Therefore, my first announcement at a scientific meeting of the discovery of elements 95 and 96, in a paper presented at an American Chemical Society symposium at Northwestern University on Friday, November 16, 1945 (published in Chemical and Engineering News, Vol. 23, page 2190, Dec. 10, 1945), described the production of element 95 by helium ions rather than by neutron bombardment.

As it actually turned out, the discovery of elements 95 and 96 was announced to the world for the first time prior to the symposium. This occurred on the Quiz Kids national radio program on Sunday, November 11, Armistice Day of 1945. I happened to be serving as a guest on this radio program and one of the kids (Richard Williams) asked me if any new chemical elements had been discovered at the Metallurgical Laboratory during the war. As the information had already been declassified for the symposium to be held the following Friday, I replied in the affirmative to his question. Following is a verbatim transcription of this question and answer:

> "Richard: Oh, another thing--have there been any other new elements discovered like plutonium and neptunium?
>
> Seaborg: Oh yes, Dick. Recently there have been two new elements discovered--elements with atomic numbers 95 and 96 out at the Metallurgical Laboratory here in Chicago. So now you'll have to tell your teachers to change the 92 elements in your schoolbook to 96 elements."

Soon after this, on December 15, 1945, I made a guest appearance on the Watson Davis program, "Adventures in Science," which was a national network (CBS) radio program at that time. Following is a verbatim transcript of a portion of that radio program:

> "Announcer: By the way, I'd like to know whether you have named these two new elements that you have discovered?
>
> Seaborg: Well, naming one of the fundamental substances of the universe is, of course, something that should be done only after careful thought. We have been faced with considerable difficulty in these cases because we have run out of planets. Naming neptunium after the planet Neptune, and plutonium after the planet Pluto, was rather logical. But so far the astronomers

haven't disclosed any planets beyond Pluto. So we'll have to go to some other method of naming.

Announcer: What's that, Dr. Seaborg?

Seaborg: This hasn't been decided yet. One possibility might be to rely on some property of these elements. We do have an idea for the naming of element 95 along these lines and may have a suggestion to offer pretty soon. And, by the way, you may be interested to know that we have received lots of suggestions. Some good and some not so good.

Announcer: Well, Dr. Seaborg, perhaps some of the listeners to "Adventures in Science" will want to make suggestions as to the naming of the new elements. Will you be willing to have them write in their suggestions?

Seaborg: Well, I don't promise to follow the suggestions. But it might be interesting to know what the public thinks about naming new chemical elements.

Announcer: Very well, Dr. Seaborg. If you want to suggest names for new elements 95 and 96, just drop a postcard to Watson Davis, Science Service, Washington 6, D.C. And to all those who write in, Mr. Davis will send them a free copy of the current issue of Chemistry Magazine which contains Dr. Seaborg's full technical paper and a new arrangement of the chemical Periodic Table. It's free for the asking. So be sure to ask for the elements 95 and 96 article and address Watson Davis, Science Service, Washington 6, D.C."

Actually, we gave a great deal of thought to the naming of elements 95 and 96. My theory that they should be chemically similar to the rare earth elements was being borne out to such an extent that we were finding it almost impossible to chemically separate them from these elements. Although we eventually succeeded, during the period of our futile efforts to do so Tom continually referred to elements 95 and 96 as "pandemonium" and "delirium."

Names were finally suggested for elements 95 and 96 in the course of a talk that I gave at the annual spring meeting of the American Chemical Society in Atlantic City on April 10, 1946 (published in Chemical and Engineering News, Vol. 24, page 1192, May 10, 1946). Element 95 was given the name "americium" (symbol Am), after the Americas, in analogy to the naming of its rare earth homologue, europium, after Europe; and for element 96 we suggested the name "curium" (symbol Cm), after Pierre and Marie Curie, in analogy to the naming of its homologue, gadolinium, after Johan Gadolin.

This, then, is the story of the discovery of the elements with the atomic numbers 95 and 96. I hope that I have succeeded in recapturing for your some of the excitement, frustrations, and satisfactions that we experienced in the course of this scientific adventure.

This work is supported by the Director, Office of Energy Research Division, of Nuclear Physics of the Office of High Energy and Nuclear Physics of the U.S. Department of Energy under Contract No. DE-AC03-76SF00098.

REFERENCES:

1. Seaborg, G. T., James, R. A. and Morgan, L. O., 'The New Element Americium (Atomic Number 95),'pp. 1525-1553 in The Transuranium Elements: Research Papers, Seaborg, G. T., Katz, J. J. and Manning, W. M., Eds., National Nuclear Energy Series, Div. IV, Vol. 14B, N.Y., N. Y. (1949)
2. Seaborg, G. T., James, R. A. and Ghiorso, A., 'The New Element Curium (Atomic Number 96)', pp. 1554-1571 in The Transuranium Elements: Research Papers, Seaborg, G. T., Katz, J. J. and Manning, Eds., National Nuclear Energy Series, Div. IV, Vol. 14B, N. Y. (1949)
3. Seaborg, G. T., The Transuranium Elements, Yale University Press (1958)
4. Seaborg, G. T., Ed., The Transuranium Elements: Products of Modern Alchemy, Dowden, Hutchinson and Ross, Inc., Stroudsburg, Penn. (1978)
5. Seaborg, G. T. , 'The Plutonium Story', pp. 1-22, in Actinides in Perspective, Edelstein, Norman M., Ed., Pergammon Press Ltd., Oxford, England (1982)
6. Ghiorso, A., 'A History of the Discovery of the Transplutonium Elements', pp. 23-56, in Actinides in Perspective, Edelstein, Norman M., Ed., Pergammon Press Ltd., Oxford, England (1982)

REMINISCENCES OF AN INSTRUMENTALIST*

A. Ghiorso
University of California
Lawrence Berkeley Laboratory
Nuclear Science
Berkeley, California 94720

ABSTRACT. Forty years ago the methods used to detect the radiations from new elements were not very sophisticated although they did suffice. It is interesting to contrast the techniques used at that time with those available today.

In thinking back over the past forty years to the time when the work on the synthesis of americium and curium was starting I was struck by the dramatic changes that have occurred in the instrumentation used for analyzing the products of these discoveries. The discoveries themselves were made possible at that time by the chemical concept that was employed so they were a triumph of chemistry as demonstrated so clearly in Glenn's talk. But instrumentation was necessary to show the nature of the radiations that were detected. In the first efforts the physical methods available to us were meager and it was necessary to enhance them in various ways to make them suitable for the searches that were undertaken.

At that time solid state detectors were not even envisioned by anyone since the transistor had not been invented yet. Available were Geiger-Mueller counters for beta rays and gaseous ionization chambers for alpha particles but these instruments merely enabled one to count the number of particles. Nothing could be determined as to the energies of such particles except at very low geometries, usually through the use of magnetic devices. At about this time nitrogen was being introduced as the gaseous medium for ion chambers and then, a little later, various gas mixtures that made possible fast electron collection. The Frisch grid method for the measurement of alpha particle energies was in the process of development in England but was not yet available in the U.S. Range measurements had been used by Europeans as a substitute for energy measurements but suffered from the disadvantage of low efficiency because it was necessary to use a low geometry to obtain a precise end point for the range. See Fig. 1.

It occurred to me that we could achieve our objective of making range measurements and still have a relatively high efficiency by using 2π geometry. In this case a range curve would be obtained with the same

N. M. Edelstein et al. (eds.), Americium and Curium Chemistry and Technology, 19–24.

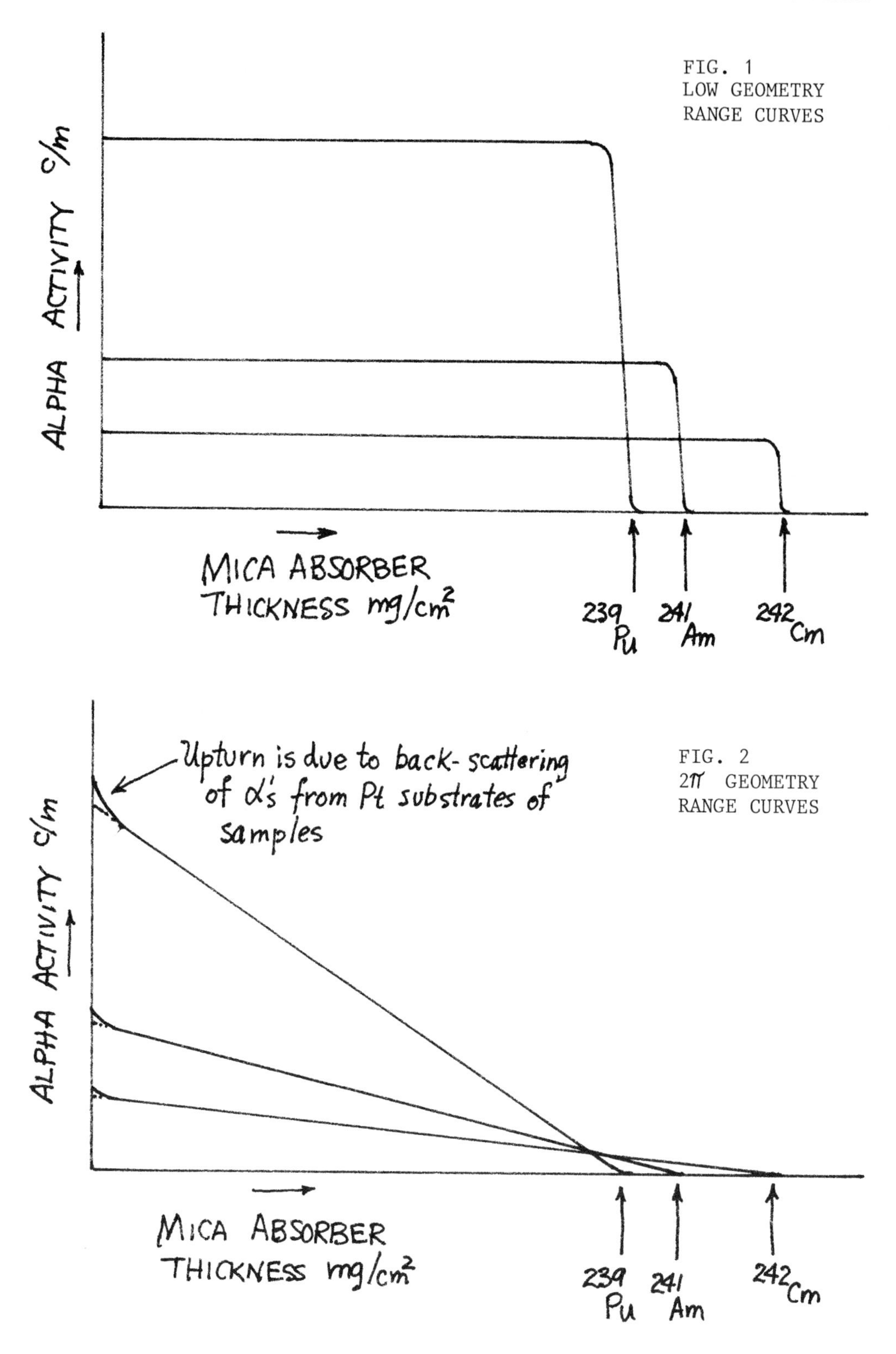
FIG. 1
LOW GEOMETRY
RANGE CURVES
ALPHA ACTIVITY c/m
MICA ABSORBER
THICKNESS mg/cm²
239 Pu
241 Am
242 Cm
FIG. 2
2π GEOMETRY
RANGE CURVES
Upturn is due to back-scattering of α's from Pt substrates of samples
ALPHA ACTIVITY c/m
MICA ABSORBER
THICKNESS mg/cm²
239 Pu
241 Am
242 Cm

end point but it would be necessary to use very flat absorbers in contact with the sample to be analyzed. The range curve would then have the shape shown in Fig. 2. The only convenient absorbers readily available to us were split sheets of mica that were carefully weighed. These were laid over the samples and the alpha particles that penetrated all the way through into the sensitive volume of the ion chamber were the ones that were counted. One of the problems that we encountered was that the surface of the mica could charge up and nullify the collecting potential for the ions created by the alphas. This was overcome by wiping the top surface with an acidic layer to make them conducting. Our chambers were operated in air and they needed a large potential (~1500 volts) to collect the ions. The collection time was quite long so that the pulse rise time was very slow and thus the required bandwidth of the amplifiers extended down to relatively low frequencies. The net result was that the chambers made very sensitive condensor microphones and tended to pick up the ambient noise in the room in spite of acoustic shielding. On sensitive counts we would watch the oscilloscope screens to make sure that our counts were genuine.

But the method did work and we were able to show that ^{242}Cm and then ^{241}Am had distinctive ranges that separated them from ^{239}Pu and thus gave them characteristic physical identities as well as new chemical identities. Needless to say this increased our confidence that two new elements had indeed been discovered.

Before we got to that point, however, Ralph James and I did several experiments at the Washington University cyclotron in St. Louis which, in retrospect, were very foolhardy. The idea was to bombard a ^{239}Pu layer with a high intensity deuteron beam and then examine the target shortly thereafter for long range alpha particle activity - alphas that would penetrate a mica layer thick enough to stop the intense ^{239}Pu alpha radiation from getting into a fast nitrogen-filled ion chamber. The hazards were two-fold. The uncovered target was a slurry of perhaps 10 milligrams of plutonium oxide evaporated to dryness onto a thick water-cooled copper plate. I don't know which was worse - being exposed to Pu from a target handled with short tongs and no hood or being exposed to the very intense beta-gamma activity from the deuteron bombardment of the copper target holder! And all of this done with no survey meters!

Of course we didn't find anything but many months later we realized that the ^{241}Am was already there, slowly growing in from the small amounts of ^{241}Pu that was present in the plutonium as a result of its production from uranium in a sea of reactor neutrons.

Speaking of these crude methods reminds me of a funny incident that took place around this time. After we had found element 96 and then 95 we obtained more and more samples of plutonium from the Clinton pile that had been exposed to increasing fluxes of neutrons to look for new isotopes. Since we knew that ^{241}Pu was a long-lived beta emitter we wanted to find ^{242}Pu. We decided that the most sensitive way to detect it would be to look for an enhanced spontaneous fission rate since we knew that ^{240}Pu had a surprisingly short fission half-life. One of the chemists made up a beautiful sample by electrolysis which was about a milligram per cm^2 thick. It was on a platinum plate about three inches in diameter and this thin plate was cemented onto a thicker brass plate for rigidity. I put the precious sample into the fission counter and watched for the big kicks in the oscilloscope that would signal spontaneous fission events.

After only fifteen minutes there was an event and I went running down the hall to alert people to the possibility that we were onto something most unusual. Then there was another big kick and then another! No doubt about it, the spontaneous fission half-life must be even shorter than we had expected. Since very little was known about spontaneous fission at that time we found this new information very exciting. The big kicks continued to come in for several hours. Then I began to realize that something was wrong for the counts seemed to come in at surprisingly regular intervals. So I timed them and found that the interval was indeed always about fifteen minutes! The oscilloscope showed nice big kicks, just the way spontaneous fission events always looked, so the counter seemed to be in working order. What could be wrong? Then I had an idea. The intense alpha activity might be charging up the platinum plate insulated from the thicker plate by the layer of cement until it reached a potential where it would discharge. This discharge would be picked up and register as a fission and then the plate would begin the cycle all over again. I shorted the two plates together and the big kicks went away.

I remember these early years as being very intense. There was an awful lot to learn and I had a very poor background from which to start. But the work was so fascinating that I had plenty of incentive to plod away at it. I remember going to the library and reading about the early experiments in radioactivity by Madame Curie and by Ernest Rutherford and his group. Their work was even more difficult since at first they did not know of the existence of isotopes. With their even more primitive equipment they used marvelous ingenuity and brilliance to unravel gradually the various alpha decay series in all their splendor. In the course of our investigations it became necessary for me to check some of their measurements using various alpha recoil techniques. I thought that with our somewhat better counter systems surely I would find that some of their results would be in error. On the contrary I found that they were always right. That was also true for the results of the earlier Seaborg group measurements which I repeated. Gradually I formulated the Ghiorso Uncertainty Principle - first experiments are likely to be correct if there are enough compensating errors!

In spite of this feeling I found that Ralph James was a very cautious fellow. I remember how he would agonize over the possibility that there was a monstrous error and somehow the assignments of the new activities were wrong. He would dream up some very unlikely scenarios which would change the picture completely - and this many months after he and Glenn had deduced the correct interpretations.

But we learned and gradually introduced new methods to the discoveries of new elements. Some we invented ourselves and others we borrowed from other experimental studies. The use of Frisch-grid ion chambers made a big difference along with the introduction of multi-channel pulse height analyzers. These tools enabled us to work with fewer and fewer atoms as long as there was time enough to do specific chemical separations.

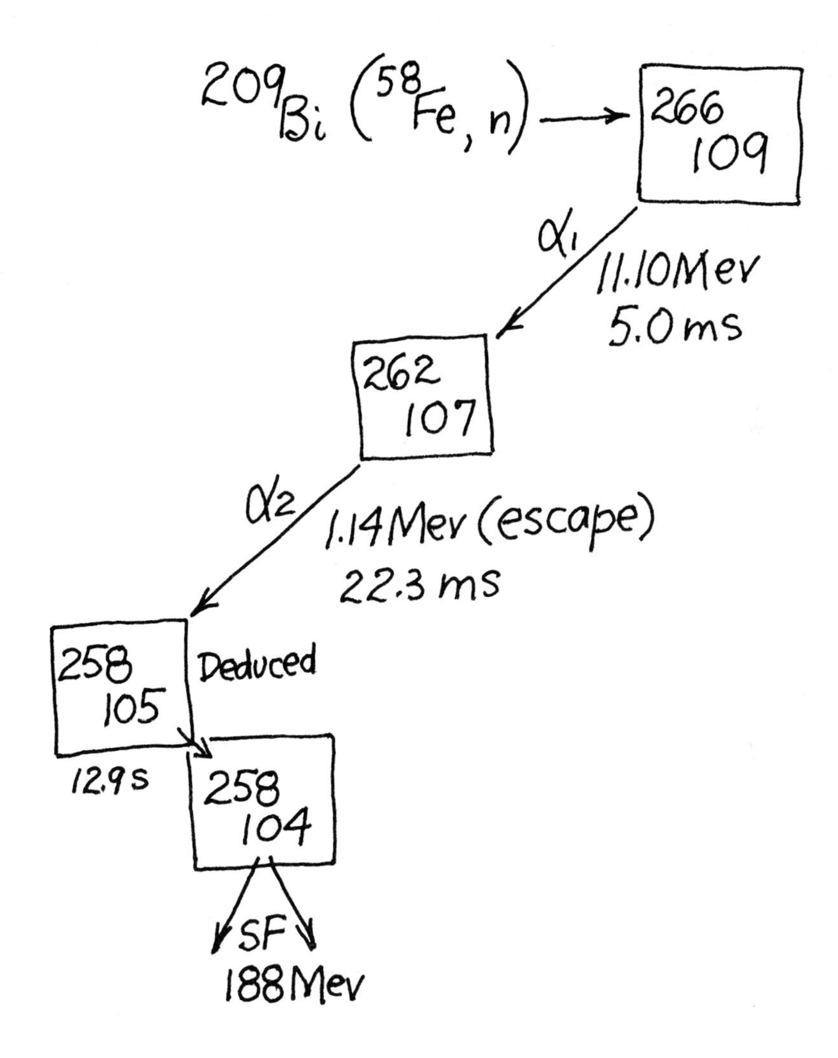

G. Münzenberg et al. 1982

FIG. 3

The big breakthrough came when we introduced the recoil technique for separating element 101 and other reaction products from highly radio-active targets. This was followed in due course in the early 60's by the use of the helium gas jet to transport these recoils to remote counting stations employing the just-invented solid-state detectors and this in turn led to the extremely powerful genetic method. With this latter technique we were able to identify particular nuclides by demonstrating alpha-alpha correlations for elements 104, 105, and 106 between unknown parents and known daughters and grand-daughters. Only a dozen or so alpha particle decays correlated in energy and time were necessary to identify these new elements. This technique was extremely important because the available half lives had become too short for conventional chemical identification.

The ultimate was finally reached by Peter Armbruster's group at GSI in 1982 when they identified one single atom of element 109 during a two week period of bombardment of ^{209}Bi by ^{58}Fe ions (Fig. 3). Their velocity separator, SHIP, was used to separate a recoiling atom of $^{266}109$ from the intense beam and implant it at a measured position in a silicon detecting crystal with a measured recoil energy. At this particular position they identified a high energy alpha particle emitted 5 milliseconds later from that atom. At the same position 22 milliseconds later they observed only part of the alpha energy from its daughter, $^{262}107$, because the alpha particle from that decay had escaped out through the face of the detecting crystal. Finally 13 seconds after that a spontaneous fission was emitted from the same place. This last member of the chain was from $^{258}104$, a known 13-ms SF emitter following electron-capture decay of $^{258}105$. Great confidence can be ascribed to this experiment since it was accomplished in a setting with a unique genetic chain and essentially no background.

I have been fortunate to participate in a dozen steps of this long string of new element syntheses and to observe first hand the profound changes that have taken place in the instrumentation essential for their identification. Each time that a new element has been discovered has meant that a new set of problems has been solved and thus has set the stage for the next step. The discoveries of curium and americium were the beginning steps for the transplutonium elements and so in some ways were the most exciting of all.

*This work was supported by the Director, Office of Energy Research, Office of High Energy and Nuclear Physics, Nuclear Science Division, U.S. Department of Energy under Contract No. DE-AC03-76SF00098.

AMERICIUM, ITS EARLY HISTORY AND GRAM-SCALE SEPARATION

R.A. Penneman (Consultant)
Los Alamos National Laboratory
Group INC-4, MS C-346
Los Alamos, New Mexico USA 87545

ABSTRACT. For this 40th Anniversary Discovery Celebration, The early contributions of the americium group at Los Alamos, established in 1948, are reviewed. They include the first major isolation of americium from impure plutonium, and significant expansion of americium chemistry.

The discovery of americium was made in 1944, as just recounted in the first paper of this session by one of the discoverers, Glenn T. Seaborg. The discovery was made at the Metallurgical Laboratory of the University of Chicago.

The Metallurgical Laboratory (later it became the Argonne National Laboratory) had been established as a result of wartime urgency in response to December 1941 urgings of Nobel Laureate Arthur Compton. Compton was an influential figure in American physics in those days, and was successful in achieving consolidation and expansion at Chicago of the atomic work begun at Columbia and Berkeley.

Thus, Fermi, Seaborg and others came to Chicago. The pace of discovery in those days was extraordinary! At the end of the first year, in December 1942, Fermi demonstrated the graphite-moderated reactor, fueled by natural uranium. Future reactors would generate neutrons and heavy elements on a production scale, surpassing by orders of magnitude the tiny amounts possible with the cyclotron. Within the next two years, Seaborg and his team had painstakingly collected enough plutonium to establish its separation schemes, and found two new elements, americium and curium.

I was but indirectly associated with Seaborg in those exciting days. As a young chemist just hired (July) at the Metalurgical Laboratory, I had been sent in August 1942 to Beverly, Massachusetts as part of a team to aid in producing the needed uranium metal. Later, on return to Chicago, I and another from Milton Burton's team made the

N. M. Edelstein et al. (eds.), Americium and Curium Chemistry and Technology, 25–33.

first measurements of the energy stored in graphite as a result of neutron displacement of carbon atoms when graphite was used as a neutron moderator in a reactor.

Part of the new knowledge being generated was in the province of physics. However, many of our present ideas of chemistry and of the extension of the Periodic Table of Elements were also being formulated at that time. Indeed, it was not until late June, 1944 that Burris Cunningham was able to furnish W. H. Zachariasen with a sample of neptunium oxide thought to be the dioxide. By x-ray diffraction analysis, Zachariasen confirmed neptunium dioxide, and wrote a memo formally suggesting that a <u>new series of elements, which he called Thorides</u>, were in hand, <u>having a persistent valence of four.</u> He based his arguments on the stability of the dioxides, on the progressve diminution in ionic radius from thorium(IV), uranium(IV), neptunium(IV) to plutonium(IV) and the similarity of the radii of tetravalent cerium and uranium.

Seaborg had long thought about the properties of the anticipated new series of elements and their properties. In July 1944, Seaborg dictated a memo in which he suggested that <u>the elements following actinium be placed in a new series called Actinides</u>. As we know, the linchpin of his successful proposal was the placement of the then unknown element of number 96 (curium) at the series midpoint, the predicted increasing stability of the trivalent state, and divalency at atomic number 102. Seaborg has mentioned earlier today the importance of this latter chemistry in the discovery of americium. As we shall see, however, americium is saved from the monotony of trivalent chemistry by its greater number of valence states and hence displays an americium chemistry of extraordinary variety.

In the period following the discovery of americium, neutron irradiation of a small amount of plutonium and separation of the americium resulting from the beta decay of mass 241 plutonium permitted Burris Cunningham to explore the chemistry of americium on the microgram scale. This remarkable work was described in 1949 (1). In those studies, tetravalent americium was demonstrated in the dioxide, pentavalent americium as the potassium carbonate salt of unknown composition. Of course, trivalent americium was demonstrated to be a very stable aqueous ion. This was the status of americium chemistry when the Los Alamos work began. All of it was found to be correct.

However, from the early tracer studies, and reinforced by the location of americium in the new actinide series as a homolog of europium, it was believed (erroneously) that americium also possessed an attainable aqueous divalent state. Establishing that this was <u>not</u> the case, was the single major change that the Los Alamos investigators were to make in the early work. Other major contributions were to follow soon after.

First of all, the building had to be outfitted for the upcoming separations work and the research anticipated to follow. The warehouse-type building did have two great advantages: first, it was already fitted with a exhaust system suitable for handling alpha activity, and second, it was contiguous to buildings where plutonium work was still being carried on. Eventually, we were to be given the concentrated residues from plutonium processing, and thus had a continuing source of americium-rich residues from which to isolate americium. I use the term americium-rich in a relative sense. Certainly americium was present in those residues which were often the color of used motor oil, saturated in salt and had a specific gravity above 1.5--it was just that americium was a such minor impurity!

When I came to Los Alamos in the fall of 1947 after finishing my graduate studies at the U. of Illinois, the great exodus after war's end had taken place. Thus, I could take over an empty building wing no longer needed for plutonium work and was able to set out to recover americium from a potential major source. The source was a stock of plutonium metal turnings that had been accumulated from machining fuel rods for the Clementine reactor.

After a year of so in installing equipment and glove boxes to handle the expected americium, we began to dissolve plutonium turnings on about a 250 gram scale and to precipitate the plutonium peroxide, leaving americium in the filtrate. As noted in a recent paper from Los Alamos, modern-day americium separators have a much easier time of it. Certainly in that early day plutonium and its residues the americium content was lower, and the impurity content higher than that observed today. It is often found now that just one precipitation of americium hydroxide from the plutonium peroxide supernatants results in americium of greater than 95% purity(2).

It did stretch our intellectual muscles to finally isolate americium from solutions that once had several hundred times as much lanthanum as americium. The abundant calcium was not difficult to remove since its hydroxide is so much more soluble than that of americium. That merely required us to wash the impure hydroxides with ammonium nitrate. The ammonium ion tied up the hydroxide, thus solubilizing the calcium while maintaining the pH high enough to keep the americium hydroxide in the precipitate. Removing the lanthanum was another matter. That removal on a large scale was accomplished by ion exchange. Also, John Hermann, in a thesis directed by Penneman, showed that homogeneous generation of oxalate enriched americium over lanthanum in the precipitate.

Los Alamos work began to focus on the new americium chemistry itself by 1950. It followed the pioneer work on americium which Cunningham, Werner and Perlman had provided in the half-dozen years following its discovery. We had an enormous advantage over the earlier workers in that, once we had it isolated, we could work with americium on the many milligram scale, utilizing spectrophotometry to ascertain americium valence states in millimolar solutions. Thus, we could rather quickly dispel the notion that americium had an attainable divalent state by observing that the trivalent absorption persisted in spite of reducing agents that yielded both europium(II) and samarium(II). It was later demonstrated by Asprey that treatment of the solid trichlorides of americium and of samarium with hot hydrogen reduced samarium to its dichloride, but did not reduce americium trichloride.

Finally having accumulated a sufficient stock of purified americium enabled the work to proceed without having to stop after every experiment to repurify--a task which plagued the earlier work. Indeed, spectrophotometry of americium in solutions as well as in the solids was invaluable. Advances in the art of high resolution spectrophotometers provided another advantage, since the absorption lines of some americium valence states are so extremely narrow. It was especially helpful when recording spectrophotometers were later available to aid in kinetic studies involving Am(V) disproportionation.

Research on the isolated americium was well enough advanced through 1950 that, at the fall American Chemical Society meeting in Chicago, the Los Alamos group could report the following significant advances:

1. The discovery of hexavalent americium,

2. Its similarity to hexavalent uranium in their sodium acetate compounds,

3. The oxidation of Am(III) to Am(VI) by peroxydisulfate,

4. The measurement of the reversible couple, Am(V)/Am(VI),

5. The achievement of Am(VI) by high acid disproportionation of Am(V),

6. Preliminary values of the second order rate constant for the disproportionation of Am(V).

7. The failure to observe an absorption feature which was ascribable to aqueous Am(IV).

The reaction involving the disproportionation of Am(V) was shown to be second order in Am(V). This reaction thus implied the formation of both Am(IV) and Am(VI):

2 Am(V) = Am(VI) + Am(IV)

Although Am(VI) was observed, only the absorption spectra of the other two known species, Am(III) and Am(V) were ever present. This evidence suggested that Am(IV) was rapidly removed by other reactions such as disproportionation, by its reduction by solvent water, or by oxidation of Am(V) to Am(VI), a reaction found later to predominate.

Our search for soluble tetravalent americium was finally to be successful, but not until 1960 and only then as a highly complexed anion. Consideration about the value of the Am(IV)/ (III) couple being somewhat over two volts persuaded me to attempt the alkaline oxidation of americium trihydrdoxide in the attempt to make the then unknown tetrahydroxide, reasoning that if its solubility product were sufficiently less, say by 10expl6, this one volt would move the oxidation potential of the tetravalent state into the region of chemical stability. Treatment of the trihydroxide with hot sodium hypochlorite did yield a black, insoluble material. When dissolved in acid it did not yield Am(IV) but only the other known, soluble americium species. Conservation of oxidation number proved that the black compound was indeed the hydroxide or hydrous oxide of tetravalent americium. But what to do about this very insoluble material?

Recall the earlier mention of ammonium ion to bind the hydroxide when we had to remove calcium hydroxide from the impure americium trihydroxide slurry? I decided to treat the black, insoluble compound (hydroxide) of tetravalent americium with a concentrated solution of ammonium fluoride. The ammonium ion was supposed to take up the freed hydroxide and the fluoride ion, having the same charge and size, was then to replace the hydroxide on the americium to form a fluoride species of tetravalent americium.

Imagine the agreeable surprise when a rose-colored solution resulted! The solution gave an absorption spectrum different from those of Am(III), Am(V) or Am(VI), and could be treated to yield these other valence states in solution. _The elusive soluble state of americium had finally been found_! My original idea had been to convert the black "hydroxide" of Am(IV) into a solid fluoride, and we were later able to crystallize out the compound of americium tetrafluoride with four molecules of ammonium fluoride, yielding an eight coordinated anionic species. Of course, Larry Asprey and I were happy to accept the _bonus_ of a _soluble species of tetravalent americium_.

In a parallel study, Asprey had obtained a crystalline product by allowing a solution of Am(V) in concentrated potassium fluoride to stand some days. The solid, crystalline substance which gradually formed gave an unusual absorption spectrum. Later, it turned out that the compound was dipotassium americium hexafluoride. So Asprey had obtained the first double fluoride salt of Am(IV) from solution.

The formula (or formulas as it turned out) of the potassium carbonate salt(s) of pentavalent americium had been unknown since Perlman and Werner first oxidized Am(III) in concentrated potassium carbonate solution to the insoluble, tan precipitate containing pentavalent americium. Joe Nigon and I found the formula of one of the compounds, and established that there was at least two other formula or compound types. It remained for Tom Keenan to make the series of 1:1 salts with the several alkali carbonates, each containing the dioxo americyl(V) group. He found the best route to this series, even with plutonium, was to use the bicarbonate rather than the carbonate. One especially interesting finding involved the discovery that with sodium bicarbonate or carbonate, oxidation with ozone yielded the mahogany-colored, soluble complex of hexavalent americium. On the other hand, use of potassium bicarbonate or carbonate invariably stopped at the insoluble pentavalent state.

There is an interesting story involving Zachariasen and our compound, potassium dioxo americium(V) carbonate. I showed the x-ray powder pattern to him, and he recalled a similar pattern from an unanalyzed plutonium compound shown to him years earlier. In his case, he said the plutonium compound had resulted from potassium hydroxide addition to a solution of hexavalent plutonium and allowing the basic solution to stand. By cell volume analysis he had concluded that the cell should contain one dioxy plutonyl(VI) ion, one potassium and three other oxygens, then thought to be hydroxyls, to balance the charge on the cations.

However, from our americium studies, we knew that the americium was pentavalent and that the compoumd contained carbonate. Therefore, the compound's three oxygens came from carbonate, not hydroxide, to give potassium plutonyl(V) carbonate, rather than the Pu(VI) hydroxide formula he originally asigned. Both formulas satisfied his cell requirements, but only one satisfied the chemistry. In retrospect, what had evidently happened was that carbon dioxide had gotten into the plutonium solution on standing, and that reduction by alpha radiation had produced Pu(V) from Pu(VI). After this, Zachariasen trusted us chemists a little more, although he always felt more comfortable with solids whose true structure was established by x-ray!

Regarding the disproportionation of Am(V), it remained for the availability of long-lived 243 americium isotope and Jim Coleman to elucidate the rate. The long-lived isotope reduced the radiation problem to a manageable level (recall that self-radiation from the 432 year isotope reduces aqueous valences higher than Am(III) at a rate of several percent an hour).

Larry Asprey and Tom Keenan made americium halides, and Asprey produced the metal. His observation of the fcc phase was long questioned, even though it was later found that the fcc phases of Am and Pu were miscible. Finally, work at the Kernforschungzentrum in Karlsruhe, on americium metal, confirmed his finding.

Shortly after our discovery of the hexavalent state, Steve Stephanou showed that it was possible to oxidize Am(III) to the fluoride-soluble Am(VI) state and to precipitate curium trifluoride from the Am(VI) in solution. Others used this hint in a later patent appplication for the separation of americium from the rare earths.

In the early 1950's, Lew Jones and I demonstrated, using infrared absorption spectra obtained from thin aqueous films of Molar solutions of the hexavalent ions, that the uranyl-type, linear dioxo structure persisted from uranium through neptunium, plutonium and americium. Furthermore, the monovalent, dioxo ions of neptunium(V) and of americium(V) gave similar absorption, indicating similar structures.

This technique of observing the infrared spectrum of aqueous solutions in thin films led to advances in other areas, that of observing stepwise formation of complex ions, for example cyanide complexes, by "finger-printing" the vibration spectra of the individual ions.

To the other names mentioned in the text as being invoved in the early-day work with americium should be added the name of Dale Armstrong, who was the designer of the large-scale equipment and instrumentation so useful in many of the cited studies. References to the main contributors to the first decade or so of americium work at Los Alamos are listed in references 3-28.

Finally, in closing my contribution to this 40 year post-discovery celebration, I wish to acknowledge the major contributions made by the European teams at their various laboratories, especially the work on the pure metals, chalcogenides and especialy the needed thermodynamic data.

I note as well as the considerable data produced by the Russian workers over many years. Although we were early in the use of alkaline oxidation, and did achieve both Am(IV) and Am(VI), we did not pursue it to the very successful accomplishment of such high overall actinide valences as they did.

References

1. "The First Isolation of Americium in the Form of Pure Compounds; Microgram-Scale Observations on the Chemistry of Americium," B.B. Cunningham, in The Transuranium Elements, National Nuclear Energy Series, Div. IV, Vol. 14B (McGraw-Hill Book Co.Inc., New York, 1949), Paper 19.2.

2. "Status of Americium-241 Recovery and Purification at Los Alamos Scientific Laboratory," H.D. Ramsey, D.G. Clifton, S.W. Hayter, R.A. Penneman and E.L. Christensen in Transplutonium Elements, Production and Recovery, (ACS Symposium Series #161), J.D. Navratil and W.W. Schulz, Eds. (American Chemical Society, Washington, D.C., 1981), Ch. 5. pp. 73-91.

3. Stephanou, S.E., Asprey, L.B., and Penneman, R.A., Report AECU-925, 1950.

4. Penneman, R.A. and Asprey, L.B., Report AECU 936, 1950.

5. Asprey, L.B., Penneman, R.A. and Stephanou, S.E., Report AECU-927, 1950.

6. Asprey, L.B., Stephanou, S.E. and Penneman, R.A., J. Amer. Chem. Soc., 1950., 72, 1425.

7. Asprey, L.B., Stephanou S.E. and Penneman, R.A.. ibid., 1951, 73, 5715.

8. Stephanou, S.E. and Penneman, R.A., ibid., 1952, 74, 3701.

9. Jones, L.H. and Penneman, R.A., J. Chem. Phys., 1953, 21, 542.

10. Stephanou, S.E., Nigon, J.P. and Penneman, R.A.. ibid., 1953, 21, 42.

11. Keenan, T.K., Penneman, R.A. and McInteer, B.B., ibid., 1953, 21, 1802.

12. Nigon, J.P., Penneman, R.A., Staritzky, E., Keenan, T. K. and Asprey, L.B., J. Phys. Chem., 1954, 58, 403.

13. Keenan, T.K., Penneman, R.A. and Suttle, J.F., ibid., 1955, 59, 381.

14. Hermann, J.A., Report AECD-3637, Los Alamos Scientific Laboratory, July 1954. (Thesis, U. of New Mexico)

15. Armstrong, D.E., Asprey, L.B., Coleman, J.S., Keenan, T.K., LaMar, L.E. and Penneman, R.A, Los Alamos Scientific Laboratory Report LA-1975, 1956.

16. Penneman, R.A. and Asprey, L.B., Proc. Int'l Conf. on Peaceful Uses of Atomic Energy, 1956, 7, 355.

17. Coleman, J.S., Penneman, R.A., Keenan, T.K., LaMar, L. E., Armstrong, D.E. and Asprey, L.B., J. Inorg. Nucl. Chem., 1957, 3, 327.

18. Armstrong, D.E., Asprey, L.B., Coleman, J.S., Keenan, T.K., LaMar, L.E. and Penneman, R.A., AIChE Journal, 1957, 3, 286.

19. Penneman, R.A. and Keenan T.K., "The Radiochemistry of Americium and Curium," National Academy of Science-NS-3006, 1960.

20. McWhan, D.B., Wallmann, J.C., Cunningham, B.B., Asprey, L.B., Ellinger, F.H. and Zachariasen, W.H., J. Inorg. Nucl. Chem., 1960, 15, 185.

21. Penneman, R.A., Coleman, J.S. and Keenan, T.K., ibid., 1961, 17, 138.

22. Asprey, L.B. and Penneman, R.A., J. Amer. Chem. Soc., 1961, 83, 2200.

23. Asprey, L.B., Penneman, R.A. and Kruse, F.H., Chem. and Eng. News, 1962, 40, No. 8, 39.

24. Asprey, L.B. and Penneman, R.A., Inorg. Chem., 1962, 1, 134.

25. Coleman, J.S., Keenan, T.K., Jones, L.H., Carnall, W.T. and Penneman, R.A., ibid., 1963, 2, 58.

26. Penneman, R.A., Kruse, F.H., Benz, R. and Douglass, R. M., ibid., 1963, 2, 799.

27. Asprey, L.B. and Penneman, R.A., Chem. and Eng. News, 1967, 45, No. 32, 74.

28. Penneman, R.A., Keenan, T.K. and Asprey, L.B., "Lanthanide/Actinide Chemistry," Amer. Chem. Soc., Advances in Chem. Series, 1967, 71, 248.

20 YEARS OF AMERICIUM AND CURIUM RESEARCH AT THE EUROPEAN INSTITUTE FOR TRANSURANIUM ELEMENTS

W. Müller
Commission of the European Communities, Joint Research Centre
Karlsruhe Establishment
European Institute for Transuranium Elements
Postfach 2266
D-7500 Karlsruhe
Federal Republic of Germany

ABSTRACT. Part of the European Institute for Transuranium Elements' research programme which is aimed at elucidating the role of 5f electrons in the chemical bond has been devoted to the study of americium and curium. The investigation of fundamental properties, carried out also in collaboration with foreign research institutes and institutions, has involved: separation and purification of Am and Cm in multigram quantities by combining various separation methods: ion exchange, solvent extraction, extraction chromatography, precipitation; preparation of Am and Cm metals by metallothermic reduction of oxides or thermal dissociation of intermetallics, followed by repeated sublimation for purification purposes; determination of bonding related physical properties: vapour pressures (sublimation enthalpy), electrical resistivity, magnetic susceptibility, volume or crystal structure changes under pressure. Examples of applied research and current work are presented.

The European Institute of Transuranium Elements at Karlsruhe is one of four research establishments of the Joint Research Centre of the Commission of the European Communities. The three other research establishments are located in Italy (Ispra), Belgium (Geel) and in the Netherlands (Petten).

The Institute for Transuranium Elements was built during the early sixties in order to centralize the efforts of EURATOM in the development of plutonium containing fuel for fast breeder reactors and for fundamental research on actinide elements.

Fundamental research of actinides started immediately after the construction of the hot cells with the reprocessing of transuranium samples irradiated at Idaho Falls. This first reprocessing run furnished the starting material (^{243}Am, ^{244}Cm) for early americium and curium research. Since 1965 an important part of the research programmes of the European Institute for Transuranium Elements has been devoted to the study of these two elements situated in the middle of the actinide series and characterized by the localization of 5f electrons.

N. M. Edelstein et al. (eds.), Americium and Curium Chemistry and Technology, 35–42.

Three periods, in part overlapping, can be distinguished corresponding to the following research areas:

- isolation and purification of americium and curium,
- preparation of americium and curium metals of high purity and
- investigation of bonding related properties (structural, thermodynamic, magnetic, electrical, electronic) of the elements.

Synthesis of well characterized binary compounds accompanied this work, but growth of their single crystals has just begun.

The first period started in 1965, the second one in 1970, the third period in the mid seventies. These three themes will be illustrated by a few examples, and the review will finish with a summary of present work and plans for the future.

ISOLATION AND PURIFICATION OF AMERICIUM AND CURIUM

Two sets of samples were reprocessed from aluminium-clad AmO_2-Al powder cermets irradiated in Idaho Falls and BR II/Mol, Belgium. The reprocessing schemes were based on experience gained in the United States - using ion exchange from highly concentrated lithium - salt solutions. Milligram quantities of highly purified ^{243}Am and ^{244}Cm were made available to the very few groups in Europe equipped at this time for work with α-emitting radioelements. Gram quantities of ^{242}Cm, resulting from a (relatively) short time irradiation of ^{241}Am in BR II were isolated and in part encapsulated to fuel the first European isotope battery [1].

Around 1970 we acquired a 10g-mixture of ^{243}Am and ^{244}Cm. The elements were separated by precipitating the major part of the americium as the AmO_2^+ - carbonate complex and purifying the curium fractions by a series of extraction chromatography (substituted ammonium salts, HDEHP) and ion exchange processes [2] (Fig. 1).

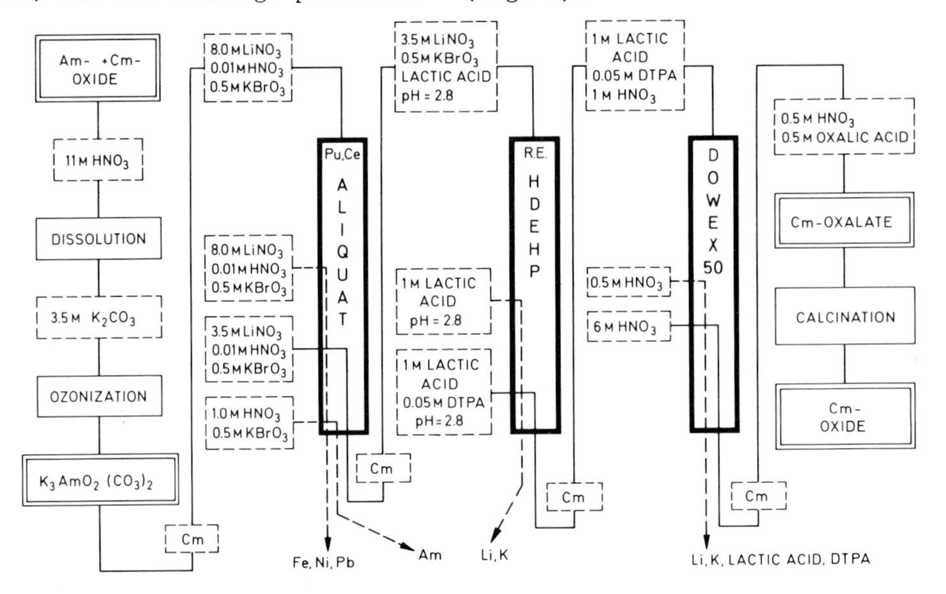

Figure 1. Purification of multigram quantities of ^{244}Cm by extraction chromatography after separation from ^{243}Am by precipitation. (Ref.2)

The resulting high purity fractions of americium and curium, after transformation into oxides, are still today the starting materials for special preparations of metals or intermetallics [3] and for the synthesis of compounds [4].

PREPARATION OF HIGH PURITY AMERICIUM AND CURIUM METALS

The reduction of anhydrous actinide halides by lithium or barium vapour very often results in metals containing non-volatile impurities and oxygen. Therefore we decided from the very beginning of our metal preparation attempts to apply methods involving volatilization of the actinide elements. Our preferred method was the reduction of prepurified oxides by electropositive, non-volatile metals followed by the distillation of the actinide metals from the reaction mixture. The principle of these methods had been developed in the case of Am metal preparation at Los Alamos [5,6], and in the case of the Cm homologue gadolinium at the Rare Earth Laboratory at Meudon/France [7]. For the preparation and manipulation of the high purity metals, a special, double glove box concept was developed [8] (Fig. 2).

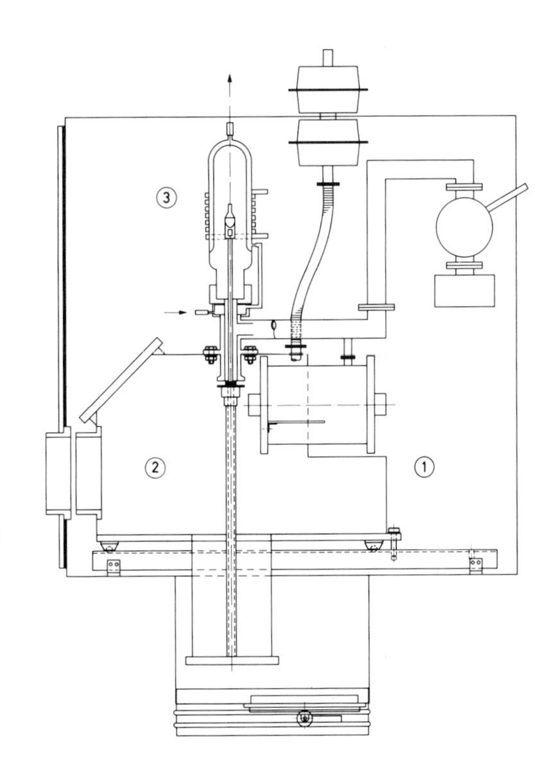

Figure 2. Scheme of the double-wall glove box for the preparation of actinide metals: (1) nitrogen box with vacuum equipment and induction furnace, (2) argon box; for changing of gloves, the argon box can be moved on rails, (3) water-cooled quartz tube. (Ref.8)

The possibility of actinide metal preparation by thermal dissociation of actinide-noble metal intermetallics was demonstrated [9], but most of the metal samples were obtained by lanthanum reduction of americium oxide [8] or thorium reduction of curium oxide [10].

INVESTIGATION OF THE PROPERTIES OF AMERICIUM AND CURIUM

Several properties of americium and curium were determined or redetermined with highly purified metals.

Americium and curium metal of known isotopic compositions and of high chemical purity were used as "spike" material for isotope dilution analysis. The half-life (and, hence, the decay constant) of ^{241}Am were redetermined by measuring the specific heat output of a metal sample [11], which had been prepared by repeated sublimation and had been fully analysed. The newly determined value of the half-life is 432 ± 0.2 years.

The electrical resistivities of both the metals [12,13] were determined between 300 and 6K. A potentiometric method was employed to measure the electrical resistivity of an americium layer of 1.89 μm thickness. Up to 20K, the resistivity varies with temperature as $T^{2.8}$, at higher temperatures deviations from the linear law were found similar to those in uranium and neptunium (Fig. 3). The resistivity of curium metal shows a sudden decrease of the slope around 60K (which is related to the magnetic transition around 52K), and at higher temperatures a behaviour similar to that of a lanthanide metal is seen.

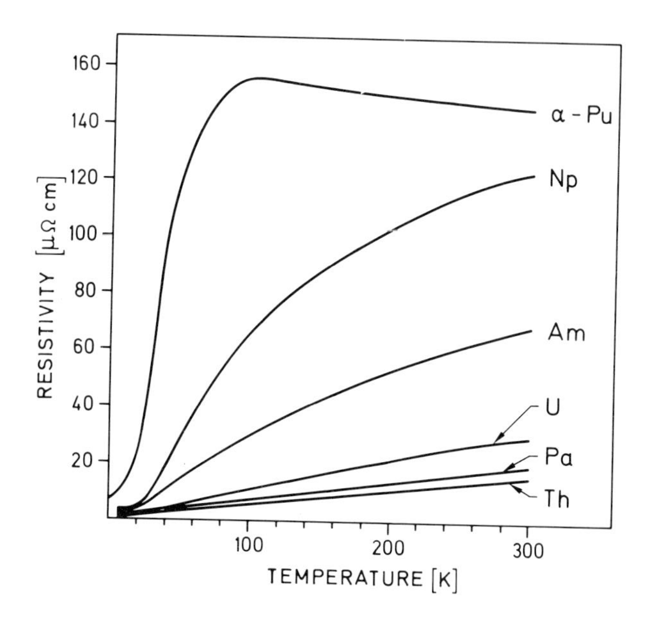

Figure 3. Electrical resistivity of actinide metals. (Ref.12)

The standard free energy of chloride formation was determined by potential measurements in molten salts [14,15]. The enthalpies of formation of Am^{3+} and of Cm^{3+} ions were determined by measuring the enthalpy of dissolution of the metals in diluted hydrochloric acid [16,17]. The heats of solution of the dhcp allotropes at room temperature in 1 molar HCl were obtained as -616.1 ± 0.8 kJ/mol^{-1} for americium, and -614.5 ± 4.5 kJ/mol for curium, respectively. From these values, the standard enthalpies of formation of the trivalent ions are deduced as -616.7 ± 1.2 kJ/mol (Am) and -615 ± 5 kJ/mol (Cm). New values for the heats of formation of various americium ions and compounds could be derived. The enthalpy of sublimation was determined by a very careful redetermination of the vapour pressures of the metals as a function of temperature [18,19] (Fig.4). The differences between the values of the two elements can be explained by taking into account the specific magnetic character of condensed americium and the transition in the gaseous state to atoms with two valence electrons, whereas Cm behaves as a trivalent, lanthanide-like metal.

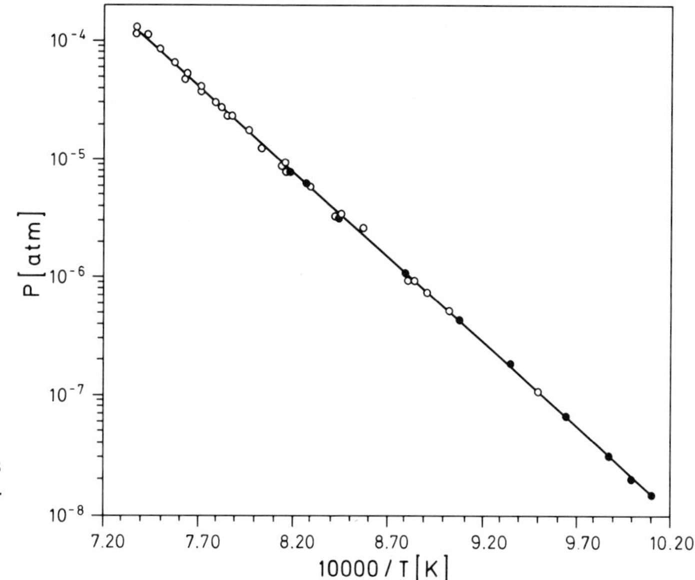

Figure 4. Vapour pressure of solid americium. Full circles and open circles correspond to different measurements series. (Ref.18)

The specific heat of americium metal has been determined [20] and the list of the thermodynamic properties of the light actinide elements is completed[21]. During a comparison of the magnetic properties of americium and curium metals, antiferromagnetic ordering of curium metal was observed [22]. The magnetic susceptibility of americium metal is almost temperature independent between room and liquid helium temperatures. Curium metal shows Curie-Weiss-paramagnetism with antiferromagnetic ordering around 52K. Its effective magnetic moment supports the hypothesis of localized 5f electrons and a metal valence of 3.

The influence of temperature [23] and of pressure [24,25,26] on the crystal structure of americium and curium metals have been investigated. The present state of the high pressure studies will be described in a later paper by U. BENEDICT [27] during this conference. Recently, the localization of 5f electrons in americium, expected and predicted by theory for a long time, could be demonstrated by X-ray and high resolution UV photoemission spectroscopy on americium metal [28]. The comparison with plutonium metal proved that the transition from delocalized to localized 5f behaviour takes place between Pu and Am (Fig. 5).

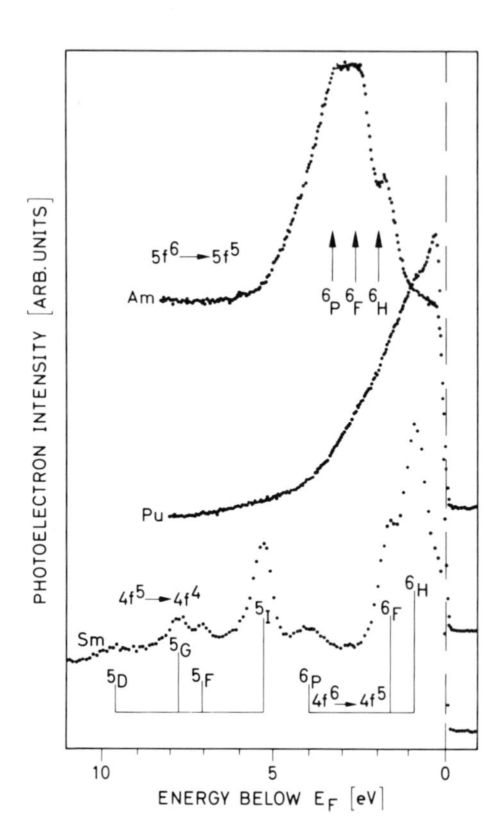

Figure 5. Conduction-band spectra of Am, Pu, and Sm metal for 40.8-eV excitation. (Ref.28)

EXAMPLES OF PRESENT DAY AMERICIUM AND CURIUM RESEARCH

- The americium and curium build-up in fuel of different reactors (LWR and FBR) has been determined in order to furnish data for the assessment of minor actinides arising in burnt nuclear fuel or fuel waste[29]. If these activities are discharged to waste they constitute a long-term radiological risk. Their recycle (especially that of ^{241}Am as the parent of the hazardous actinide nuclide ^{237}Np) will not only reduce that risk, but contribute to nuclear energy generation in a fast reactor. Therefore the properties of mixed uranium-minor actinide (MA = Am, Cm, Np) oxides are being studied from the point of view of their irradiation behaviour. In particular thermal conductivity, phase diagram, Gibb's free energy, fission product composition, and the reaction between minor actinide mixed oxides and Na [29] have been measured.
- In cooperation with a commercial fuel fabrication company kg-amounts of ^{241}Am are being purified by repeated oxalate precipitation, followed by calcination.
- Americium is sorbed from waste effluents by inorganic ion exchangers containing zirconium [30].

- ^{241}Am Mössbauer sources of high resolution are prepared by deposition of Am vapour [31].

FUTURE WORK

Future work will include the continuation and extension of high pressure techniques for both synthesis and investigation of americium and curium containing systems. The growth of single crystals that are sufficiently large for physical measurements of binary americium and curium compound systems will be attempted. The preparation and study of actinide-intermetallics or (ternary) compounds are planned.

ACKNOWLEDGEMENTS

The author thanks his numerous colleagues for their contribution to the development of our knowledge on americium and curium. Most of the results presented in this review have been obtained by (in general international) cooperation between different research institutes and institutions.

REFERENCES

1. Müller, W., Atompraxis 1969, 15, 1.
2. Buijs, K., Maino, F., Müller, W., Reul, J., Toussaint, J.Cl., J. Inorg. Chem. Supplement 1976, 209.
3. Buijs, K., Toussaint, J.Cl., Bottini, G., Intern. J. Appl. Rad. Isot. 1975, 26, 564.
4. Damien, D., Charvillat, J.P., Müller, W., Inorg. Nucl. Chem. Lett. 1975, 11, 451.
5. Johnson, K.W.R., Leary, J.A., 1964, LA-2992.
6. Wade, W.Z., Wolf, T., J. Inorg. Nucl. Chem. 1967, 29, 2577.
7. Schiffmacher, G., Trombe, F., C.R. Acad. Sci. Paris 1969, 268, 159.
8. Müller W., Fuger, J., Spirlet, J.C., J. Inorg. Nucl. Chem. Supplement 1976, 139.
9. Müller, W., Reul, J., Spirlet, J.C., Atomwirtschaft 1972, XVII,415.
10. Müller, W., Reul, J., Spirlet, J.C., Revue de Chimie Minérale 1977, 14, 212.
11. Ramthun, H., Müller, W., Intern. J. Appl. Rad. Isot. 1975, 26, 589.
12. Schenkel, R., Schmidt, H.E., Spirlet, J.C., Transplutonium Elements 1976, Müller W. and Lindner R., editors, North-Holland Publishing Company, p. 149.
13. Schenkel, R., Ph.D. Thesis, Saarbrücken, 1977.
14. Martinot, L., Spirlet, J.C., Duyckaerts, G., Müller, W., Bull. Soc. Chim. Belg. 1974, 83, 449.
15. Martinot, L., Reul, J., Duyckaerts, G., Müller, W., Bull. Soc. Chim. Belg. 1975, 84, 657.
16. Fuger, J., Spirlet, J.C., Müller, W., Inorg. Nucl. Chem. Lett. 1972, 8, 709.
17. Fuger, J., Reul, J., Müller, W., Inorg. Nucl. Chem. Lett. 1975, 11, 265.

18. Ward, J.W., Müller, W., Kramer, G.F., Transplutonium Elements 1976, Müller W., Lindner R., editors, North-Holland Publishing Company, p. 161.
19. Ward, J.W., Ohse, R.W., Reul, J., J. Chem. Phys. 1975, 62, 2366.
20. Müller, W., Schenkel, R., Schmidt, H.E., Spirlet, J.C., McElroy, D.L., Hall, R.O.A., Mortimer, M.J., J. Low Temp. Phys. 1978, 30, 561.
21. Hall, R.O.A., Lee, J.A., Mortimer, M.J., McElroy, D.L., Müller, W., J. Low Temp. Phys. 1980, 41, 397.
22. Kanellakopulos, B., Blaise, A., Fournier, J.M., Müller, W., Solid State Comm. 1975, 17, 713.
23. Sari, C., Müller, W., Benedict, U., J. Nucl. Mat. 1972/73, 45, 73.
24. Benedict, U., J. Less-Common Metals 1984, 100, 153.
25. Haire, R.G., Benedict, U., Peterson, J.R., Dufour, C. Itié, J.P., J. Less-Common Metals, in press.
26. Benedict, U., Haire, R.G., Peterson, J.R., Itié, J.P., J. Phys. F: Metal Phys. 1985, 15, L29.
27. Benedict, U., Symp. 40th anniversary of the discovery of Am and Cm, Honolulu, HI, Dec. 19, 1984.
28. Naegele, J.R., Manes, L., Spirlet, J.C., Müller, W., Phys. Rev. Lett. 1984, 52, 1834.
29. Koch L., Bartscher, W., Coquerelle, M., De Meester, R., Richter, K., Sari, C., Schmidt, H.E., to be published, Jahrestagung Kerntechnik 85, May 7-9 1985, München.
30. Crispino, E., Gerontopoulos, P., Arcangeli, G., Cao, S., Forno, M., Müller, W., Radiochim. Acta 1984, 36, 69.
31. Spirlet, J.C., pers. communication, 1984.

FRONTIERS OF CHEMISTRY FOR AMERICIUM AND CURIUM*

O. L. Keller, Jr.
Oak Ridge National Laboratory
P. O. Box X
Oak Ridge, TN 37831

ABSTRACT. The discoveries of americium and curium were made only after Seaborg had formulated his actinide concept in order to design the chemistry needed to separate them from irradiated ^{239}Pu targets. Their discoveries thus furnished the first clear-cut evidence that the series exists and justified Seaborg's bold assumption that even though Th and Pa appeared to presage a following 6d series, the pattern established by the periodic table after Cs and Ba would be repeated exactly after Fr and Ra. That is to say, a new 5f element rare earth series (the actinides) would follow Ac in the same way the 4f rare earth series (the lanthanides) follows La. The consequences of the resulting half-filled $5f^7$ shell at Cm were originally presented by Seaborg as a test of his hypothesis. Recent research is outlined that substantiates Seaborg's predictions in new and definitive ways.

Americium and curium are of historic importance in chemistry because the method used in their discovery by Seaborg, Ghiorso and coworkers furnished the first clear-cut evidence that the actinide series exists. And, as is evident in this Symposium, americium and curium are still furnishing insights into the systematics and the underlying electronic phenomena implied by the formulation of the periodic table on the basis of the actinide concept.

As everyone knows, scientists often assume the validity of some pattern when they set up their own experiments. In his first attempts to discover Am and Cm, Seaborg (1) made the reasonable assumption that the pattern of chemical properties set up by U, Np, and Pu would persist to elements 95 and 96. The nuclear reactions employed in these new element discovery experiments were the bombardment of plutonium-239 with alpha-particles to produce ^{242}Cm and with neutrons to produce ^{241}Pu which would then decay to ^{241}Am. The chemical separations of Am or Cm from the irradiated ^{239}Pu targets were then based on the idea that they

*Research sponsored by the Division of Chemical Sciences, Office of Basic Energy Sciences, U. S. Department of Energy, under contract DE-AC05-84OR21400 with Martin Marietta Energy Systems, Inc.

N. M. Edelstein et al. (eds.), Americium and Curium Chemistry and Technology, 43–49.

would be fairly readily oxidizable to the VI state to ions analogous to PuO_2^{2+}. The experiments based on this assumption did not bear fruit, and it became apparent that some new tack had to be taken.

There were also, of course, various assumptions about the untried nuclear reactions that were used in these unsuccessful attempts to produce Am and Cm from the ^{239}Pu target. But Seaborg assumed correctly that the assumptions about the chemistry were what needed changing.

In his first attempts to predict the chemistry of Am and Cm, Seaborg chose as a guide a very simple pattern - namely, a direct extrapolation from the properties of U, Np, and Pu. When this approach did not work, he then made the very bold assumption that even in this heavy element region it was possible to take the regularity set by the periodic table in the lighter elements quite literally (Figure 1). If the pattern established by the periodic table after Cs and Ba were repeated in exactly the same way after Fr and Ra, then a new 5f element rare earth series (the actinides) would follow Ac in the same way the 4f rare earth series (the lanthanides) follows La. If such a 5f rare earth series were actually following Ac - but the apparent analogy with the lanthanides was not really going to be apparent until Am and Cm - then these new undiscovered elements should have chemical properties similar to Eu and Gd rather than U, Np, and Pu. Even though we have had the opportunity to look in the back of the book and see the answer, the assumptions that led Seaborg to his new formulation of the periodic table look bold even today, and it is rare indeed that anyone has such penetrating insights on the basis of only the scanty evidence that was available at the time.

In making the assumption that the straightforward regularity of the periodic table would become apparent once again at Am and Cm, Seaborg had very little guidance from either experiment or theory (2). Th and Pa, the first two elements after Ac, behave chemically very much like Hf and Ta, the first two members of the 5d transition series. Thus it was natural to assume that Th and Pa were the first two members of the 6d transition series. On the other hand, work in the wartime Manhattan Project was helpful in showing that something new was going on in the U, Np, and Pu region because Np and Pu bear no resemblance to Re and Os, and the chemistry of U differs in important ways from the chemistry of W. Also, although theory did not offer any concrete evidence either, several computations made in the 30's and early 40's did suggest that 5f electrons could be stabilized versus 6d in the U/Pu region (2,3).

In spite of the extreme skepticism expressed by his colleagues about the possible validity of the proposed actinide concept, Seaborg went ahead and designed a new chemical scheme on the basis of it for separating Am and Cm from the irradiated ^{239}Pu targets. His tenacity paid off because he, Ghiorso, and their colleagues quickly met with success. The initial evidence for the validity of the actinide concept was thus established by nature of the chemistry employed in the new element discovery experiments for Am and Cm.

In his 1945 article (4) describing the discovery of Am and Cm and his formulation of the actinide series, Seaborg laid special stress on the testing of his hypothesis through the extra stability that would be imparted to the III oxidation states of elements 96 and 103 by the half filled $5f^7$ and filled $5f^{14}$ shells that should occur in them if actinium really is the first member of a postulated new 5f rare earth series. He mentioned several times in the article, in fact, that Cm should exhibit the III state almost exclusively because of its $5f^7$ half filled shell.

Experimental determination of the extra chemical stability of the III oxidation state of curium relative to americium and berkelium furnished important early support for the actinide concept. Of parallel interest, but much later, it was shown also by R. J. Silva and coworkers (5) that Lr has a most stable oxidation state of III as predicted by Seaborg for the last member of the actinide series. This experiment on Lr (the only chemical one that has ever been performed on this element) was particularly important since shortly before it was done the surprising discovery had been made by Maly, Sikkeland, Silva, and Ghiorso that No, the second to the last member of the series, is most stable in the divalent state (6).

Although no additional chemical experiments have been possible yet with Lr, several types of experiments have demonstrated the extra stability of the trivalent state in Cm that Seaborg predicted. A good example is the heat of sublimation of Cm over that of Am and Bk. This effect is shown from data of Kleinschmidt, Ward, and Haire (7) in Figure 2. Their value of the heat of sublimation for Lr was predicted on the basis of its trivalency, which has already been established for the trivalent ion in aqueous solution as noted above. The low heats of sublimation of Fm, Md, and No were predicted on the basis of their expected divalency. Actually, the divalency of Fm and Md has been demonstrated experimentally already by the gas chromatography results of Hubener and Zvara (8) and there can be little doubt that No is divalent. So the heat of sublimation of Lr will be a most important quantity to measure. It may actually be much lower than predicted by Kleinschmidt and coworkers because, although Lr is trivalent in aqueous solution, the electronic configuration of the metal may be different from what would be predicted on the basis of analogy with Lu. Strong relativistic effects in the Lr region will definitely stabilize the $7s^2$ closed shell more than the $6s^2$ closed shell in Lu, and they also could cause the replacement of the expected 6d electron, which is very interactive, with a $7p_{1/2}$ electron, which is relatively inert. These two effects could work in conjunction to lower the heat of sublimation of Lr substantially (9).

Fortunately, it has become possible recently to experimentally demonstrate in a direct manner the extra stability of the highly localized $5f^7$ shell in curium. In essence, the approach is to test the stability of the f shells in transplutonium metals by placing them under a high enough pressure that they delocalize into the spd bands. That is to say that the pressure is high enough to cause the f electrons to change from a localized non-bonding character into a delocalized bonding character. A recent review has been given by Benedict (10).

All four of the metals - Am, Cm, Bk, and Cf - now have been studied (11,12,13). An illustration using results from a paper by Roof, Haire and coworkers (13) is given in Figure 3. In this experiment, Am metal

PERIODIC TABLE SHOWING HEAVY ELEMENTS AS MEMBERS OF AN ACTINIDE SERIES

Arrangement by Glenn T. Seaborg

1945

1 H 1.008																1 H 1.008	2 He 4.003
3 Li 6.940	4 Be 9.02											5 B 10.82	6 C 12.010	7 N 14.008	8 O 16.000	9 F 19.00	10 Ne 20.183
11 Na 22.997	12 Mg 24.32	13 Al 26.97										13 Al 26.97	14 Si 28.06	15 P 30.98	16 S 32.06	17 Cl 35.457	18 A 39.944
19 K 39.096	20 Ca 40.08	21 Sc 45.10	22 Ti 47.90	23 V 50.95	24 Cr 52.01	25 Mn 54.93	26 Fe 55.85	27 Co 58.94	28 Ni 58.69	29 Cu 63.57	30 Zn 65.38	31 Ga 69.72	32 Ge 72.60	33 As 74.91	34 Se 78.96	35 Br 79.916	36 Kr 83.7
37 Rb 85.48	38 Sr 87.63	39 Y 88.92	40 Zr 91.22	41 Cb 92.91	42 Mo 95.95	43	44 Ru 101.7	45 Rh 102.91	46 Pd 106.7	47 Ag 107.880	48 Cd 112.41	49 In 114.76	50 Sn 118.70	51 Sb 121.76	52 Te 127.61	53 I 126.92	54 Xe 131.3
55 Cs 132.91	56 Ba 137.36	57 La 138.92 / 58-71 see La series	72 Hf 178.6	73 Ta 180.88	74 W 183.92	75 Re 186.31	76 Os 190.2	77 Ir 193.1	78 Pt 195.23	79 Au 197.2	80 Hg 200.61	81 Tl 204.39	82 Pb 207.21	83 Bi 209.00	84 Po	85	86 Rn 222
87	88 Ra	89 Ac / see Ac series	90 Th	91 Pa	92 U	93 Np	94 Pu	95	96								

LANTHANIDE SERIES	57 La 138.92	58 Ce 140.13	59 Pr 140.92	60 Nd 144.27	61	62 Sm 150.43	63 Eu 152.0	64 Gd 156.9	65 Tb 159.2	66 Dy 162.46	67 Ho 163.5	68 Er 167.2	69 Tm 169.4	70 Yb 173.04	71 Lu 174.99
ACTINIDE SERIES	89 Ac	90 Th 232.12	91 Pa 231	92 U 238.07	93 Np 237	94 Pu	95	96							

Figure 1. Reprinted with permission from Ref. 4) Copyright 1945, American Chemistry Society.

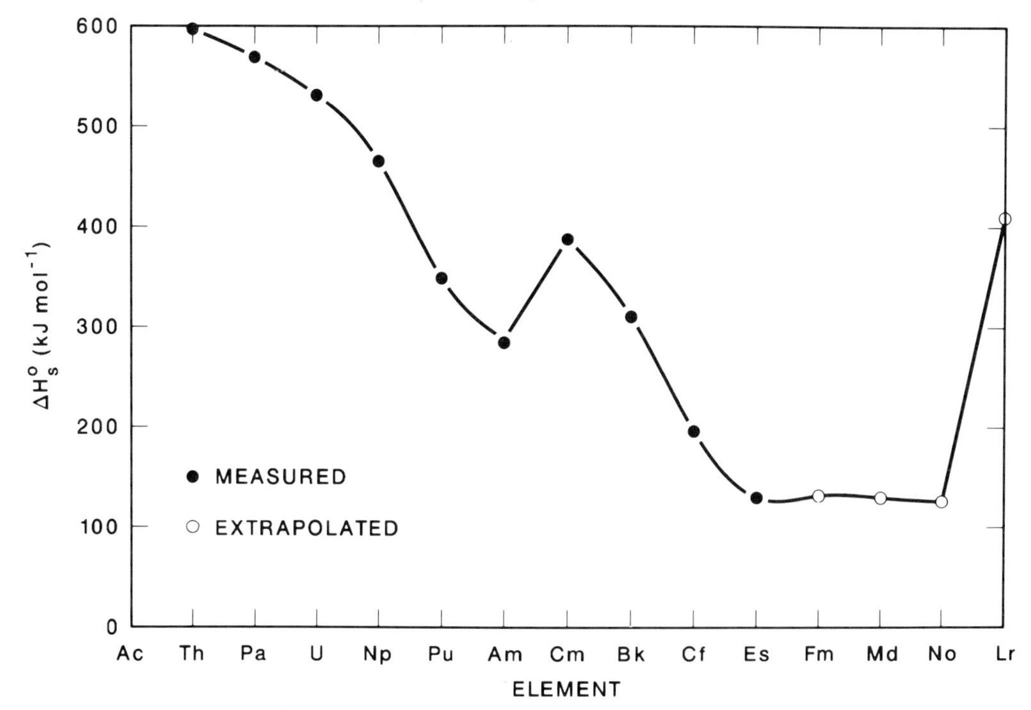

Figure 2. Heats of Sublimation of Actinide Metals (Data from Ref. 7).

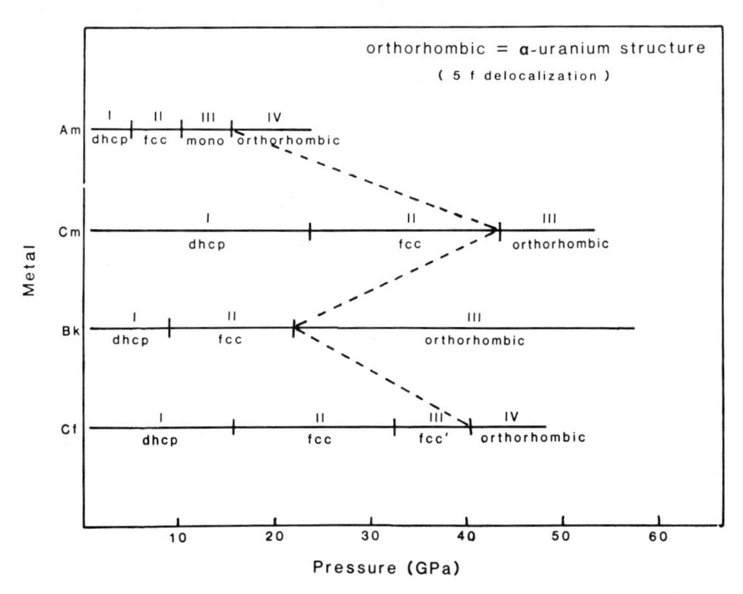

Figure 3. Phase Changes and Delocalization of f orbitals in Am metal under pressure (Ref. 13).

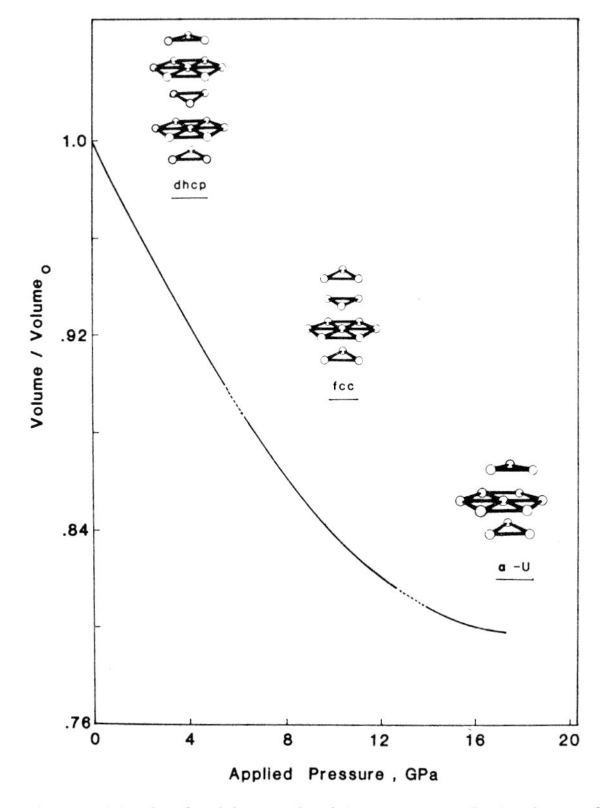

Figure 4. Half-filled shell stabilization of Cm metal shown by anamalously high pressure for f orbital delocalization (Ref. 14).

ultimately assumes the alpha uranium structure which is associated with fspd hybridization.

Thus the high pressure technique gives an ideal way to determine f orbital stability because the more stable the orbitals, the higher the pressure required to bring about their delocalization. The results for the four metals Am, Cm, Bk, and Cf are given in Figure 4 (14). Since the pressure required to delocalize the f orbitals in Cm is much higher than in Am or Bk, its extra stability is clearly shown. Thus the stability of the f shell in Cm is demonstrated directly and its position as the middle member of the actinide series is verified in a most definitive way.

Another verification of the actinide concept comes from the discovery by J. L. Smith and R. G. Haire (15) of superconductivity in Am metal. The superconductivity is allowed in Am metal because it has a nonmagnetic ground state like La, which is also a superconductor. But the important point for adding to our understanding of the systematics of the actinide series is that Am metal is trivalent and nonmagnetic because it has a unique f^6 configuration in the ground state. Even europium, americium's homologue in the lanthanide series, does not exhibit the f^6 configuration in the metal (as shown by its divalency) and is not superconducting. Smith and Haire point out also that the f electron character in Am and La must be important in allowing superconductivity since other trivalent elements that do not have available f orbitals are not superconducting. Studies of Am superconductivity under high pressures should be most enlightening in elucidating the role of f orbitals in this phenomenon, and hopefully we will be treated to the results of such experiments in the near future.

The discovery of Am and Cm was especially useful for developing an understanding of the underlying 5f electronic properties of the series because they introduced a simplicity that was not offered by U, Np, and Pu. But suppose, for example, that Am had turned out to be divalent like No! Then it would have been even more difficult to discover Am although, fortunately, Cm would still have saved the day. Actually, even with the simplicity introduced by the discovery of Am and Cm, there was still enough confusion that many scientists provisionally classified the elements Th through Cm as both sub-group elements and as a second rare earth series at the same time. One of the scientists who did that was Seaborg himself (Figure 1). But within a few years it became obvious that Seaborg was correct in placing them as a second rare earth series and the sub-group classification disappeared.

Certainly a key result of the efforts of transuranium chemists over the last 40 years has been the validation and elucidation of the actinide concept. As we move ahead into further investigations of the architecture of the periodic table through chemical studies of Lr, the first members of the transactinide series and beyond, we are fortunate to have as a guide the valuable lessons from the work of Seaborg and his coworkers in those exciting early days of the discovery of Am and Cm.

ACKNOWLEDGMENT. I thank Drs. R. G. Haire, J. K. Gibson, and U. Benedict for helpful discussions during the preparation of this article.

REFERENCES

1. Seaborg, G. T. Man-Made Transuranium Elements, Prentice-Hall, Inc., Englewood Cliffs, NY (1963).
2. van Spronsen, J. S. The Periodic System of Chemical Elements, Elsevier, Amsterdam (1969).
3. Seaborg, G. T., Katz, J. J., and Manning, W. M. The Transuranium Elements, Part II, McGraw-Hill Book Company, Inc., New York (1949), pp. 1492-1524.
4. Seaborg, G. T. Chem. Eng. News, 23, 2190 (1945).
5. Silva, R., Sikkeland, T., Nurmia, M., and Ghiorso, A. Inorg. Nucl. Chem. Letters, 6, 733 (1970).
6. Maly, J., Sikkeland, T., Silva, R., and Ghiorso, A. Science, 160, 1114 (1968).
7. Kleinschmidt, P. D., Ward, J. W., and Haire, R. G. Proc. Symp. on High Temp. Materials, Chem II, Vol. 83-7, Z. A. Munir and D. Cubicciotti, eds., The Electrochemical Society, Inc., Pennington, NJ (1983), pp. 23-31.
8. Hubener, S. and Zvara, I. Radiochemica Acta, 31, 89 (1982).
9. Keller, Jr., O. L. Paper presented at the International Conference on Nuclear and Radiochemistry, Lindau, FRG, October 1984. To be published in Radiochemica Acta.
10. Benedict, U. J. Less-Common Metals, 100, 153 (1984).
11. Benedict, U., Peterson, J. R., Haire, R. G., and Dufour, C. J. Phys. F: Met. Phys., 14, L43 (1984).
12. Benedict, U., Haire, R. G., Peterson, J. R., and Itie, J. P. J. Phys. F: Met. Phys. (to be published).
13. Roof, R. B., Haire, R. G., Schiferl, D., Schwalbe, L. A., Kmetko, E. A., and Smith, J. L. Science, 207, 1353 (1980).
14. Haire, R. G., Benedict, U., Peterson, J. R., Dufour, C., and Itie, J. P. J. Less-Common Metals (in press, 1985).
15. Smith, J. L. and Haire, R. G. Science, 200, 535 (1978).

Part II

Solution Chemistry and Analytical Procedures

STUDIES OF AMERICIUM AND CURIUM SOLUTION CHEMISTRY IN THE USSR

B.F.Myasoedov and N.Yu.Kremliakova
V.I.Vernadsky Institute of Geochemistry and
Analytical Chemistry USSR Academy of Sciences
19 Kosygin street, 11797 Moscow, USSR

ABSTRACT. The historical review of the fundamental achievements on Americium and Curium properties study in the USSR is presented in the paper. The main attention is payed to the properties of these elements in the highest oxidation states. Methods of obtaining, kinetics of redox processes, mechanism of transformation and stability of Am(IV), Am(V), Am(VI), Cm(IV) and Cm(VI) are discussed. The values of redox potentials for a number of reactions are presented. Americium behaviour in the highest oxidation states during extraction, sorption and precipitation, as well as the new methods of this element determination are discussed in conclusion.

The chemical properties of americium and curium in solutions were well studied for forty years since the time of their discovery by G.T.Seaborg, L.O.Morgan, R.A.James and A.Chiorso (1,2). Numerous data concerning the methods of americium and curium preparation and stability in different oxidation states, the kinetics of redox reactions, the hydrolysis and complexation and the composition of forming compounds have been accumulated. On the base of fundamental knowledge of the chemical properties of americium and curium a lot of simple and effective methods of isolation, concentrating and separation with the use of solvent extraction, sorption, precipitation and coprecipitation have been suggested. Highly sensitive and selective methods for the determination of americium and curium in different materials, both technological and natural have been developed. The investigations of the chemical properties of these elements are carried out in many countries, but especially widely and actively in the USA, USSR, France, GFR and England. In the present review we would like to give briefly the main results of

N. M. Edelstein et al. (eds.), Americium and Curium Chemistry and Technology, 53–79.

americium and curium chemical properties investigations, carried out in the USSR. But the authors have not attempted to give an exhaustive analysis of all of the Soviet literature concerning this question. The works of the fifties-seventies are quite fully summarized in reviews and monographs (3-6).

The period of investigation of the chemical properties of americium and curium in our country can be divided conditionally into two stages. The first (till the seventies) mainly dealt with the study of chemical properties of americium and curium in the trivalent oxidation state and their behaviour during the processes of solvent extraction, sorption and coprecipitation. The second stage, starting at the beginning of the seventies, gave its principal attention to the investigation of americium and curium chemical properties in "unusual", chiefly, the highest oxidation states and the utilization of the latter in new separation and isolation methods.

1. CHEMICAL PROPERTIES OF TRIVALENT AMERICIUM AND CURIUM

1.1. Complex Formation

The study of americium and curium complexation with inorganic and organic ligands has been performed in the Soviet Union since 1951. Spectrophotometry, methods of solubility, electromigration, solvent extraction, ion exchange and others have been used for this purpose. The composition of simple and complex americium and curium compounds, isolated from solutions has been also studied by different methods. The complex formation with inorganic ligands, such as perchlorate (7), chloride (7,8), nitrate (9,10), thiocyanate (11), sulfate (9), carbonate (7), phosphate (12,13), as well as trimetaphosphate (14) has been studied. Americium (III) and curium (III) hydroxides are precipitated by ammonium hydroxide and alkali (15) as pink gelatinous compounds. $Am(OH)_3$ solubility in 0.3 M NH_4OH solution is equal to 2.4 mg Am/l at 20°C (16) and that of Cm - 10 mg Cm/l (15). Yakovlev and collaborators (17) prepared americium sulfate $Am_2(SO_4)_3 \cdot 5\ H_2O$, and the double sulfates of americium with potassium, rubidium, cesium and thallium. They studied phase diagrammes, absorption spectra and solubilities of these compounds. The double sulfate of curium with potassium has been described in a paper (15). Its solubility varies from 10 to 20 mg Cm/l depending on acidity. Curium carbonate has been shown to dissolve completely in 40% K_2CO_3 solution (15).

The complexation and the composition of americium and curium ions with organic ligands, important in different chemical analytical procedures, including the isolation and determination of these elements, have been studied in detail. In acetate solutions Am(III) and Cm(III) form complexes of the ML^{2+} and ML_2^+ types. The stability of CmL_2^+ complexes is higher than that of americium (18). Great attention has been given to Am(III) and Cm(III) complexation with oxalate-ions. These complexes have been studied by ion-exchange (9,19,20), solvent extraction (21,22) and electromigration (23). The composition of these complex compounds, their stability and solubility in different media have been determined (16). Lebedev and coworkers (24) determined the composition of americium oxalate $Am_2(C_2O_4)_3 \cdot 9H_2O$. In tartaric acid solutions with pH 2-9 americium and curium form, like rare earths, complexes of the ML^+ and ML_2^- types (21,22,25) and with citrate ions - ML^0, ML_2^{3-} and $M(HL)_3^0$ (23). Stability constants of americium and curium with some α-oxycarboxylic acids, widely used in ion exchange separation methods, such as with lactate (26,27) and α-oxybutyrate-ions (of the ML^{2+}, ML_2^+ and ML_3^0 types) (28-30) have been determined. Trivalent americium and curium form the most stable complexes with polyaminopolyacetic acids. The formation of ML^0 and ML^{2-} complexes with nitriletriacetic acid in the pH region of 2 to 6 has been established. Their stability constants differ little from the analogous rare earths (21, 22,31). Americium and curium form with ethylenediaminetetracetic acid (EDTA) complexes mainly of the ML^- type (20-22,32). These complexes exist in solution in a pH range of 1-3,3. At high acidities (pH<2) they form protonated complexes MHL^0. Americium and curium form complexes with 1,2-diaminocyclohexanetetraacetate - (21,22,33) and diethylenetriaminepentaacetate (DTPA)-ions (34) that are more stable than these with EDTA. The stability constant of americium and curium complexes with DTPA of the MHL-type, determined by the electromigration method, is equal to $2 \cdot 10^{14}$ (34). Complexation of americium and curium with phosphororganic ligands has been studied by a number of workers (35-39).

1.2. Isolation and Separation Methods

Precipitation,solvent extraction, sorption and electrochemical methods all have been developed to separate and isolate trivalent americium and curium.

1.2.1. <u>Precipitation</u>. In early work in order to isolate microquantities of americium and curium coprecipitation with the following precipites has been widely studied: lanthanum fluoride (40), potassium sulfate (41,43), double sulfate of potassium and lanthanum (44-46), oxalates of calcium, uranium (16) and lanthanum (47) as well as some organic coprecipitants (48-50).

1.2.2. Solvent Extraction. Extractants of different classes such as organophosphorous compounds, high-molecular-weight amines, beta-diketones and others, as well as their mixtures (synergistic extraction) have been used for extraction and separation of americium and curium. The application of neutral organophosphorous compounds have been studied most. Trialkylphosphates $(RO)_3PO$, trialkylphosphonates $(RO)_2RPO$, trialkylphosphinates $(RO)R_2PO$ and trialkylphosphine oxides R_3PO belong to this class. According to the data of the papers (51,52) the extractability of Am and Cm decreases in the order $R_3PO > (RO)R_2PO > (RO)_2RPO > (RO)_3PO$. The effect of structure and the nature of substituents at the phosphorous atoms (51,53,54) the concentrations of acids and salting out agents in the aqueous phase (55,58), and the extractant concentration (40) on americium and curium extraction efficiency have been thoroughly studied. Different authors (55,57,58,59) proved that the composition of extracted americium and curium species with any of four kinds of reagents is $Me(NO_3)_3 \cdot 3A$. Tributylphosphate (TBP) and phosphine oxides are often used in practice. More recently bidentate organophosphorous compounds have been intensively studied for americium and curium extraction from high acidity solutions in the presence of large amounts of salting out agents (60,61). As for other organophosphorous compounds, dialkylphosphoric acids, particularly di-2-ethylhexylphosphoric acid (HDEHP), have been widely employed as extractants for americium and curium. Kosyakov and coworkers (62) have shown the efficiency of americium extraction from 0.5 M HNO_3 by HDEHP solutions to be inversely proportional to solvent polarity. The efficiency also increases with the elongation of the normal alkyl radical. For back extraction of americium from the organic phase, they suggested the use of 2-4 M HNO_3 or K_2CO_3 solutions (62). Americium and curium separation from some rare-earth fission elements takes place during extraction from concentrated LiCl solutions (62,63). Among the extractants of the other class, used for TPE isolation, including trivalent Am and Cm, tertiary amines have been the most investigated. It has been shown in the reference (64) that Am and Cm extraction by TOA solutions in cyclohexane is greatly influenced by the nitric acid concentration and the nature of the salting out agent and diluent. Chudinov and coworkers (65) calculate the separation factor of Am and Cm in the system of methyldioctylamine-nitrate in xylol-$LiNO_3$ to be close to 3 and to depend weakly on amine and $LiNO_3$ concentrations. Hydrochloric acid solutions have proven to be useful for Am and Cm separation from rare earths. Thus the effect of chloride-ion, H^+-ions and metal ion concentrations on the separation factor of Cm and Eu have been studied in detail (63). The largest separation factor has been obtained in the system of 20% tridecylamine in xylol - 11.2 M LiCl solution. The diluent nature considerably affects americium and curium isolation, but has practically no effect on the rare earths.

Beta-Diketones have also been used for americium and curium isolation and their following determination. 1-phenyl-3-methyl-4-benzoyl-pyrazolone-5 (PMBP) is the most thoroughly studied example in the works of the soviet authors (66). It extracts a number of elements, including curium (67) from more acid media than thenoyltrifluoroacetone (TTA). The presence of salts greatly influences americium extraction by this reagent (68). The usage of PMBP allows the attainment of a high degree of americium separation from plutonium, based on the extraction of the latter by 0.1 M PMBP solution in benzene from 3 M HNO_3 solution (69). The efficiences of americium and curium extraction by beta-diketones increase strongly in the presence of organic phase donor active additives such as TPB and TOPO (synergistic extraction). The most attention has been paid to americium and curium extraction in the systems: TTA, PMBP-TBP, TOPO (67,70). In PMBP-TBP system the composition of extracted curium compound has been found to be $Cm(PMBP)_3 \cdot 2TBP$ (70). The americium extraction constant in such a system is two fold greater than in TTA-TBP system (71). The diluent nature influences the synergistic effect: americium and curium extraction constants and adducts stability constants decrease with the increasing of diluent polarity (71).

1.2.3. Sorption. For americium and curium isolation and separation practically all known methods have been studied: ion exchange on organic resins, extraction chromatography, partition and thin-layer chromatography, and sorption on inorganic and chelate resins. Cation exchangers of different types have been used for the separation of macroamounts of americium and curium by ammonium α-oxybutyrate solutions (15); for group concentration of TPE in $HClO_4$ solutions (72), for TPE and REE separation in HCl solutions (63) or in the presence of alcohols (73); for americium and curium separation from La and Pu in 1 M HNO_3 solution (10,72). The application of anion exchangers and aqueous-organic solutions as eluents gives new possibilities for the separation of elements with close chemical properties (75-81). Americium and curium separation proves to be the best on anion exchangers with small DVB (~4%) content and in 0.5 M HNO_3 solution in the presence of 80% of CH_3OH. Their concentration is best made on the resins with high DVB content (8%) and from $\geq$0.5 M HNO_3 solutions containing $\geq$90% alcohol (78). By sorption on anion exchangers from 0.5-1 M HNO_3 in 90% alcohol Am and Cm can be separated from other elements, which are poorly sorbed under these conditions: elements with $Z < 56$, as well as Hf, Ta, Re, U, Np(V). The separation from fission rare earths occurs during resin washing with 0.5 M NH_4NCS solution in 80% alcohol (76,79,80), and the separation of the americium and curium remaining on the resin is carried out with 0.5 M HNO_3 in the presence of 85% of alcohol (81). The system anion exchanger-alcohol solution in nitric acid can be successfully used for the isolation of americium, curium and the other TPE in a highly pure state (82,83). Introducing α-hydroxyisobutyric acid in water-alcohol solutions of nitric acid considerably increases the separation factor of Am-Cm pair to ~ 6 (85, 86).

There are some references, summarized in the review (87), concerning the application of inorganic sorbents for americium and curium separation and isolation from solutions of inorganic acids and their salts. In order to improve the selectivity of sorbents, different chelate groups are introduced into them. Myasoedov and Molochnikova (88) have studied the possibility of the application of chelate resins on the base of aminopolystyrene and arsenazo I for americium and curium isolation from HCl, HNO_3 and H_2SO_4 solutions. This sorbent has proven to be very useful for the separation of these elements in inorganic acids solutions and for their separation from europium. The possibility of the separation of americium and curium tracer quantities from uranium and plutonium by thin-layer chromatography on silicagel with TBP has been studied in the paper (89).

1.2.4. Electrochemical Methods. These methods are not widely used for americium and curium separation. They are usually employed for electrochemical isolation of these metals, for making samples for radiometric measurements and for the investigation of their states in aqueous solutions (13,23,30, 32-34,39). In paper (90) the authors have suggested a method for separation of tracer amounts of americium and curium, based on their selective sorption by a platinum anode.

1.3. Methods of Determination

Spectrophotometric, radiometric and emission spectrometry methods have been developed for the detection and quantitative determination of americium and curium in the early papers. The first colored reaction of Am and Cm with the use of arsenazo III has been described in paper (91). The spectrophotometric method based on this reaction allows the determination of these elements down to concentrations of 0.02 mkg/ml (92). Absorption spectra of americium in different oxidation states in acid solutions and the use of characteristic absorption bands for the quantitative determination of these elements have been thoroughly studied in papers (7,16,93,94). This method has been widely used for the investigation of Am(V) disproportionation in perchloric (16,98), nitric (16,95) and sulfuric acids (7,16,95,96), as well as Am(V) self reduction in solutions of perchloric (7,16,94,95), sulfuric (16,95), nitric (16, 95,97) and hydrochloric (7) acids.

In order to determine Am in samples of air, waste waters and other samples in the presence of Pu and U, an α-radiometric method with a preliminary ion exchange isolation of americium has been developed (98). The determination bf ^{243}Am by the γ-rays of the daughter ^{239}Np, is carried out by the extraction of the latter with a mixture of PMBP and TBP (99).

The emission spectrum of americium has been well studied. Some characteristic bands have been suggested for the quantitative determination of this element (100).

2. CHEMICAL PROPERTIES OF AMERICIUM AND CURIUM IN THE HIGHEST OXIDATION STATES

Transplutonium elements are known to be characterized by a larger number of oxidation states than the analogous 4f-elements-lanthanides (Table 1). In aqueous solutions most of TPE are in the trivalent oxidation state in the absence of oxidizing or reducing agents. However Am,Cm, Bk and Cf can exist in the highest oxidations states, while Md can be rather easily reduced to Md^{2+}. The most stable oxidation state for No in aqueous solutions is +2. Recently the existence of Am(VII) (101), Cm(VI) (102), Cf(V) (103) and Cf(IV) (104) and Md(I) (105), under different conditions has been proved. As mentioned earlier, the main emphasis in studies of americium and curium properties in the second period in the USSR has been given to the investigation of their unusual oxidation states and to their use in methods of isolation, separation and determination.

2.1. Oxidation State+4

Both americium and curium can be prepared in the +4 oxidation state. Attempts to obtain americium and curium in the tetravalent state in aqueous solutions have long been unsuccessful due to the very high values of the potentials of the Me(IV)/Me(III) pair, which for these elements are equal to 2.44 and ~3V respectively. In acids or in carbonates of alkaline metals, Am(V) or Am(VI) have been

TABLE I. Oxidation states and standard redox potentials of TPE

Oxidation states	Am	Cm	Bk	Cf	Es	Fm	Md	No	Lr
+7	±								
+6	$\overset{+}{+1.7}$	±							
+5	$\overset{+}{+1.1}$								
+4	$\overset{+}{+2.45}$	$\overset{\pm}{+3.1}$	$\overset{+}{+1.67}$	$\overset{+\,\pm}{+3.1}$	$\overset{\pm}{+4.5}$				
+3	$\overset{+}{-2.6}$	$\overset{+}{-3.8}$	$\overset{+}{-2.8}$	$\overset{+}{-1.7}$	$\overset{+}{-1.2}$	$\overset{+}{-0.65}$	$\overset{+}{-0.15}$	$\overset{+}{+1.45}$	+
+2	±	±	±	±	±	±	+	+	
+1							±		

\+ – the most stable oxidation state in aqueous solutions.
\+ – relatively stable oxidation state.
± – unstable oxidation state.

Redox potentials are in volts (versus normal hydrogen electrode) and belong to neighbouring (in vertical) couple of ions.

obtained by the action of oxidizing agents on Am(III) solutions. And upon dissolving americium dioxide in acids the mixture of Am(III) and Am(VI) is obtained. Only in strong complexing media in which an essential decrease of redox potentials is observed has it been possible to develop reliable methods for americium and curium stabilization in the tetravalent oxidation state.

Asprey and Penneman were the first (106) to obtain Am(IV) in aqueous solutions by dissolving $Am(OH)_4$ in saturated ammonium fluoride solution. Over the last few years a number of convenient practical methods for oxidizing Am(III) to Am(IV) in 8-15 M H_3PO_4 have been developed. Electrochemical and chemical (107,108) (by the mixture of Ag_3PO_4 and $(NH_4)_2S_2O_8$ (108-112) methods have been used. Am(IV) stability in phosphoric acid solutions has been studied in detail (109,110,112,113). In 8-15 M H_3PO_4 Am(IV) reduces to Am(III) and in 2-7 M H_3PO_4 Am(VI) forms together with Am(III). Besides the characteristic bands at 742 and 920 nm in the absorption spectrum of Am(IV) in H_3PO_4 solution there is an intense band in the UV area ($\mathcal{E}_{333}$=1942, $\mathcal{E}_{370}$=1085±20 $m^{-1}cm^{-1}$). This band is used for spectrophotometric determination of americium in solutions containing curium (108).

It has recently been proven (114), that Am(IV) can be obtained in acetonitrilic solutions, containing 0.3-2,0 M H_3PO_4. With +1.4V anode potential with respect to mercury-sulfate electrode, americium can be completely oxidized to Am(IV) in 30 minutes. Oxidation occurs both in pure acetonitrilic solutions and in aqueous-acetonitrilic with water content up to 50%. On standing Am(IV) gradually reduces to Am(III) according to the first order reaction with a half-life of 5.6 hours. Am(IV) disproportionation is not observed under these conditions. Values of the formal redox potential of the Am(IV)/Am(III) couple in 0.3-2.0 M H_3PO_4 solutions in acetonitrile (1.3-1.1V with respect to the mercury-sulfate electrode) and in solutions containing 0.5 M H_3PO_4 up to 20% of water have been measured. The possibility of quantitative oxidation of americium to Am(IV) in H_3PO_4 solutions in acetonitrile has been used for the coulometric determination of americium in microgram amounts by Am(IV)/Am(III) couple with an accuracy of 1-2%.

Americium (III) oxidation to Am(IV) can also be carried out in solutions of unsaturated heteropolytungstates, $K_7PW_{11}O_{39}$ or $K_{10}P_2W_{17}O_{16}$. Significant changing of the redox potential of the Me(IV)/Me(III) couple occurs due to the strong complex formation of tetravalent actinides ions with unsaturated phosphotungsten acids. The maximum changes for Am,Bk,Ce and plutonium take place at pH 4-5 and are equal to 0.9-1.0V relative to E_0 for the corresponding couple in 1 M $HClO_4$ (115,116). Oxidation of

Am(III) to Am(IV) in solutions of unsaturated heteropolytungstates is carried out by persulfate ions with heat (115, 117-121), or electrochemically (115,120). Kylyako et al.(112) have studied in detail the kinetics of electrochemical oxidation of americium. Quantitative oxidation of americium takes place at the anode potential of +1.77V. Under these conditions Am(IV) is stable in the presence of oxidizing agents for several days (117,119); and without oxidants it is reduced at a slow rate (about 1% per hour) (115,120) by radiolysis products and by water. It also has been proved that americium (III) can be oxidized to Am(IV) without heating by a mixture of $AgNO_3$ with $(NH_4)_2S_2O_8$ in 0.1-6 M HNO_3 solutions containing a four-fold excess of $K_{10}P_2W_{17}O_{61}$ with respect to Am (123). In such solutions at HNO_3 concentration $\leqslant 3$ M, Am(IV) is stable for 5-7 days. Obtaining stable Am(IV) in phosphotungstate solutions undoubtedly provides an opportunity to use such a system for analytical purposes.

In dilute solutions of $Am(ClO_4)_3$ containing N_2O, $S_2O_8^{2-}$ or XeO_3, exposed to γ-rays or pulses of electrons, a non-stable Am(IV) appears, which transforms very quickly ($\sim 10^{-3}$ sec) to Am(V) and Am(VI) (124-126). As has been mentioned above, tetravalent americium is stable in solutions of potassium phosphotungstate for several days (123). This stability has allowed recently the establishment for the first time of the conditions for tetravalent americium extraction (127). Am(IV) is quantitatively extracted from 0.5-1 N H_2SO_4 solutions by 3% dioctylamine (DOA) solution in dichlorethane (DCE) in the presence of heteropolyanions, which on passing to the organic phase, stabilize tetravalent americium (Fig.1).

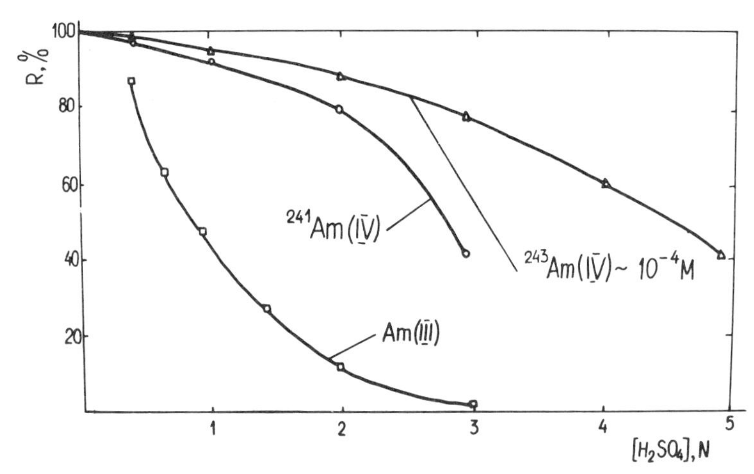

Fig.1 Americium extraction by dioctylamine from H_2SO_4 solutions in the presence of 10^{-3} M $K_{10}P_2W_{17}O_{61}$. 1-10 M Am(IV); 2- ^{241}Am(IV); 3-10 M Am(III).

TABLE II. Selective coulometric determination of americium

Couple	Composition of the solutions	Oxidation time, Hrs	Detection limits,	Relative standard deviation
Am(VI)/Am(V)	2M H_2PO_4+0.1M $HClO_4$	1.5	5	0.015
Am(IV)/(III)	0.5M H_3PO_4 in CH_3CN	0.5	5	0.02
Am(IV)/Am(III)	0.1M $HClO_4$+$6 \cdot 10^{-3}$ PW	0.25	6-70	0.03
Am(IV)/Am(III)	1-2M Na_2CO_3 (pH=10)	0.5	5	0.025

The presence of Am(IV) in the organic phase has been proved spectrophotometrically (Fig.2).

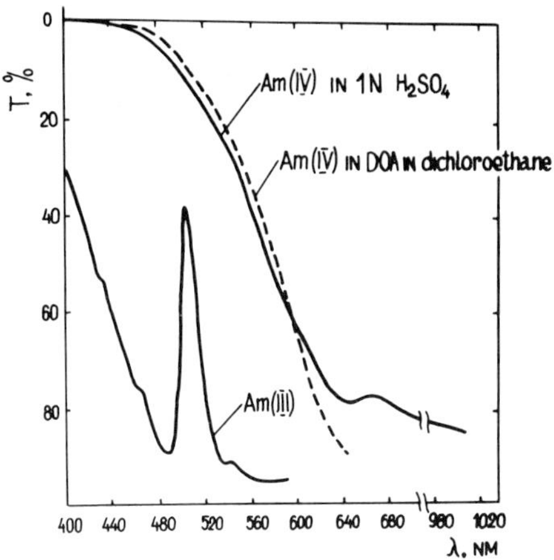

Figure 2 Absorption spectra of 10^{-3} M Am(III) and Am(IV). 1-Am(III); 2-Am(IV) in 1 N H_2SO_4 in the presence of 10^{-2} M $K_{10}P_2W_{17}O_{61}$; 3-Am(IV) in inorganic extract (DOA in DCE).

The period of half-reduction of Am(IV) in the organic phase is equal to 2.5 hours. The existence of reversible electrochemical couples for Am(VI)/Am(V) and Am(IV)/Am(III) has permitted the development of coulometric methods for americium determination with a high sensitivity and accuracy (Table 2). The possibility of electrochemical oxidation of Am(III) up to Am(IV) and Am(VI), and a high stability of the higher oxidation states of americium under certain conditions underlie these methods. Recently the stable Am(IV) has been shown to exist not only in 10-15 M phosphoric acid solutions but to be formed during electrochemical oxidation of Am(III) in 0.3-2.0 M H_3PO_4 solutions in acetonitrile. The available equipment allows the determination of 5-10 μg of americium with an error of 1-2%.

It is possible to determine 5 to 100 μg of americium by the Am(IV)/Am(III) couple in 0.1 M $HClO_4$ in the presence of 6.10^{-3} M potassium phosphotungstate (128). In this solution Am(III) is completely oxidized to Am(IV) in 15 minutes under a potential of 1.7V. Am(IV) is very stable in the presence of phosphotungstate-ions and so the 10-fold excess does not prevent americium determination. Many elements oxidizable under these conditions (Pu,Ce and others) also do not prevent this determination, since they are not reduced at the potential of 1.17V. The measurement error does not exceed 3% even in the presence of 100-fold quantities of other metals.

Coulometric determination of americium by the

Am(IV)/Am(III) couple can be carried out in an acetonitrile solution of phosphoric acid where Am(III) has been established to be oxidized rapidly and reversibly electrochemically to Am(IV) (114). Microgramm amounts of americium can be determined with a measurement error of 1-2% in 0.5 M H_3PO_4 solutions in acetonitrile.

Curium in the trivalent oxidation state is very stable in aqueous solutions and in most cases it is not oxidized under the conditions that Am(III) oxidation occurs. This can be explained by the high values assumed by the oxidation potentials of the Cm(IV)/Cm(III) couple, which are about 0.6-0.7V higher than that of Am(IV)/Am(III) one. Tetravalent Cm was originally obtained by dissolving CmF_4 in a concentrated solution of cesium and ammonium fluoride (129). However, curium (III) can be oxidized to Cm(IV) by potassium persulfate in $K_{10}P_2W_{17}O_{61}$ solution on heating (115,117), though the stability of Cm(IV) under such conditions is not high. It completely reduces in 1.5-3 hours. The solution absorption spectrum of Cm(IV) and the data on its reduction are presented on Fig.3. The Cm(IV) absorption spectrum in solution is characterized by a broad-band with maxima at 520,748 and 932 nm (115).

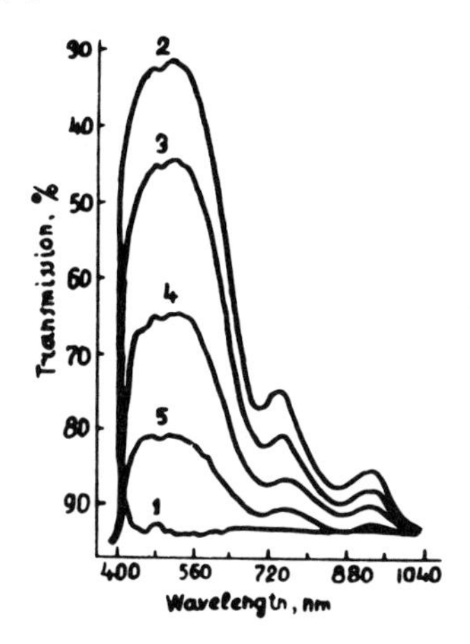

Figure 3. Absorption spectra of Cm(III) (1) and Cm(IV) (2-5) in solution of potassium phosphotungstate. C =4.06x10^{-4} M, 25°C. Time after oxidation, 2-0.5 min; 3-60 min; 4-110 min; 5-140 min.

2.2. Pentavalent Americium

Several methods of Am(V) preparation in solution are known. Recently convenient methods have been suggested of electrochemically obtaining Am(V) by Am(VI) reduction in 1-10 M NaOH at an anode potential of 0.2V (130), in 1 M K_2CO_3 at potential of 0.54V (131) and in dilute

$HClO_4+H_3PO_4$ solution (132) at potential 1.27V. Am(VI) is quantitatively reduced to americium (V) in 30 minutes. Am(V) is stable in dilute H_3PO_4 and reduced slowly to Am(III) by radiolysis products at a rate of about 0.2% hour. In weak acid solutions Am(V) is generally more stable than Am(VI) (133-135). With increasing acidity Am(V) stability falls due to the increase of the disproportionation rate (the same effect is observed with increase of PO_4^{3-}-ion concentration (136)). In solutions of many acids Am(V) is immediately oxidized by Am(IV) to Am(VI) (137). In $K_{10}P_2W_{17}O_{61}$ solutions this reaction is not complete and it is possible to observe the existence of americium in four oxidation states - III,IV,V and VI (138, 139). The reaction between Am(V) and Am(III) in phosphotungstate solutions has been studied. It leads to the formation of Am(IV) (140). In 5 M NaOH Am(III) can be completely oxidized electrochemically to Am(V) (at an anode potential of +1V) and then to Am(VI) (130).

Americium oxidation by persulfate-ions on heating in carbonate solutions results in Am(V) and Am(VI). The kinetics of this process has been studied in detail (141). Am(III) in 1 M K_2CO_3 at a potential of 1V transforms it into Am(VI), which reduces electrochemically at 0.9V to Am(V) (130). Partial oxidation of Am(III) to Am(V) and Am(VI) in 3-5 M K_2CO_3 under strong γ-irradiation has been also observed (142).

Effective extraction methods for americium separation from other TPE, based on the usage of pentavalent americium, have been developed. These methods are especially useful for the isolation of tracer quantities of americium for its further determination in solutions of complex composition (143). Am(V) can be extracted from aqueous solutions by PMBP (144), HDEHP (145) and ammonium pyrrolidinedithiocarbaminate (PDTC) (146). The extraction of trivalent actinide-ions can be suppressed with the help of potassium phosphotungstate (PT) as is seen from Fig.4. The data of Table 3 show that Am(V) is well extracted by all named extractants from the solutions with pH 5. The authors of papers (147,145) were the first who extracted macroamounts of Am(V) using PMBP (147) and HDEHP (145). They also obtained absorption spectra of Am(V) in the organic phase and data on its stability under these conditions. The separation factor of americium and curium using Am(V) extraction by PMBP or HDEHP is equal to $(3\div6)\cdot10^3$ (145,147). Trivalent TPE, on the contrary, can be extracted with high distribution coefficients by extractant mixtures (PMBP with TOPO, TTA with TOPO or picrolonic acid with sulfoxide) from weak acid solution, while Am(V) remains in the aqueous phase under these conditions. A high degree of Am(V) separation from TPE also takes place (132,143,148). Effective separation

TABLE III. Am(V) isolation from TPE

Aqueous phase composition	Organic phase composition	distribution ratio		Am(V)/Me(III)
		Am(V)	Me(III)	
0.01M NH_4NO_3, pH=5 (acetate buffer)	0.06M NH_4PDTC in isopentanol-ethanol-mixture	30	0.02	$1.5 \cdot 10^3$
0.01M NH_4NO_3, pH=5 (acetate buffer+ 0.001M PW)	0.05M PMBP in isopentanol	12.7	0.002	$6.4 \cdot 10^3$
0.1M NH_4NO_3, pH=5 (acetate buffer+0.001M PW)	0.5M HDEHP in octane	30	0.01	$3.0 \cdot 10^3$
0.1M $HClO_4$+0.01M H_3PO_4	0.05M PMBP+0.025M TOPO in cyclohexane	0.02	900	$4.5 \cdot 10^{4*}$
0.1M HNO_3	0.16M PA in methyl isobutyl ketone	0.26	194	$7.3 \cdot 10^{2*}$

*Me(III)/Am(V)

of americium (V) and Cm(III) can be performed by extraction chromatography with HDEHP as the stationary phase (149). Americium purification from curium is equal to 5.10^3 and that of curium from americium is some what worse and equal to ~60. The utilization of zirconium phosphate has proven to be very promising for Am(V) separation from curium (150-152).

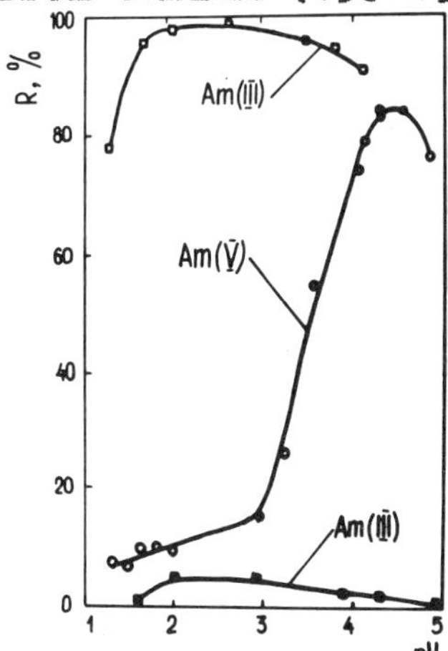

Figure 4. Effect of pH on Am(III) (1,3) and Am(V) (2) extraction by 0.5 M HDEHP solution in octane from acetate solutions (□ and o) and from 10^{-3} M solution of PW (■ and ●).

2.3. Hexavalent Oxidation State

Americium and curium are both known in oxidation state (VI). They exist in solution in the form of the oxygenated MeO_2^{2+}-ion. The potential of the Am(VI)/Am(III) couple in acidic solution is equal to 1.69V, and that of Am(VI)/Am(V) couple is equal to 1.60V. Many methods of Am(III) oxidation to Am(VI) have been described (3).

The following methods of obtaining Am(VI) have been recently developed: in phosphoric acid solutions - electrochemically (107,108) and by a mixture of Ag_3PO_4+ $(NH_4)_2S_2O_8$ (108,110,153); in $HClO_4$ solutions with a small admixture of H_3PO_4 - electrochemically (132); in nitric acid solution with a mixture of $AgNO_3$+$(NH_4)_2S_2O_8$ in the presence of $K_{10}P_2W_{17}O_{61}$ (123) or by $Na_2S_2O_8$ with heating (139); in 5 M NaOH solutions - electrochemically (130); and in acetate solution - electrochemically (154).

The conditions of Am(VI) preparation in H_3PO_4 and HNO_3 solutions in the presence of $K_{10}P_2W_{17}O_{61}$ have been studied in detail. Quantitative americium oxidation to Am(VI) takes place in 0.1-3 M H_3PO_4 solutions by Ag_3PO_4+ $(NH_4)_2S_2O_8$ mixture at room temperature (153). Electrochemical Am(III) oxidation at an anode potential of about

2.0V in 0.1 M $HClO_4$ solution in the presence of 10^{-5}-2 M H_3PO_4 is a convenient method for Am(VI) preparation in solution. The yield of Am(VI) in nitric acid solutions in the presence of phosphotungstate-ions during the oxidation by persulfate-ions depends on the pH of the solution and on $P_2W_{17}O_{61}$-ion concentration. Quantitative americium oxidation to Am(VI) takes place in solutions with pH 1 and with an excess of $P_2W_{17}O_{61}$-ions (139).

Investigations of Am(VI) stability in aqueous solution of different content have shown (132-135,139,153,155, 156) americium to reduce gradually to Am(V) due to its own α-radiation. Am(VI) is the most stable state in H_3PO_4 solutions (156) and in the presence of $K_{10}P_2W_{17}O_{61}$ (123,139). In the latter case it practically doesn't reduce during several days. Thus the rate of ^{243}Am(VI) reduction in $6\cdot10^{-3}$M $K_{10}P_2W_{17}O_{61}$ solutions is equal to 20% for 400 hours, with Am(V), Am(IV) and Am(III) being the products of its reduction (139).

Americium oxidation to Am(VI) is used as a method of separation from other elements and its determination. Hexavalent americium is extracted by HDEHP solutions (149,157,158) and by a mixture of HDEHP with TOPO (158). Americium (VI) extraction by 0.5 M HDEHP solution in benzene occurs at americium concentrations $\geqslant$ 2 mg/ml (149). Hexavalent americium is rather stable in the organic phase and this permits the measurement of the americium (VI) absorption spectrum and the study of the mechanism of its reduction (157). The mixture of PMBP with TOPO (143) in cyclohexane has proven to be very effective for americium and curium separation. Americium is first oxidized to the hexavalent state. During the extraction process it is reduced to the nonextractable pentavalent state. Curium and other trivalent elements pass into the organic phase with high distribution coefficients (Fig.5). The method provides quantitative americium isolation and a high degree of americium purification from curium ($>10^3$) and fission products (143). It has been used for the isolation of macroamounts of americium from irradiated plutonium, as well as for final purification of americium samples from curium and radioactive isotopes of rare-earth elements. Curium content in the final americium sample thus purified is equal to 10^{-6} % (by weight).

Efficient separation of Am(VI) from Cm(III) can also be achieved by extraction chromatography using HDEHP as the stationary phase (149) on zirconium phosphate (150) and on calcined calcium fluoride powder. The separation factor of Am(VI) from Cm(III) during quantitative americium isolation is equal to 2.10^2 and that of Cm from Am - 10^2 (148).

A coulometric method of americium determination by the Am(VI)/Am(V) couple using a 2 M H_3PO_4+0.14 M $HClO_4$

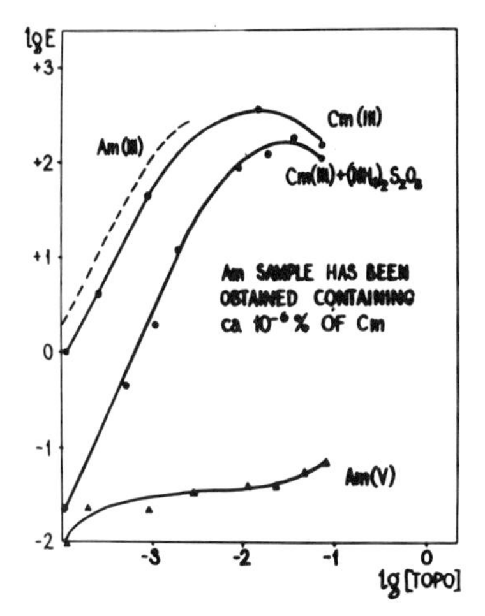

Figure 5. Am(V) and Cm(III) extraction by mixture of 0.05 M PMBP with TOPO in cyclohexane. 1-Cm(III) from 0.1 M HNO_3; 2-Cm(III) from 0.1 M HNO_3 in the presence of 50 mg/ml $(NH_4)_2S_2O_8$; 3-Am after oxidation from 0.1 M HNO_3, containing 50 mg/ml $(NH_4)_2S_2O_8$.

mixture as the electrolyte has been developed (159). The method allows the determination of microgram amounts of americium in the presence of great quantities of curium and plutonium. The method of spectrophotometric titration has been recently suggested for Am(VI) determination in the presence of both Am(V) and Am(III) (160).

2.3.1. Curium (VI). Until recently attempts to prepare hexavalent curium have failed. Krot was the first to predict theoretically the possibility of curium existence in the penta- and hexavalent state (161).

According to theoretical estimates (162), the standard oxidation potential of Cm(VI)/Cm(V) couple is not very high (+1.54) and Cm(VI) can exist in aqueous solutions and will be more stable than Am(IV) and Cm(V) (163). However its direct preparation by Cm(III) oxidation is difficult due to the high potential of Cm(IV)/Cm(III) couple. Cm(VI) (as CmO_2^{2+}) has been shown to appear following the β-decay of $^{242}AmO_2^{+}$ in solid $K_3AmO_2(CO_3)_2$ and to exist in solution after americium carbonate is dissolved in 0.1 M $NaHCO_3$.

2.4. Heptavalent Americium

The first attempts to obtain Am(VII) were carried out immediately after the preparations of Np(VII) and Pu(VII) (101). Alkaline solutions were used in order to obtain Am(VII) (164). O^--ion radicals and ozone in 3 M NaOH were utilized for Am(VI) oxidation. Absorption spectra of Am(VII), obtained under such conditions, have a characteristic absorption maximum at 740 nm with $\mathcal{E}\sim 330$ (Fig.6). Am(VII) isn't stable in alkaline so-

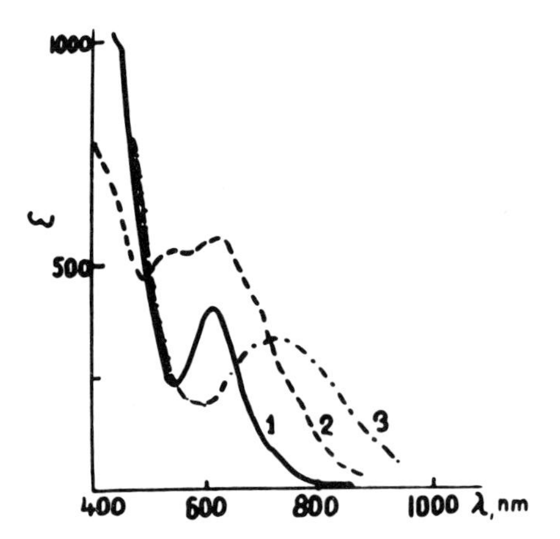

Figure 6. Absorption spectra of heptavalent actinides in alkaline solution. 1-Np; 2-Pu; 3-Am.

lutions and the time for half-reduction to Am(VI) at 0°C is equal to 8 min. Hydrogen peroxide, hydrazine, hydroxylamine sulfite-, and ferrothyanide-ions, Np(VI) and Pu(VI) transform Am(VII) to hexavalent or lower oxidation states (165). The potential of the Am(VII)/Am(VI) couple has been estimated from the linear dependence of electron spectra of Np(VII), Pu(VII) and Am(VII) on their potentials. It proves to be equal to 1.05V in 1 M NaOH. The potential of the Am(VII)/Am(VI) couple decreases with alkaline concentration and reaches the potential of Am(VI)/Am(V) couple. So Am(VII) can be obtained in 12-18 M NaOH solutions owing to Am(VI) disproportionation (166).

3. CONCLUSION

In closing this brief review of the main achievements in the study of the chemical properties of americium and curium ium and their behaviour during extraction and sorption separation processes, we would like to underline some trends of further investigations in these directions. First of all it is necessary to investigate the chemical properties of these elements in their rare oxidation states, including the composition and the structure of their complexes in solutions. It is also necessary to search for and to use new complexing media, such as carbonate solutions, to stabilize the highest oxidation states. As for the extraction,it is attractive to use multiphase systems including aqueous ones and to study the kinetics of interphase equilibria. Further investigation should be carried out of chemical properties of transplutonium elements in unusual oxidation states. The application of new extractants and sorbents are likely to provide soon effective isolation and separation methods for radioactive elements and reliable techniques for their control in radiochemical industries.

4. REFERENCES

1. Seaborg, G.T.; James, R.A.; Morgan, L.O. The Transuranium Elements. National Nuclear Energy Series, Div. IV, vol.14B, Paper 22.1. New-York - Toronto - London, McGraw-Hill Book, Company Inc.,1949.
2. Seaborg, G.T.; James, R.A.; Ghiorso, A. The Transuranium Elements. National Nuclear Energy Series, Div.IV, vol.14B, Paper 22.2. New-York - Toronto - London, McGraw-Hill Book, Company Inc.,1949.
3. Myasoedov, B.F.; Guseva, L.I.; Lebedev, I.A.; Milyukova, M.S.; Chmutova, M.K. "Analytical Chemistry of Transplutonium Elements", John Wiley and Sons, New-York - and Israel Program for Scientific Translations, Ierusalem, 1974.
4. Myasoedov, B.F.; Lebedev, I.A. Radiochimica Acta, 1983, 32, 55.
5. Lebedev, I.A.; Myasoedov, B.F. Radiokhimiya, 1982, 24, 700.
6. Myasoedov, B.F.; Lebedev, I.A. Critical Review, 1985.
7. Yakovlev, G.N.; Kosyakov, V.N. Materialy Mezhdunarodnoy Konferentsii po Mirnomu Ispolzovaniyu Atomnoy energii (Geneva, 1955), Moskova, Goskhimizdat, 1958, 7, 445.
8. Barbanel, Yu.A.; Mikhaylova, N.K. Radiokhimiya, 1969, 11, 595.
9. Lebedev, I.A.; Pirozhkov, S.B.; Yakovlev, G.N. Radiokhimiya, 1960, 2, 549.
10. Starik, I.E.; Ginzburg, F.Z. Radiokhimiya, 1961, 3, 685.
11. Lebedev, I.A.; Yakovlev, G.N. Radiokhimiya, 1962, 4, 304.
12. Borisov, M.S.; Elesin, A.A.; Lebedev, I.A.; Filimonov, V.T.; Yakovlev, G.N. Radiokhimiya, 1966, 8, 42.
13. Moskvin, A.I. Radiokhimiya, 1969, 11, 458.
14. Elesin, A.A.; Lebedev, I.A.;Piskunov, E.M.; Yakovlev, G.N. Radiokhimiya, 1967, 9, 161.
15. Dedov, V.B.; Volkov, V.V.; Gvozdev, B.A.; Ermakov, V.A.; Lebedev, I.A.; Razbitnoy, V.M.; Trukhlyaev, P.S.; Chuburkov, Yu.T.; Yakovlev, G.N. Radiokhimiya, 1965, 7, 453.
16. Yakovlev, G.N.; Kosyakov, V.N. Trudy Vtoroy Mezdunarodnoy Konferentsii po Mirnomu Ispolzovaniyu Atomnoy

Energii (Geneva, 1958), Moskova, Atomizdat, 1959, 4, 147.
17. Yakovlev, G.N.; Gorbenko-Germanov, D.S.; Lenkova, R.A.; Razbitnoy, V.M.; Kazansky, K.S. Zhurn. obshch. khimii, 1958, 28, 2624.
18. Lebedev, I.A.; Yakovlev, G.N. Sb "Soosazhdenie i adsorbtsiys radioactivnykh elementov", Moskova-Leningrad, Nauka Publishers, 1965, 183.
19. Gel'man, A.D.; Moskvin, A.I.; Zaytsev, M.M.; Mefodieva, M.P. "Kompleksnye soedineniya transuranovykh elementov", Moskova, Izdatel'stvo AN SSSR, 1961.
20. Moskvin, A.I.; Khalturin, G.V.; Gel'man, A.D. Radiokhimiya, 1965, 7, 670.
21. Stary, I. Radiokhimiya, 1966, 8, 504.
22. Stary, I. Radiokhimiya, 1966, 8, 509.
23. Stepanov, A.V.; Makarova, T.P. Radiokhimiya, 1965, 7, 670.
24. Lebedev, I.A.; Pirozhkov, S.V.; Razbitnoy, V.M.; Yakovlev, G.N. Radiokhimiya, 1960, 2, 351.
25. Moskvin, A.I.; Khalturin, G.V.; Gel'man, A.D. Radiokhimiya, 1962, 4, 163.
26. Ermakov, V.A.; Stary, I. Radiokhimiya, 1967, 9, 197.
27. Lebedev, I.A.; Yakovlev, G.N. Radiokhimiya, 3, 455.
28. Gvozdev, B.A.; Chuburkov, Yu.T. Radiokhimiya, 1965, 7, 419.
29. Dedov, V.B.; Lebedev, I.A.; Ryzhov, M.N.; Trukhlyaev, P.S.; Yakovlev, G.N. Radiokhimiya, 1961, 3, 701.
30. Lebedev, I.A.; Shalinets, A.B. Sb "Khimiya transuranovykh i oskolochnykh elementov", Leningrad, Nauka Publishers, 1967, 140.
31. Gedeonov, L.I.; Lebedev, I.A.; Stepanov, A.V.; Shalinets, A.B.; Yakovlev, G.N. Sb. "Khimiya transuranovykh i oskolochnykh elementov", Leningrad, Nauka Publishers, 1967, 102.
32. Lebedev, I.A.; Maksimova, A.M.; Stepanov, A.V.; Shalinets, A.B. Radiokhimiya, 1967, 9, 707.
33. Stepanov, A.V.; Makarova, T.P.; Maksimova, A.M.; Shalinets, A.B. Radiokhimiya, 1967, 9, 710.
34. Lebedev, I.A.; Filimonov, V.T.; Shalinets, A.B.; Yakovlev, G.N. Radiokhimiya, 1968, 10, 93.
35. Borisov, M.S.; Elesin, A.A.; Lebedev. I.A.; Filimonov, V.T.; Yakovlev, G.N.; Piskunov, E.M. Radiokhimiya, 1967, 9, 166.
36. Elesin, A.A.; Zaytsev, A.A. Preprint NIIAR P-65, Melekess, 1970.
37. Elesin, A.A.; Zaytsev, A.A.; Karaseva, V.A.; Nazarova, I.I.; Petukhova, I.V. Preprint NIIAR P-64, Melekess, 1969.
38. Shalinets, A.B.; Borobieva, V.V. Radiokhimiya, 1968, 10, 102.

39. Lebedev, I.A.; Shalinets, A.B. Radiokhimiya, 1968, 10. 233.
40. Skobelev, N.K.; Gvozdev, B.A.; Druin, V.N. Atomnaya energiya, 1968, 24, 65.
41. Grebenshchikova, V.I.; Bobrova, V.N. Radiokhimiya, 1961, 3, 544.
42. Grebenshchikova, V.I.,Bryzgalova, R.V. Zhurn. analitich. khimii, 1958, 3, 36.
43. Khlopin, V.G. Trudy G.R.I.,Leningrad-Moskova, GONTI, 1938, 4, 34.
44. Grebenshchikova, V.I.;Chernyavskaya, N.B. Zhurn. neorgan. khimii, 1959, 4, 941.
45. Grebenshchikova, V.I.; Chernyavskaya, N.B. Radiokhimiya, 1961, 3, 650.
46. Grebenshchikova, V.I.; Chernyavskaya, N.B. Radiokhimiya, 1962, 4, 232.
47. Grebenshchikova, V.I.; Bryzgalova, R.V. Radiokhimiya, 1960, 2, 152.
48. Akimova, T.G. Radiokhimiya, 1967, 9, 206.
49. Kuznetsov, V.I.; Akimova, T.G. Kontsentrirovanie aktinoidov soosazhdeniem s organicheskimi soosaditelyami, Moskova, Atomizdat, 1968.
50. Kuznetsov, V.I.; Akimova, T.G.; Eliseeva, O.P. Sb "Radiokhimicheskie metody opredeleniya microelementov", Moskova-Leningrad, Nauka Publishers, 1965, 44.
51. Gureev, E.S.; Dedov, V.B.; Karpacheva, S.M.; Lebedev, I.A,; Ryzhov, M.N.; Trukhlyaev, P.S.; Shvetsov, I.K.; Yakovlev, G.N. Third United Nations. Intern. Conf. on the Peaceful Uses of At. Energy (Geneva, 1964), Report 28p/p/348.
52. Zemlyanukhin, V.I.; Savoskina, G.P.; Pushlenkov, M.F. Radiokhimiya, 1963, 6, 714.
53. Zemlyanukhin, V.I.; Savoskina, G.P.; Pushlenkov, M.F. Radiokhimiya, 1963, 6, 694.
54. Laskorin, B.N.; Skorovarov, D.I.; Filippov, E.A.; Volodin, N.I. Radiokhimiya, 1976, 18, 373.
55. Zemlyanukhin, V.I.; Savoskina, G.P.; Radiokhimiya, 1961, 3, 411.
56. Rozen, A.M.; Khorkhorina, L.P. Zhurn. reorgan. khimii, 1957, 2, 1956.
57. Mc-Kay. Materialy Mezdumarodnoy Konferentsii po Mirnomu Ispolzovaniyu Atomnoy Energii (Geneva), 1955, Moskova, Goskhimizdat, 1958, 7, 388.
58. Zemlyanukhin, V.I.; Savoskina, G.P.; Pushlenkov, M.F. Radiokhimiya, 1962, 4, 655.
59. Zemlyanukhin, V.I.; Savoskina, G.P.; Pushlenkov, M.F. Radiokhimiya, 1962, 4, 570.
60. Chmutova, M.K.; Kochetkova, N.E.; Myasoedov, B.F.; J. Inorg. Nucl. Chem.,1980, 42, 897.

61. Myasoedov, B.F.; Chmutova, M.K.; Karalova, Z.K. 117th ACS National Meeting Actinide Separations. Honolulu, Hawaii, April 3-5, 1979. ACS Symposium Series 117, Washington, 1980, 101.
62. Gureev, E.O.; Kosyakov, V.N.; Yakovlev, G.N. Radiokhimiya, 1964, 6, 655.
63. Vdovenko, V.M.; Krivokhvatsky, A.S.; Kovalskaya, M.P.; Smirnova, E.A.; Kulikov, S.M.; Alekseeva, N.A.; Belov, L.M. Third United Nations, Intern. Conf. on the Peaceful Uses of At. Energy (Geneva) 1964, Report 28/p/345.
64. Miluykova, M.S.; Myasoedov, B.F.; Ryzhova, L.V. Zhurn. analitich. khimii, 1972, 27, 67.
65. Chudinov, E.G.; Pirozhkov, S.V.; Stepanchikov, V.I. Preprint I.A.E. 1753, Moskova, 1968.
66. Zolotov, Yu.A.; Kuzmin, N.M. "Ekstraktsiya metallov atsilpirazolonami", Moskova, Nauka Publishers, 1977.
67. Chmutova, M.K.; Kochetkova, N.E. Zhurn. analitich. khimii, 1969, 24, 1757.
68. Chmutova, M.K.; Pribylova, G.A.; Kochetkova, N.E. Xhurn. analitich. khimii, 1970, 25, 2357.
69. Chmutova, M.K.; Paley, P.N.; Zolotov, Yu.A. Zhurn. analitich. khimii, 1968, 23, 1476.
70. Chmutova, M.K.; Kochetkova, N.E. Zhurn. analitich. khimii, 1970, 25, 710.
71. Kochetkova, N.E.; Chmutova, M.K.; Myasoedov, B.F. Zhurn. analitich. khimii, 1971, 27, 678.
72. Guseva, L.I.; Tikhomirova, G.S. Sb. "Khimiya transuranovykh i oskolochnykh elementov", Leningrad, Nauka Publishers, 1967, 149.
73. Guseva, L.I.; Tikhomirova, G.S. Radiokhimiya, 1972, 14, 252.
74. Bochkarev, V.A.; Voevodin, E.N. Radiokhimiya, 1965,
75. Guseva, L.I.; Grigorieva, S.I.; Tikhomirova, G.S. Radiokhimiya, 1971, 13, 778.
76. Guseva, L.I.; Tikhomirova, G.S. Radiokhimiya, 1968, 10, 246.
77. Guseva, L.I.; Tikhomirova, G.S. Radiokhimiya, 1970, 12, 771.
78. Guseva, L.I.; Tikhomirova, G.S.; Grigorieva, S.I. Radiokhimiya, 1973, 15, 397.
79. Moskvin, L.N.; Novikov, V.T. Radiokhimiya, 1972, 14, 106.
80. Guseva, L.I.; Tikhomirova, G.S. Radiokhimiya, 1974, 16, 152.
81. Lebedev, I.A.; Myasoedov, B.F.; Guseva, L.I. J. Radioanal. Chem.,1971, 21, 259.
82. Guseva, L.I.; Tikhomirova, G.S. Radiokhimiya, 1973, 15, 401.
83. Guseva, L.I.; Myasoedov, B.F.; Tikhomirova, G.S. et al. Radioanal. Chem.,1973, 13, 293.
84. Guseva, L.I.; Tikhomirova, G.S. Radiokhimiya, 1977, 19, 188.

85. Guseva, L.I.; Tikhomirova, G.S. Radiokhimiya, 1978, 20, 809.
86. Guseva, L.I.; Tikhomirova, G.S. Radioanal. Chem., 1979, 52, 369.
87. Barsukova, K.V.; Myasoedov, B.F. Radiokhimiya, 1981, 23, 489.
88. Myasoedov, B.F.; Molochnikova, N.P. J. Radioanal. Chem.,1970, 6, 67.
89. Volynets, M.P.; Guseva, L.I. Zhurn. analitich. khimii, 1968, 23, 947.
90. Samartseva, A.G. Radiokhimiya, 1966, 8, 269.
91. Myasoedov, B.F.; Milyukova, M.S.; Ryzhova, L.V. Radiochem. Radioanal. Letters, 1970, 5, 19.
92. Myaso edov, B.F.; Milyukova, M.S.; Ryzhova, L.V. Zhurn. analitich. khimii, 1972, 27, 1860.
93. Zaytsev, A.A.; Kosyakov, V.N.; Rykov, A.G.; Sobolev, Yu.P.; Yakovlev, G.N. Sb. "Isotopy i izlucheniya v khimii", Moskova, Izdatelstvo AN SSSR, 1958, 326.
94. Zaytsev, A.A.; Kosyakov, V.N.; Rykov, A.G.; Sobolev, Yu.P.; Yakovlev, G.N. Radiokhimiya, 1960, 2, 348.
95. Zaytsev, A.A.; Kosyakov, V.N.; Rykov, A.G.; Sobolev, Yu.P.; Yakovlev, G.N. Radiokhimiya, 1960, 2, 339.
96. Zaytsev, A.A.; Kosyakov, V.N.; Rykov, A.G.; Sobolev, Yu.P.; Yakovlev, G.N. Atomnaya Energiya, 1959, 7,69.
97. Ermakov, V.A.; Timofeev, G.A.; Rykov, A.G.; Yakovlev, G.N. Preprint NIIAR P-71, Melekess, 1970.
98. Vorobieva, A.M.; Fomicheva, V.I. Radiokhimiya, 1965, 7, 728.
99. Lebedev, I.A.; Myasoedov, B.F.; Paley, P.N. J. Radioanal. Chem.,1970, 5, 61.
100. Oganov, M.N.; Striganov, A.R.; Sobolev, Yu.P. Optika i spektroskopiya, 1956, 1, 965.
101. Krot, N.N.; Gel'man, A.D.; Mefodieva, M.P. et al. "Semivalentnoe sostoyanie neptuniya, plutoniya, ameritsiya", Moskova, Nauka Publishers, 1977.
101. Peretrukhin, V.F.; Erin, E.A.; Dzyubenko, V.I. et al. DAN SSSR, 1978, 242, 1359.
103. Kosyakov, V.N.; Erin, E.A.; Vityutnev, V.M. et al. Radiokhimiya, 1982, 24, 551.
104. Kosyakov, V.N.; Timofeev, G.A.; Erin, E.A. et al. Radiokhimiya, 1977, 19, 82.
105. Mikheev, N.B.; Kamenskaya, A.N.; Spitsin, V.I. et al. Radiokhimiya, 1981, 23, 736.
106. Asprey, L.B.; Penneman, R.A. J. Am. Chem. Soc.,1961, 83, 2200.
107. Myasoedov, B.F.; Mikhaylov, V.M.; Lebedev, I.A. et al. Radiochem. Radioanal. Letters, 1973, 14, 17.
108. Myasoedov, B.F.; Lebedev, I.A.; Milyukova, M.S. Rev. Chim. Miner.,1977, 14, 160.
109. Myasoedov, B.F.; Milyukova, M.S.; Litvina, M.N. Radiochem. Radioanal. Letters, 1976, 25, 33.

110. Milyukova, M.S.; Litvina, M.N.; Myasoedov, B.F. Radiokhimiya, 1977, 19, 349.
111. Lebedev, I.A.; Mazur, Yu.F.; Myasoedov, B.F. Radiokhimiya, 1978, 20, 207.
112. Litvina, M.N.; Malikov, D.A.; Milyukova, M.S.; Myasoedov, B.F. Radiokhimiya, 1980, 22, 653.
113. Lebedev, I.A.; Frenkel, V.Ya; Myasoedov, B.F. Radiokhimiya, 1977, 19, 570.
114. Kulyako, Yu.M.; Perevalov, S.A.; Frenkel, V.Ya.; Lebedev, I.A.; Maysoedov, B.F. Radiochem. Radioanal. Letters, 1983, 59, 235.
115. Kosyakov, V.N.; Timofeev, G.A.; Erin, E.A.; Andreev, V.I.; Kopytov, V.V.; Simakin, G.A. Radiokhimiya, 1977, 19, 511.
116. Baranov, A.A.; Simakin, G.A.; Kosyakov, V.N.; Erin, E.A.; Kopytov, V.V.; Timofeev, G.A.; Rykov, A.G. Radiokhimiya, 1981, 21, 127.
117. Saprykin, A.S.; Shilov, V.P.; Spitsin, V.I.; Krot, N.N. DAN SSSR, 1976, 226, 853.
118. Saprykin, A.S.; Spitsin, V.I.; Krot, N.N. DAN SSSR, 1976, 231, 150.
119. Erin, E.A.; Kopytov, V.V.; Rykov, A.G.; Kosyakov, V.N. Radiokhimiya, 1981, 21, 63.
120. Shilov, V.P.; Bukhtiyarova, T.N.; Spitsin, V.I.; Krot, N.N. Radiokhimiya, 1977, 19, 565.
121. Erin, E.A.; Kopytov, V.V.; Vasiliev, V.Ya.; Rykov, A.G. Radiokhimiya, 1981, 23, 727.
122. Kulyako, Yu.M.; Lebedev, I.A.; Frenkel, V.Ya.; Trofimov, T.I.; Myasoedov, B.F. Radiokhimiya, 1981, 23, 937.
123. Milyukova, M.S.; Litvina, M.N.; Myasoedov, B.F. Radiochem. Radioanal. Letters, 1980, 44, 259.
124. Nikolaevsky, V.B.; Shilov, V.P.; Krot, N.N.; Pikaev, A.K. Radiokhimiya, 1976, 18, 368.
125. Pikaev, A.K.; Shilov, V.P.; Nikolaevsky, V.B. et al. Radiokhimiya, 1977, 19, 720.
126. Pikaev, A.K.; Shilov, V.P.; Spitsin, V.I. DAN SSSR, 1977, 232, 387.
127. Myasoedov, B.F.; Milyukova, M.S.; Kuzovkina, E.V. et al. DAN SSSR, 1984.
128. Kulyako, Yu.M.; Trofimov, T.I.; Frenkel, V.Ya. et al. Zhurn. analitich. khimii, 1981, 36, 2343.
129. Keenan, T.K. J. Am. Chem. Soc.,1963, 83, 3719.
130. Peretrukhin, V.F.; Nikolaevsky, V.B.; Shilov, V.P. Radiokhimiya, 1974, 16, 833.
131. Simakin, G.A.; Volkov, Yu.F.; Visyashcheva, G.I. et al. Radiokhimiya, 1974, 16, 859.
132. Frenkel, V.Ya.; Kulyako, Yu.M.; Lebedev, I.A.; Myasoedov, B.F. Zhurn. analitich. khimii, 1979, 34, 330.
133. Myasoedov, B.F.; Lebedev, I.A.; Frenkel, V.Ya.;

Vyatkina, I.I. Radiokhimiya, 1975, 16, 822.
134. Vladimirova, M.V.; Ryabova, A.A.; Kulikov, I.A.; Milovanova, A.S. Radiokhimiya, 1977, 19, 725.
135. Ermakov, V.A.; Frolov, A.A.; Rykov, A.G. Radiokhimiya, 1979, 21, 615.
136. Frenkel, V.Ya.; Lebedev, I.A.; Tikhonov, M.F. et al. Radiokhimiya, 1981, 23, 846.
137. Frenkel, V.Ya.; Lebedev, I.A.; Kulyako, Yu.M.; Myasoedov, B.F. Radiokhimiya, 1979, 21, 836.
138. Kulyako, Yu.M.; Lebedev, I.A.; Frenkel, V.Ya. et al. Radiokhimiya, 1981, 23, 839.
139. Erin, E.A.; Kopytov, V.V.; Rykov, A.G.; Kosyakov, V.N. Radiokhimiya, 1979, 21, 63.
140. Erin, E.A.; Vityutnev, V.M.; Kopytov, V.V.; Vasiliev, V.Ya. Radiokhimiya, 1982, 24, 179.
141. Chistyakov, V.M.; Ermakov, V.A.; Mokrousov, A.D.; Rykov, A.G. Radiokhimiya, 1974, 16, 810.
142. Osipov, S.V.; Andreychuk, N.N.; Vasiliev, V.Ya.; Rykov, A.G. Radiokhimiya, 1977, 19, 522.
143. Myasoedov, B.F.; Molochnikova, N.P.; Lebedev, I.A. Zhurn. analitich. khimii, 1971, 26, 1984.
144. Myasoedov, B.F.; Molochnikova, N.P. Radiochem. Radioanal. Letters, 1974, 18, 33.
145. Molochnikova, N.P.; Myasoedov, B.F.; Frenkel, V.Ya.; Radiochem. Radioanal. Letters, 1983, 59, 25.
146. Myasoedov, B.F.; Molochnikova, N.P.; Davydov, A.V. Radiokhimiya, 1979, 21, 400.
147. Molochnikova, N.P.; Frenkel, V.Ya.; Myasoedov, B.F.; Lebedev, I.A. Radiokhimiya, 1982, 24, 303.
148. Myasoedov, B.F.; Molochmikova, N.P.; Kuvatov, Yu.G.; Nikitin, Yu.E. Radiokhimiya, 1981, 23, 43.
149. Kosyakov, V.N.; Yakovlev, G.N.; Kazakova, G.M. Radiokhimiya, 1979, 21, 262.
150. Shafiev, A.I.; Efremov, Yu.V.; Nikolaev, V.M.; Yakovlev, G.N. Radiokhimiya, 1971, 13, 129.
151. Shafiev, A.I.; Efremov, Yu.V.; Yakovlev, G.N. Radiokhimiya, 1973, 15, 265.
152. Shafiev, A.I.; Efremov, Yu.V.; Yakovlev, G.N. Radiokhimiya, 1975, 17, 498.
153. Milyukova, M.S.; Litvina, M.N.; Myasoedov, B.F.; Radiochem. Radioanal. Letters, 1980, 42, 21.
154. Fedosev, A.M.; Krot, N.N. Radiokhimiya, 1982, 24,
155. Kulikov, I.A.; Frenkel, V.Ya.; Vladimirova, M.V. et al. Radiokhimiya, 1979, 21, 839.
156. Frenkel, V.Ya.; Lebedev, I.A.; Myasoedov, B.F. Radiokhimiya, 1980, 22, 75.
157. Timofeev, G.A.; Levakov, B.I.; Vladimirova, N.V. Radiokhimiya, 1975, 17, 124.
158. Timofeev, G.A.; Levakov, B.I.; Andreev, V.I. Radiokhimiya, 1977, 19, 525.
159. Frenkel, V.Ya.; Kulyako, Yu.M.; Lebedev, I.A. et al.

Zhurn. analitich. khimii, 1980, 35, 1759.
160. Kornilov, A.S.; Chistyakov, V.M.; Frolov, A.A.; Vasiliev, V.Ya.; Rykov, A.G. Radiokhimiya, 1981, 23, 731.
161. Krot, N.N. Radiokhimiya, 1975, 17, 677.
162. Spitsin, V.I.; Ionova, G.V. Radiokhimiya, 1978, 20, 328.
163. Ionova, G.V.; Spitsin, V.I. DAN SSSR, 1978, 241, 390.
164. Krot, N.N.; Shilov, V.P.; Nikolaevsky, V.B.; Pikaev, A.K.; Gel'man, A.D.; Spitsin, V.I. DAN SSSR, 1974, 217, 589.
165. Shilov, V.P.; Radiokhimiya, 1976, 18, 659.
166. Nikolaevsky, V.B.; Shilov, V.P.; Krot, N.N.; Peretrukhin, V.F. Radiokhimiya, 1975, 17, 431.

COULOMETRIC DETERMINATION OF AMERICIUM AND CURIUM FOR THE PREPARATION OF REFERENCE SOLUTIONS

P. Brossard and S. Lafontan
CEA - Service CQ
B.P. 12
91680 Bruyères-le-Châtel
France

ABSTRACT. An improvement of the method for the determination of americium by constant - current coulometry is proposed in order to prepare standardized samples used in alpha-spectrometry and mass spectrometry. The use of a microcomputer for automatic data acquisition and their numerical treatment is described. This method gives a precision of about 0,2 % for 100 micrograms of americium (relative standard deviation of the mean, 95 % confidence). The same technique is also suitable to prepare ^{244}Cm reference solutions.

I. INTRODUCTION

During the past few years a lot of work has been done focusing on the preparation of actinide reference samples to be used in alpha-spectrometry. Any method to measure the concentration of a standardized solution must be, besides accurate and reproducible, absolute. For example the well-known EDTA-complexometry, which is quite accurate, cannot be used for reference materials since the result depends on the EDTA concentration of the reageants. (1)

To achieve such a goal in the case of americium, the potential imposed coulometry, currently used for plutonium and uranium, was first proposed (2), (3). Unfortunately, the reproductibility was not very good since the americium-IV is dismuted in Am III and Am V. Only recently, a method which meets the required conditions was described (4) (5). The idea was to combine the EDTA complexation of americium with the constant - current coulometry. Since it gives very good results, the method is now widely used. However for each determination one needs a fairly large amount of americium and this appears to be a serious problem since pure americium solutions of high isotopic quality are not largely available.

The present work was carried out in order to improve this method, i.e., lowering the quantities needed for one determination.

In principle, the method is quite simple. The coulometric determination is based on the electrochemical equilibrium between mercury-EDTA complex and americium :

N. M. Edelstein et al. (eds.), Americium and Curium Chemistry and Technology, 81–87.

$$Hg\ EDTA^{2-} + Am^{3+} \longrightarrow Am\ EDTA^{-} + Hg^{2+} \quad (1)$$

The free mercury ions are removed by cathodic reduction on a mercury-pool electrode. The number of coulombs needed for a total elimination of the ions Hg^{2+} is then directly proportional to the amount of americium previously added. The detection of the end-point is done by amperometry with two polarized indicator electrodes (dead-stop-end-point).

In order to minimize the random error on the end-point determination, these are automatically calculated by a microcomputer which has previously recorded the titration curves.

2. EXPERIMENTAL

2.1. Apparatus

The device used for the coulometry of americium has been described in a previous paper (4). It mainly consists of a coulometric cell (figure 1) connected to an electrochemical buret (wich delivers and integrates the reduction current) and an ampèrometry detection unit. The later imposes a low potential (20 mV) between the two droplet-mercury electrodes and the current passing through is measured. This current is directly related to the concentration of the free mercury ions in the solution.

2.2. Reagents

2.2.1. Acetic buffer : For one liter of solution, 2.20 grams of mercury are dissolved in nitric acid. Then one adds 3.7 grams of Na_2EDTA and 57 ml of pure acetic acid. The pH is then adjusted to about 5 with ammonia.

2.2.2. Solution of americium : The solution of the desired isotope is prepared from the oxide and purified by reverse phase chromatography on HDEHP-gas chrom-Q column. Then the solution is treated with oxalic acid. The oxalate is calcined into an oxide which is finally dissolved in nitric acid to the desired concentration.

2.3. Procedure

One pours in the cell about 8 cc of mercury and 6 cc of "acetic buffer". The auxiliary compartment is filled with 1M ammonium nitrate. The nitrogen pressure is set to the pressure in this compartment. Then a first reduction of the free Hg^{2+} ions present in the solution is performed. The current measured between the two detection electrodes is recorded as a function of the delivered number of coulombs. When the end point has been detected the reaction is stopped and an accurately weighted amount of the americium solution is introduced into the cell. The reduction is then repeated and the second end-point determination is recorded. This procedure leads to titration curves like those presented on figure 2. One can compute the unknown concentration of americium, C, with the relation :

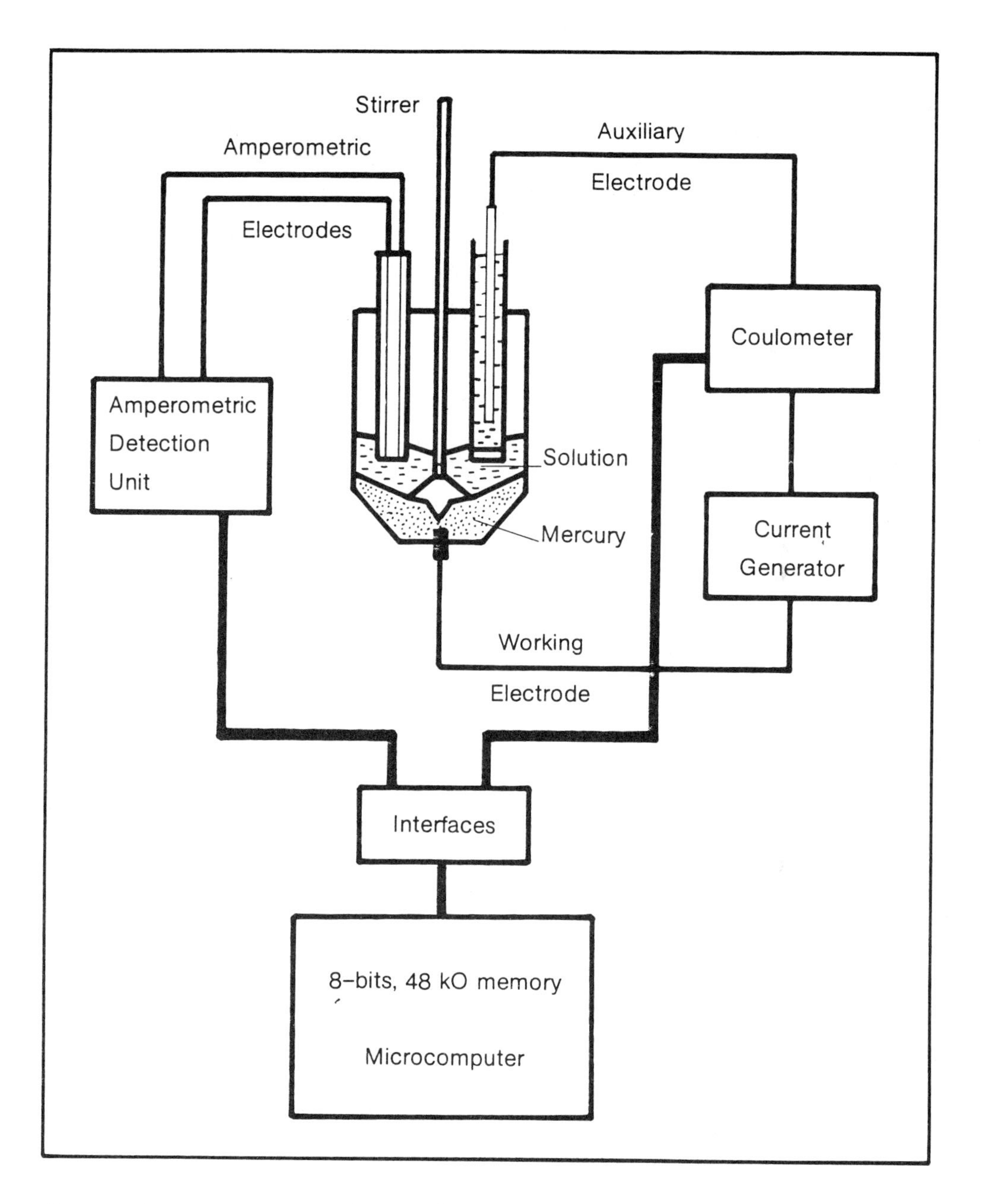

Figure 1. Apparatus used for the coulometric determination of americium and curium.

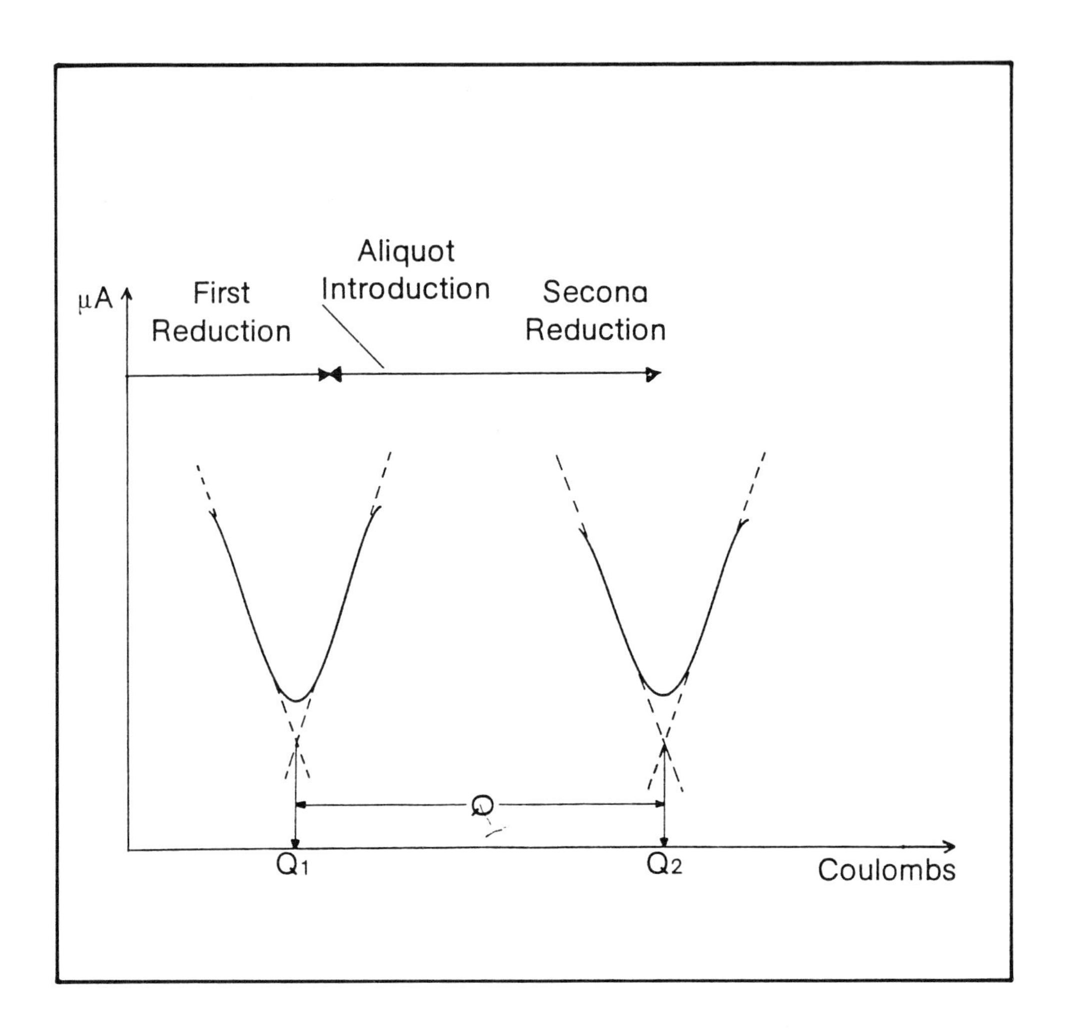

Figure 2. Titration curves obtained during the constant-current coulometry of americium and curium.

Method	Conc. found (mg/g)	Std. Deviation (%)	Difference with the theoretical value (%)
Hand-determination	3.186	0.50	0.2
Polynomial fitting (fifth degree)	3.204	1.07	0.76
Polynomial fitting (eighth degree)	3.191	0.34	0.36
Fast Fourier Transform	3.183	0.23	0.1

Table I. Comparison between the numerical methods in the case of the titration of cadmium (C_0=3.1797 ± 0.006 mg/g, 8 determinations, aliquot weight: ≈100 µg).

Weight of the aliquot	60 - 120 g
Number of determinations	8
Measured concentration of americium (mg/g)	2.009 ± 0.004
Alpha-spectrometry value (mg/g)	2.01 ± 0.02

Table II. Typical results obtained in the coulometric titration of americium and comparison with alpha-spectrometry.

$$C = \frac{Q\ M}{2mF} \qquad (2)$$

where M is the atomic weight, m the weight of the aliquot and F the Faraday constant (96 484.57 C/Mole), Q is the difference between the number of coulombs given by the second end-point (Q_2) and the first one (Q_1).

2.4. Determination of the end-point

The problems arise with the determination of the end-points. The electrical noise cannot be reduced to zero without affecting the reproductibility. Then, because of the presence of oscillations along the curve, the minimum cannot be read directly. The method to obtain it consists in drawing two tangents at the inflexion points on each side of the curve (dashed lines on the figure 2). If the curve is symmetric, the intersection point will give the minimum. Unfortunatelly, this does not occur very often. Even more important is the fact that two operators will never draw exactly the same tangents and thefore will not found the same result. This discrepancy will become more important if one lowers the amount of americium used to make the determination. The reason is that the current measured by the amperometric detection unit is, somehow, proportionnal to the concentration of Hg^{2+}. Therefore the height of the amperometric wave is decreasing with the decreasing weight of the aliquot, so the determination becomes more difficult.

To avoid these difficulties the idea is to record the titration curves into the memory of a microcomputer and to obtain the minimum by numerical means. This is possible since both the coulometer and the amperometric detector have BCD-coded outputs. Such a connection represents only a slight modification to the apparatus but has great advantages.

Figure 1 gives a schematic view of the modified apparatus. To perform the automatic data acquisition and the calculations we used an Apple II microcomputer with 48 kØ of RAM-memory. The interfacing is done by three parallel interfaces of 16 bits each.

Once the experimental data points are recorded one has to fit them with a smooth curve in order to determine the minimum. Several methods are then available ; for this kind of curves the most popular ones are the polynomial approximation and the Fourier Transform.

For the first one we may need up to the eighth degree polynomials. For the second one, the frequencies contained in the signal (digitized in a power of two number of points) are computed by the Fast Fourier Transform. Then the higher frequencies are set to a zero value and the inverse transform is computed to get a smoothed titration curve.

3. RESULTS AND DISCUSSION

In order to see which of the two methods is the best according to our problem, we first performed a set of runs with a cadmium solution instead of americium ; the procedure was exactly the same. The cadmium solution was prepared from a reference sample of metal dissolved in nitric acid, all dilutions being done by weighing. The exact concentration of the solution was 3.1797 mg/g (± 0.02 %). Eight runs were carried out with aliquots of about 100 µg.

The results are presented in table I. Values directly obtained by hand-determination of the minimum are also given.

One can easily see that we did, indeed, improve our results by using the microcomputer. The polynomial fitting is good only if the polynomial is of the eighth degree. We also concluded that, the Fourier Transform appears to be the most suitable method for smoothing such titration curves. The explanation comes from the fact that the electrical noise in the signal represents the most disturbing factor.

These results gives us confidence in choosing the Fast Fourier Transform. However the computationnal time was quite long (about 15 minutes for a 512 points signal). To overcome this disadvantage the program was written down in assembly language. This leads to affordable times (less than 2 min.).

Finally we performed measurements of americium solutions. The aliquot weights were ranging from 60 to 120 µg (in americium). The concentration of the solution was also determined by alpha-spectrometry. The results are given in table II. If we are in good agreement with alpha-spectrometry, we also get an excellent reproducibility. In fact, comparing with results previously published (4), we managed to keep the accuracy and the reproducibility at very low values even though we used ten times less material. The next step is to prepare reference solution of curium 244. Runs are underway and the first results indicate a good reproducibility with 80 µg curium244 aliquot weights.

4. CONCLUSION

The use of a microcomputer appears to be of a great help for the preparation of reference solutions of americium and curium by coulometry. It allows the amount of material used for the measurement to be lowered without affecting the quality of the coulometric determination. One can reach a precision of about 0,2 % with only 100 µg of material for each determination. Associated with α-spectrometry, such a method will certainly permit the preparation of certified solutions.

5. REFERENCES

(1) Buijs K. and W. Bartscher, Anal. Chim. Acta, 88, 403 (1977)
(2) Koehly G., Anal. Chim. Acta, 33, 410 (1965)
(3) Stokely J.R. and W.D. Shults, Anal. Chim. Acta, 45, 417 (1969)
(4) Bergey C. and L. Fouchard, Talanta, 26, 445 (1979)
(5) Bergey C., Mikrochimica Acta (Wien), 2, 207 (1981)

A STUDY ON THE STABILITY OF AMERICIUM(V) AND AMERICIUM(VI) IN NITRATE MEDIA

M. Hara and S. Suzuki
The Research Institute for Iron, Steel and Other Metals,
Tohoku University
Katahira 2-1-1, Sendai 980
Japan

ABSTRACT. The oxidation of Am(III) to Am(VI) with sodium bismuth trioxide was investigated in a nitric acid solution, and a procedure was developed to oxidize Am(III) at low concentrations completely. The oxidant was added to a nitric acid solution of Am(III), and the mixture was mechanically stirred; the Am(VI) solution free from the excess oxidant was obtained by filtering the mixture. An Am(V) solution was prepared by passing the Am(VI) solution through a column of Celite 545 containing bis(2-ethylhexyl)hydrogenphosphate, which reduced Am(VI) to Am(V). The stability of Am(V) and Am(VI) at concentrations lower than 10^{-6} M in nitric acids and in solutions containing ozone gas, fluoride, or phosphate ions was investigated by means of the coprecipitation technique with $BiPO_4$ and ThF_4. It was found that ^{241}Am(V) and ^{241}Am(VI) were gradually reduced to Am(III) in nitric acids at low acidities and high temperatures, but the reduction was effectively inhibited by the presence of ozone gas or fluoride ions. On the other hand, ^{243}Am(V) and ^{243}Am(VI) were found to be rather stable in nitric acids. But, they were readily reduced to Am(III) in the phosphate solutions in the same manner as ^{241}Am(V) and ^{241}Am(VI).

INTRODUCTION

Most investigations on the chemical properties of Am(V) and Am(VI) have been carried out by the use of an alpha-emitter ^{241}Am($t_{1/2}$ = 433 y); its alpha-decay leads to the radiolytic reduction of Am(V) and Am(VI).(1,2) When the long-lived isotope ^{243}Am($t_{1/2}$ = 7400 y) is used, the radiolytic reduction is not so remarkable as that in ^{241}Am solutions.(3) On the other hand, Am(V) disproportionates to Am(VI) and Am(IV) in moderately concentrated acids.(3-5) The radiolytic reduction and disproportionation have been investigated by means of spectrophotometry using ^{241}Am and ^{243}Am solutions at concentrations of 10^{-3} M or above.

The ^{241}Am and ^{243}Am solutions diluted to about 10^{-7} and 10^{-6} M, respectively, can be studied by radiometry. In dilute solutions, Am(V) may be free from any disproportionation, while the radiolytic reduction of ^{241}Am(V) and ^{241}Am(VI) may proceed, as suggested by the rate laws.(1-

N. M. Edelstein et al. (eds.), Americium and Curium Chemistry and Technology, 89–104.

5). Am(V) and Am(VI) may also be reduced by small amounts of impurities.

The information concerning the stability of Am(V) and Am(VI) at low concentrations is rather poor. Only a little information was obtained from previous works(6,7) in which the stability of Am(V) in several media was studied by means of solvent extraction with 2-thenoyltrifluoroacetone and of column chromatography using bis(2-ethylhexyl)hydrogenphosphate, HDEHP, as an extractant, where the rapid reduction of Am(VI) by HDEHP was observed.

When the solutions of Am(V) and Am(VI) were submitted to the coprecipitation method with some fluorides and phosphates at 0 - 5 ^{o}C, no change of oxidation states of americium was observed.(8,9) However, this finding can not guarantee a high stability of Am(V) and Am(VI) in fluoride and phosphate solutions at room temperature or above.

An acidic solution of Am(VI) at low concentrations has been conventionally prepared by the oxidation of Am(III) with silver-catalysed ammonium persulfate,(9) but this method has disadvantages as described below. First, the acidity of the solution must be lower than 0.2 M because persulfate decomposes via an acid-catalysed path(10) to produce peroxymonosulfuric acid which would rapidly reduce any Am(VI) formed. Second, the resulting Am(VI) solution suffers contamination from the excess oxidant and its reduction products. Such disadvantages would make it more difficult to investigate the chemical properties of Am(VI) such as its stability, complex formation, and redox reactions.

By the way, sodium bismuth trioxide was previously used for the determination of manganese. This method is known to be fatally flawed by the presence of Co(II) and Ce(III) because they are oxidized to Co(III) and Ce(IV) with the reagent.(11) The standard potential for the Bi(III)-Bi(V) couple must be quite enough to oxidize Am(III) to Am(VI) with Bi(V), since the potentials for the Am(III)-Am(VI), the Co(II)-Co(III), and the Ce(III)-Ce(IV) couples are 1.69,(4) 1.95,(12) and 1.74 volts, (13) respectively. The low solubility of sodium bismuth trioxide in acid solutions is an advantage since the reagent not dissolved can be easily removed from the reacting solution, although the oxidant dissolved in the solution can oxidize only a small amount of materials.

In this work, the oxidation of Am(III) with sodium bismuth trioxide was investigated at low concentrations of americium, and a study in nitric acid solutions has resulted in successful developments.(14) The oxidation reaction was followed by the coprecipitation method with $BiPO_4$. The stability of Am(V) and Am(VI) in nitric acid solutions and in solutions containing ozone gas, fluoride or phosphate ions was also investigated.(15) The change of americium oxidation state with time was followed by the coprecipitation method with $BiPO_4$ and ThF_4.

MATERIALS AND METHODS

Dissolution of oxidant

Sodium bismuth trioxide powder, min. 80 wt%, was added to the nitric acid solution in a glass test tube, and the mixture was vigorously stirred. The mixture was then filtered with a glass fibre filter-paper,

Whatman GF/A, or a column packed with quartz glass wool, A grade(1-5 μm). The concentration of Bi(V) and total bismuth(Bi(III) + Bi(V)) in the filtrate were determined by iodometry and chelatometry, respectively. Potassium iodide, and solutions of sodium thiosulfate and starch were used for iodometry; sodium dihydrogenphosphate was used to hinder a yellow coloration resulting from the formation of bismuth iodide. Disodium ethylenediaminetetraacetate and xylenol orange were used as a chelating agent and an indicator, respectively.

Oxidation of Am(III) with sodium bismuth trioxide

The powdered oxidant was added to a nitric acid solution of ^{241}Am(III) or ^{243}Am(III) in a glass test tube, and the mixture was vigorously stirred. Then, an aliquot of the mixture was analysed by the coprecipitation test with $BiPO_4$. In some cases, the undissolved oxidant was filtered off from the mixture and an aliquot of the filtrate was also analysed by the coprecipitation test.

Coprecipitation test

The test was carried out under such conditions that Am(III) was quantitatively carried down, while most of Am(V) and Am(VI) remained in the solution.(8) $BiPO_4$ and ThF_4 were precipitated at 0 °C from the americium solution to which nitrate solutions of Bi(III) and Th(IV) were added, respectively, followed by the addition of excess precipitants under the conditions described below:

Precipitate	$BiPO_4$	ThF_4
Volume of the mixture	5 ml	3 ml
Acidity of the mixture	0.5 M	0.1 M
Amount of metal ion	Bi^{3+} 10 mg	Th^{4+} 1 mg
Precipitant	NaH_2PO_4	NH_4F
Time of mixing	7 min	10 min

The mixture was either filtered by a glass fibre filter-paper or centrifuged at 6000 rpm for 5 min. The fraction of americium carried, Y, was determined as usual by measuring the gamma-activity in the mixture and the filtrate(or the supernatant). The gamma-ray assay was performed by the use of a flat NaI(Tl) scintillation probe connected to a scaler or by the use of a Ge(Li) detector connected to a multi-channel pulse-height analyzer.

Preparation of Am(V) and Am(VI) solutions

Ten mg/ml of sodium bismuth trioxide was added to a 0.1 M HNO_3 solution of Am(III), and the mixture was mechanically stirred for 50 min at room temperature and then for 10 min at 0 °C. An Am(VI) solution was obtained by filtration of the mixture with a glass fibre filter-paper.

Am(V) solution was prepared at 0 °C by passing the Am(VI) solution through a column of Celite 545 containing HDEHP, which rapidly reduced

Am(VI) to Am(V). Solutions of nitric acid and ammonium fluoride or sodium dihydrogenphosphate were added to the Am(V) and Am(VI) solutions making ^{241}Am and ^{243}Am concentrations up to about 10^{-7} and 10^{-6} M, respectively. A reference Am(VI) solution was prepared as usual by the oxidation of Am(III) with silver-catalysed ammonium persulfate.(9)

Stability of Am(V) and Am(VI)

The Am(V) and Am(VI) solutions were maintained in a thermostat at a temperature kept constant to ± 0.5 °C. Appropriate aliquots of the stock solutions were analysed by the coprecipitation test at controlled time intervals.

RESULTS AND DISCUSSION

Dissolution of oxidant

The concentration of total bismuth ions in a nitric acid solution was found to be almost independent when the amounts of the oxidant used was in the range of 2 - 20 mg/ml. Solutions containing more than 20 mg/ml oxidant could not be effectively stirred. Thus the amount of oxidant was adjusted to 10 mg/ml in the following experiments.

It was observed that the concentration of Bi(V) was apparently constant in a nitric acid solution when mixed with the oxidant, but increased with increasing acidity and temperature, as shown in TABLE I. However, the concentration of Bi(III) (equal to the difference between the concentrations of total bismuth and Bi(V)) increased rapidly with time of mixing at 30 °C, especially at high acidities, while it remained at low levels at 0 °C, as shown in Figure 1.

TABLE I. The concentration of Bi(V) in the nitric acid solution mixed with $NaBiO_3$.

$[HNO_3]$	$[Bi(V)]$,* 10^{-3} M	
M	0 °C	30 °C
0.25	0.04 ± 0.01	0.03 ± 0.01
0.5	0.05 ± 0.01	0.05 ± 0.02
1.0	0.13 ± 0.01	0.16 ± 0.01
2.0	0.17 ± 0.04	0.26 ± 0.03
4.0	0.47 ± 0.05	0.74 ± 0.04

* The mean of the four values measured every 15 min of mixing.

The concentration of Bi(V) in the solution free from the undissolved oxidant did decrease more rapidly with time at 30 °C, especially at high acidities, as shown in Figure 2. That is, the increase in the concentration of Bi(III) seemed to be closely correlated with the

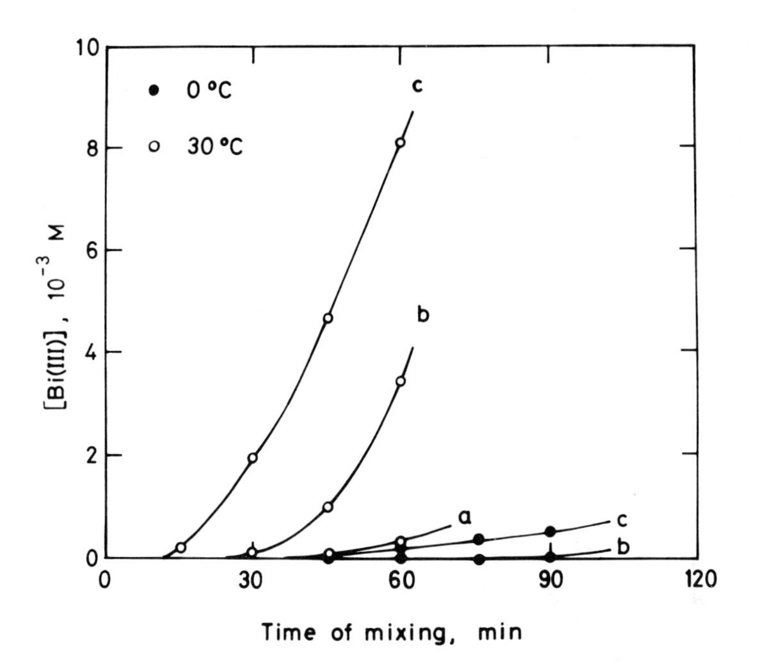

Figure 1. The concentration of Bi(III) in 1 M(a), 2 M(b), and 4 M(c) HNO_3 mixed with $NaBiO_3$ at 0 and 30 °C.

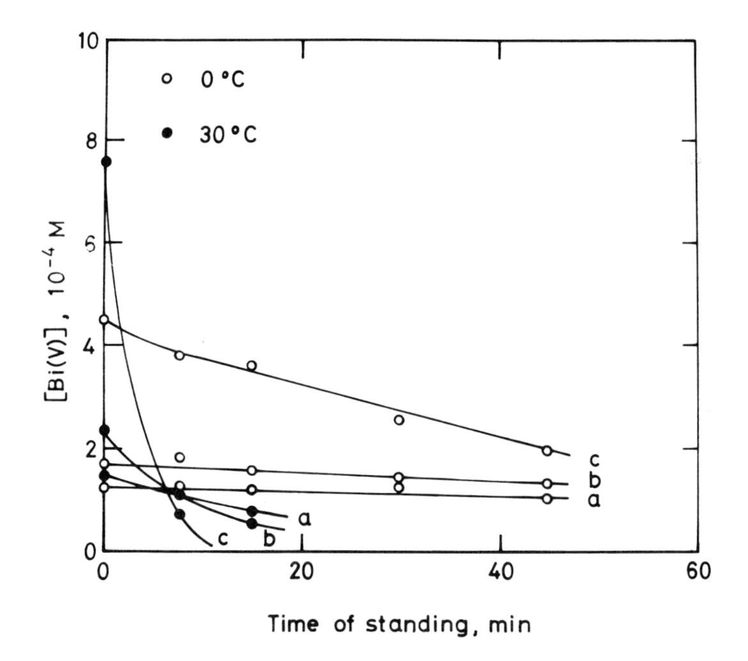

Figure 2. The concentration of Bi(V) in 1 M(a), 2 M(b), and 4 M(c) HNO_3 free from undissolved $NaBiO_3$ at 0 and 30 °C.

decrease in the concentration of Bi(V).

The chemical species of Bi(V) generated by the dissolution of oxidant powder may be neutral species such as $NaBiO_3$ or $HBiO_3$. The concentration of Bi(V) seemed to be constant at a constant acidity and temperature because the reduction of Bi(V) to Bi(III) with water or hydrogen ion was apparently balanced by the dissolution of the oxidant, irrespective of the acidity and the temperature of the solution. This situation would be the same even if low concentrations of americium were present.

Oxidation of Am(III) with sodium bismuth trioxide

Figures 3 and 4 indicate the results of the coprecipitation test which was carried out on the americium solution mixed with the oxidant. With lower acidity and/or higher temperature, the fraction of americium carried decreased rapidly with the reaction time to an apparently constant value. This value was in good agreement with the value which was obtained from the reference Am(VI) solution. Therefore, it was concluded that the oxidation of Am(III) to Am(VI) in solution was completed when the value given above was found in the coprecipitation test.

When the mole fraction of each oxidation state of americium in the reacting solution is expressed as f_X(X = III, V, and VI), Y, which is obtained by the coprecipitation test, may be given by an equation:(8)

$$Y = f_{III}Y_{III} + f_{V}Y_{V} + f_{VI}Y_{VI} \quad (1)$$

where Y_X(X = III, V, and VI) means the fraction of Am(X) carried, which can be determined by the coprecipitation test on corresponding single-component solution. The values for Y_X were determined as follows: Y_{III} = 0.99 ± 0.01, Y_V = 0.25 ± 0.03, and Y_{VI} = 0.05 ± 0.02 by the $BiPO_4$ coprecipitation test, and Y_{III} = 0.99 ± 0.01, Y_V = 0.12 ± 0.02, and Y_{VI} = 0.30 ± 0.02 by the ThF_4 coprecipitation test.

In the reacting solutions containing excess oxidant Am(V) could be ignored because the standard potentials for the Am(III)-Am(V) and Am(V)-Am(VI) couples are 1.74 and 1.60 volts,(4) respectively, in comparison with the potentials for the Am(III)-Am(VI) and Bi(III)-Bi(V) couples, as discussed before. That is, Am(V) would be readily oxidized to Am(VI) even if initially present. Then, Eq. 1 may be rewritten as:

$$Y = f_{III}Y_{III} + f_{VI}Y_{VI}. \quad (2)$$

Using the Y_{III} and Y_{VI} given above, and the Y shown in Figures 3 and 4, the f_{VI} could be deduced by solving Eq. 2 because $f_{III} + f_{VI} = 1$. It was also plotted in each figure. It is clear that the oxidation reaction of Am(III) with $NaBiO_3$ proceeded more rapidly at lower acidities and/or at higher temperatures. The reaction time needed for half or complete oxidation of Am(III) is given in TABLE II for several cases. The rate of oxidation does not depend so much on the concentration of Bi(V) as on the acidity and the temperature. This finding explains qualitatively the oxidation process, for which kinetic and stoichiometric analysis has not been given.

The recommended procedure for the preparation of an Am(VI) solution

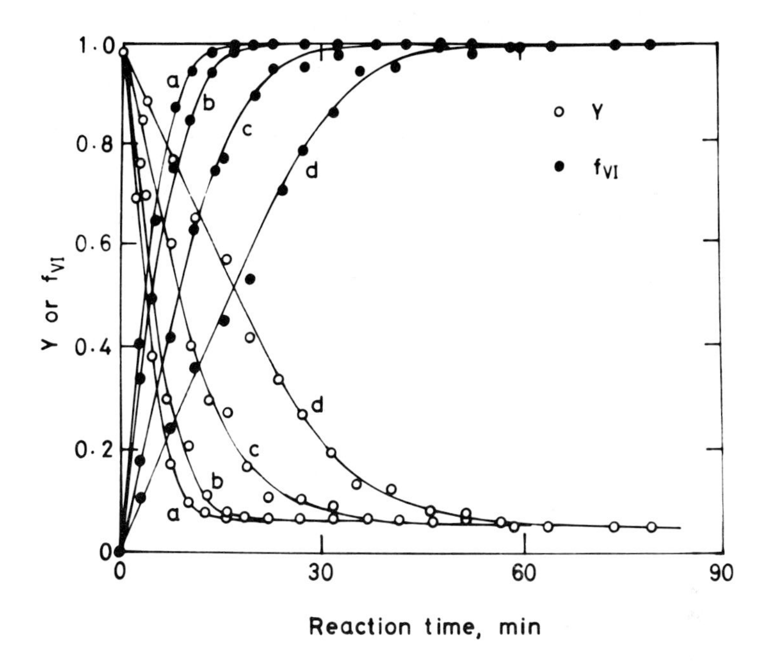

Figure 3. Oxidation of Am(III) with $NaBiO_3$ in 0.35 M(a), 0.5 M(b), 0.7 M(c), and 1.0 M(d) HNO_3 at 0 °C.

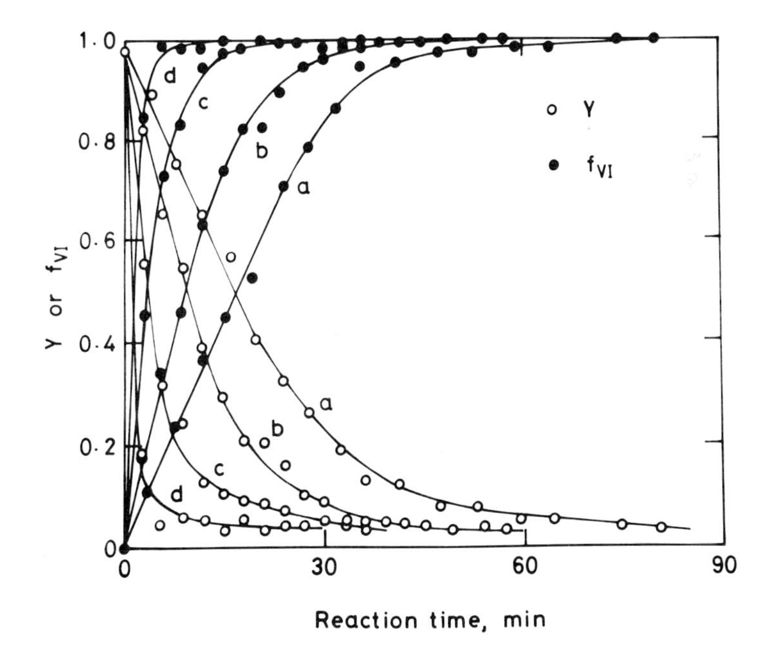

Figure 4. Oxidation of Am(III) with $NaBiO_3$ in 1.0 M HNO_3 at 0 °C(a), 10 °C(b), 20 °C(c), and 30 °C(d).

is as follows: the solution resulting from the reaction of Am(III) and $NaBiO_3$, and when allowed enough time as given in TABLE II, must be allowed to stand at as low a temperature as possible until the undissolved oxidant is filtered off. The acidity of the solution is maintained at 0.2 - 2 M in order to avoid both the contamination by a large amount of Bi(III) and the precipitation of its hydrolysis products.(16)

TABLE II. The time required for the half and complete oxidation of Am(III) to Am(VI) with $NaBiO_3$

$[HNO_3]$, M	Temperature, °C	Time for half oxidation, min	Time for complete oxidation, min
0.5	0	4	30
0.5	10	<3	25
0.5	20	<2	20
0.5	30	<1	5
1.0	0	16	50
1.0	10	9	40
1.0	20	<3	22
1.0	30	<2	12
2.0	20	15	40
2.0	30	<4	18

Stability of Am(V) and Am(VI)

In HNO_3 solutions. The $BiPO_4$ coprecipitation test was carried out on the nitric acid solution of $^{241}Am(V)$ and $^{241}Am(VI)$. The results for the 0.5 M HNO_3 solutions at 0 - 40 °C and for the (0.125 - 2 M) HNO_3 solutions at 30 °C are shown in Figures 5 and 6, respectively. The measured value for Y increased with time of storage, especially at high temperatures and at low acidities. Dependence of the increase in Y on acidity was found only for the Am(VI) solutions at the temperatues higher than 20 °C. The increase in Y may be caused by the reduction of Am(V) and Am(VI) to Am(III) in the stock solutions, because Am(III) is more effectively carried by $BiPO_4$.

The rate law for the radiolytic reduction of Am(V) can be expressed by an empirical equation:

$$-\frac{d[Am(V)]}{dt} = \frac{d[Am(III)]}{dt} = k_1[Am]_T, \qquad (3)$$

where k_1 is an apparent rate constant and $[Am]_T$ means the total americium concentration. The decrease in Am(VI) concentration with time can be expressed by an equation:

$$-\frac{d[Am(VI)]}{dt} = k_2[Am]_T, \qquad (4)$$

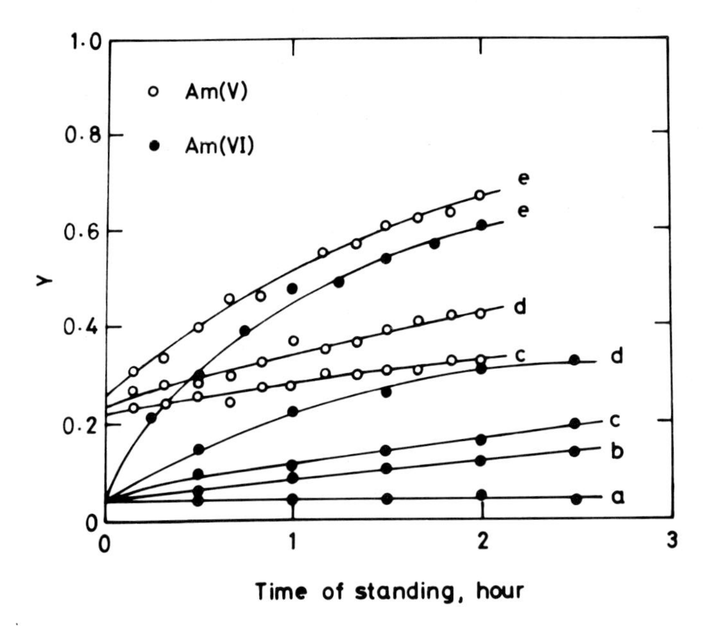

Figure 5. Result of the $BiPO_4$ coprecipitation test carried out on the 0.5 M HNO_3 solutions of $^{241}Am(V)$ and $^{241}Am(VI)$ stocked at 0 °C(a), 10 °C(b), 20 °C(c), 30 °C(d), and 40 °C(e).

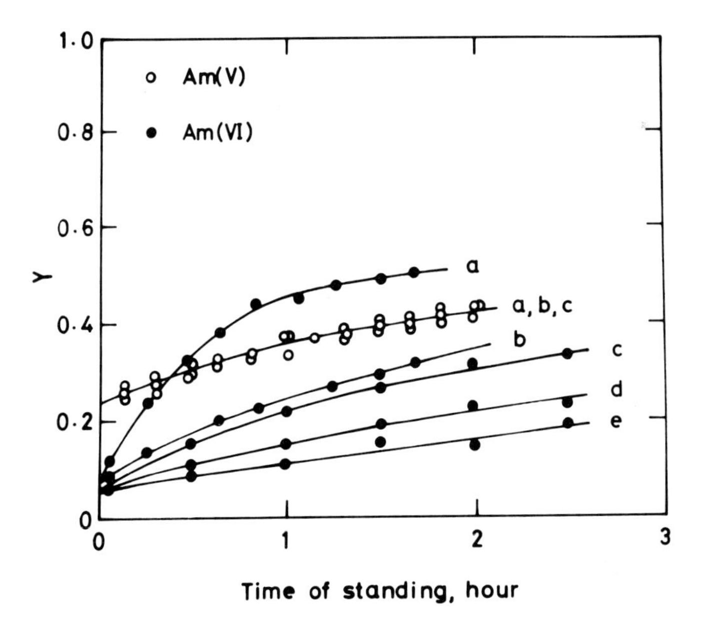

Figure 6. Result of the $BiPO_4$ coprecipitation test carried out on the 0.125 M(a), 0.25 M(b), 0.5 M(c), 1.0 M(d), and 2.0 M(e) HNO_3 solutions of $^{241}Am(V)$ and $^{241}Am(VI)$ stocked at 30 °C.

where k_2 is an apparent rate constant. Values for k_1 and k_2 have been estimated to be 0.01 and 0.03 h^{-1}, respectively, in dilute acid solutions.(1,2) If no reactions other than radiolytic reduction proceed, the growth of Am(V) in the Am(VI) solution can be expressed by an equation:

$$\frac{d[Am(V)]}{dt} = (k_2 - k_1)[Am]_T. \quad (5)$$

If the radiolytic reduction predominantly proceeds in the stock solutions of ^{241}Am, Eq. 1 can be expressed by the following equations, using the solutions of Eqs. 3 - 5:
for the case of ^{241}Am(V) solutions,

$$Y = Y_V + (Y_{III} - Y_V)k_1t, \quad (6)$$

and for the case of ^{241}Am(VI) solutions,

$$Y = Y_{VI} + \{(Y_{III} - Y_{VI})k_1 + (Y_V - Y_{VI})k_2\}t. \quad (7)$$

Unfortunately, the linear relations expected by Eqs. 6 and 7 were valid only when the stock solutions were at high acidities and at low temperatures, as shown in Figures 5 and 6. It is clear that reactions other than radiolytic reduction also occurred, especially at low acidities and at high temperatures. They are assumed to be first-order reactions such as Am(VI) ---> Am(V) and Am(V) ---> Am(III), while the radiolytic reduction of Am(V) and Am(VI) is zero-order reaction with respect to Am(V) and Am(VI).

The rate law for the above-mentioned reactions can be expressed as:
in the case of ^{241}Am(V) solutions,

$$-\frac{d[Am(V)]}{dt} = \frac{d[Am(III)]}{dt} = k_1'[Am(V)], \quad (8)$$

and in the case of ^{241}Am(VI) solutions,

$$-\frac{d[Am(VI)]}{dt} = k_2'[Am(VI)], \text{ and} \quad (9)$$

$$\frac{d[Am(V)]}{dt} = k_2'[Am(VI)] - k_1'[Am(V)], \quad (10)$$

where k_1' and k_2' are apparent rate constants. If these first-order reactions predominantly proceed in the stock solutions, Eq. 1 can be written as follows, using the solutions of Eqs. 8 - 10:
for the case of ^{241}Am(V) solutions,

$$Y_{III} - Y = (Y_{III} - Y_V)\exp(-k_1't), \quad (11)$$

and for the case of ^{241}Am(VI) solutions,

$$Y_{III} - Y = C\cdot\exp(-k_1't) + (Y_{III} - Y_{VI} - C)\exp(-k_2't), \quad (12)$$

where $C = (Y_{III} - Y_V)k_2'(k_2' - k_1')^{-1}$.

As Y_{III} is nearly equal to unity, Figures 7 and 8 illustrate the plot of log(1 - Y) vs t, using the data shown in Figures 5 and 6. The relationship expected by Eqs. 11 and 12 was confirmed for the stock solutions of low acidities and at high temperatures. Thus, it is reasonable to assume that the radiolytic reduction and first-order reactions compete in the dilute ^{241}Am solutions.

The values for k_1' and k_2' were estimated by means of curve-fitting techniques; the results are shown in TABLE III. The values for k_1' and k_2' fluctuated, and seem to depend upon whether the nitric acid solution (used to prepare the stock solutions) was freshly prepared by dilution of conc. nitric acid or was allowed to age slightly. Therefore, some material which induced first-order reactions was probably introduced into the stock solutions from the nitric acid.

TABLE III. Apparent rate constants for the reduction of ^{241}Am(V) and ^{241}Am(VI) in HNO_3 solutions

[HNO_3],	Temperature,	Apparent rate constant, h^{-1}		
M	°C	k_1' a)	k_1' b)	k_2' b)
0.125	30	0.14 ± 0.01	0.14 ± 0.02	2.37 ± 0.11
0.25	30	0.12 ± 0.01	0.14 ± 0.02	2.07 ± 0.23
0.5	30	0.13 ± 0.01	0.09 ± 0.02	0.69 ± 0.23
0.5	20	0.06 ± 0.01	-	-
0.5	40	0.40 ± 0.01	0.30 ± 0.05	2.00 ± 0.43

a) Results for Am(V) solutions.
b) Results for Am(VI) solutions.

When the 0.05 M HNO_3 solutions of ^{243}Am(V) and ^{243}Am(VI) were investigated, no change in Y with time of storage was observed even on stock solutions at 30 °C. This finding may be explained only by the large difference in emission rate of alpha-particles between ^{241}Am and ^{243}Am. Therefore, first-order reactions are considered to be the radio-induced reactions which may be caused by a product resulting from the interaction between the radiolysis product and the material introduced from the nitric acid.

This assumption was supported by a finding for the ^{241}Am(V) and ^{241}Am(VI) solutions containing ozone gas; Y was constant for the time of storage during which the stimulative smell of ozone gas was present in the stock solution vessel. It can be concluded that ozone gas effectively eliminates any reducible materials as chlorine gas does.(4)

In HNO_3-NH_4F solution. The coprecipitation test with ThF_4 was carried out on the 0.05 M HNO_3-(0.05 and 0.5 M) NH_4F solutions of ^{241}Am(V) and

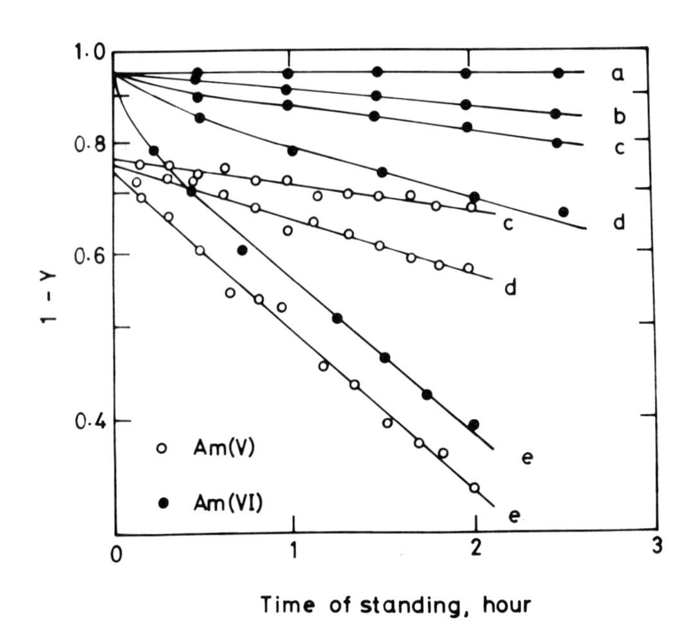

Figure 7. The plot of log(1 - Y) vs t for 0.5 M HNO_3 solutions of $^{241}Am(V)$ and $^{241}Am(VI)$ stocked at 0 °C(a), 10 °C(b), 20 °C(c), 30 °C(d), and 40 °C(e).

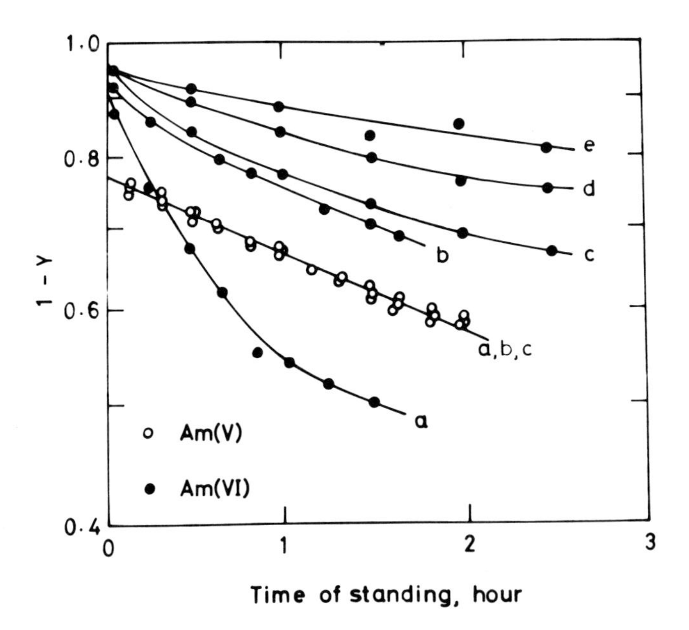

Figure 8. The plot of log(1 - Y) vs t for 0.125 M(a), 0.25 M(b), 0.5 M(c), 1.0 M(d), and 2.0 M(e) HNO_3 solutions of $^{241}Am(V)$ and $^{241}Am(VI)$ stocked at 30 °C.

^{241}Am(VI) at 30 °C, and the result is shown in Figure 9. The Y for the stock solutions containing the fluoride increased more slowly than for the case without fluoride, and no change in Y was observed for the 0.5 M NH_4F solutions. The stock solutions of acidity higher than 0.05 M called for high fluoride concentrations to minimize the variation in Y.

This finding can be explained by the suggestion that ^{241}Am(V) and ^{241}Am(VI) become free from all reduction reactions if they form fluoro complexes, as in the case of the acetate(7) and the chloride solutions (17) of ^{241}Am(V). The excellent stability of ^{243}Am(V) and ^{243}Am(VI) in the ozonized fluoride solutions was also confirmed as expected.

In HNO_3-NaH_2PO_4 ozonized solutions. The coprecipitation test with $BiPO_4$ was carried out on the 0.05 M HNO_3-(0 - 2 M) NaH_2PO_4 and on the (0.025 - 0.2 M) HNO_3-0.5 M NaH_2PO_4 solutions of ^{241}Am(V) and ^{241}Am(VI) at 30 °C. The Y for these ozonized phosphate solutions increased rapidly with time of storage, especially for the solutions of low acidities and high phosphate concentrations. The plot of log(1 - Y) vs t is illustrated in Figures 10 and 11; a good linearity was obtained in most cases.

The same result was obtained both for the 0.05 M HNO_3-0.5 M NaH_2PO_4 solutions of ^{241}Am(V) and ^{241}Am(VI) at 0 - 40 °C and for the (0.05 - 0.5 M) HNO_3-1 M NaH_2PO_4 solutions of ^{243}Am(V) and ^{243}Am(VI) at 30 °C. Evidently, the variation in Y with time for the phosphate solutions of ^{241}Am(VI) and ^{243}Am(VI) can be explained neither by Eq. 7 nor by Eq. 12.

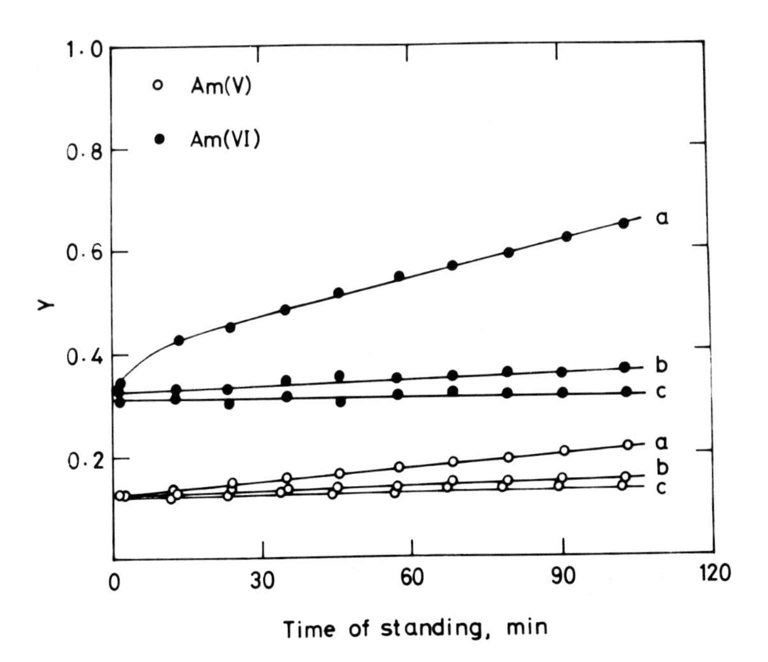

Figure. 9. Result of the ThF_4 coprecipitation test carried out on the 0.5 M HNO_3-(0 M(a), 0.05 M(b), and 0.5 M(c)) NH_4F solutions of ^{241}Am(V) and ^{241}Am(VI) stocked at 30 °C.

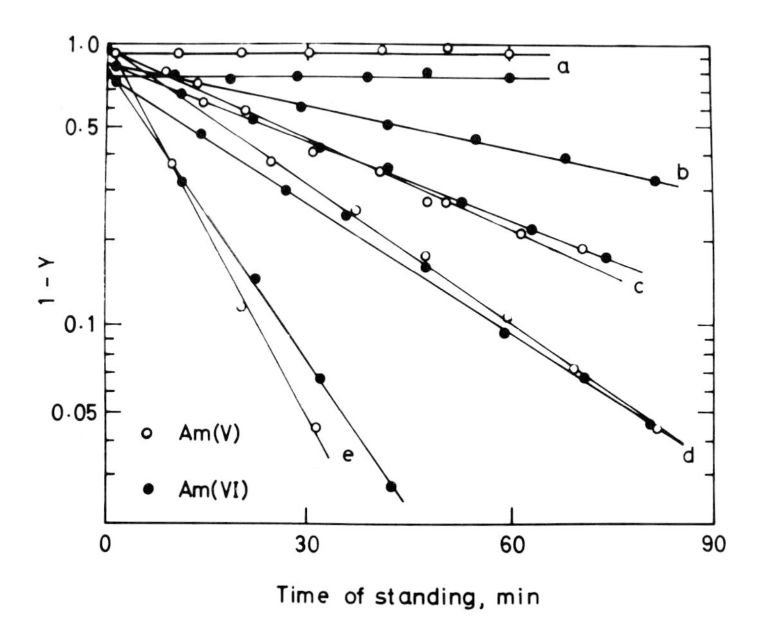

Figure 10. The plot for log(1 - Y) vs t for 0.05 M HNO_3-(0 M(a), 0.1 M(b), 0.5 M(c), 1.0 M(d), and 2.0 M(e)) NaH_2PO_4 solutions of $^{241}Am(V)$ and $^{241}Am(VI)$ stocked at 30 °C.

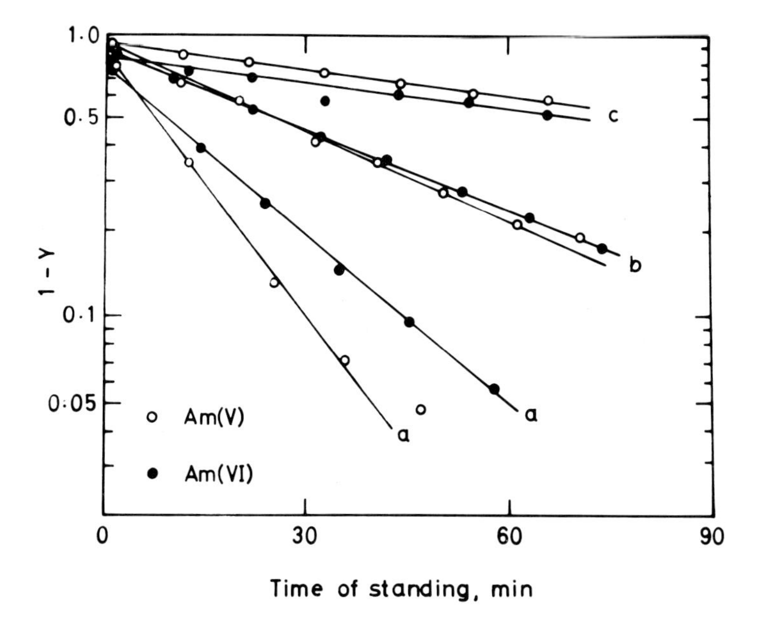

Figure 11. The plot for log(1 - Y) vs t for (0.025 M(a), 0.05 M(b), and 0.2 M(c)) HNO_3-0.5 M NaH_2PO_4 solutions of $^{241}Am(V)$ and $^{241}Am(VI)$ stocked at 30 °C.

It is known that Am(VI) at such high concentrations as 2×10^{-3} M (^{243}Am 85 % and ^{241}Am 15 %) is readily reduced to Am(III) in phosphoric acids.(18) It may also be assumed that Am(V) and Am(VI) at low concentrations are reduced to Am(III) in the ozonized phosphate solutions, and that the reduction rates are given by the following equations:

$$-\frac{d[Am(V)]}{dt} = \frac{d[Am(III)]}{dt} = k_1^*[Am(V)], \text{ and} \qquad (13)$$

$$-\frac{d[Am(VI)]}{dt} = \frac{d[Am(III)]}{dt} = k_2^*[Am(VI)], \qquad (14)$$

where k_1^* and k_2^* are apparent rate constants. If the change of oxidation states of americium in the ozonized phosphate solutions is predominantly caused by these reactions, Y can be expressed as follows, using the solutions of Eqs. 13 and 14:
for the case of Am(V) solutions,

$$Y_{III} - Y = (Y_{III} - Y_V)\exp(-k_1^*t), \qquad (15)$$

and for the case of Am(VI) solutions,

$$Y_{III} - Y = (Y_{III} - Y_{VI})\exp(-k_2^*t). \qquad (16)$$

As Y_{III} is nearly equal to unity, Eqs. 15 and 16 can explain the linearity obtained by the plot of log(1 - Y) vs t, as shown in Figures 10 and 11. Therefore, it seems reasonable to assume that Am(V) and Am(VI) are reduced to Am(III) in the ozonized phosphate solutions.

Values for k_1^* and k_2^* were obtained from the slope of the straight lines. They were found to be larger for the stock solutions at low acidities, at high phosphate concentrations, and at high temperatures, as summarized in TABLE IV. The reduction of ^{243}Am(V) and ^{243}Am(VI) proceeded more rapidly than in the solutions of ^{241}Am because the ^{243}Am concentration was about ten times higher than the ^{241}Am concentration.

The available information concerning the behavior of americium in phosphoric acid(18) and the phosphato complex formation of neptunium and plutonium(19,20) suggests that Am(V) and Am(VI) may form the phosphato complexes such as $AmO_2H_2PO_4$ and $AmO_2H_2PO_4^+$, respectively in the phosphate solutions investigated. Therefore, the rapid reduction of Am(V) and Am(VI) might result from an intramolecular electron transfer in the phosphato complexes formed.

References

1. Hall, G.R.; Markin, T. L.; J. Inorg. Nucl. Chem., 1957, **4**, 296.
2. Yakovlev, G. N.; Kosyakov, V. N.; Proc. 1st Int. Conf. On the Peaceful Uses of Atomic Energy, Geneva(1955), 1956, **7**, 363.
3. Coleman, J. S.; Inorg. Chem., 1963, **2**, 53.

TABLE IV. Apparent rate constants for the reduction of Am(V) and Am(VI) in HNO_3-NaH_2PO_4 ozonized solutions.

$[HNO_3]$,	$[NaH_2PO_4]$,	Temperature,	Apparent rate constant, h^{-1}	
M	M	°C	k_1* a)	k_2* a)
0.05	0	30	-0.02 ± 0.03	-0.05 ± 0.02
0.05	0.1	30	-	0.67 ± 0.01
0.05	0.5	30	1.44 ± 0.04	1.28 ± 0.02
0.05	1.0	30	2.32 ± 0.05	2.13 ± 0.04
0.05	2.0	30	6.10 ± 0.04	4.78 ± 0.04
0.05	0.5	0	-0.06 ± 0.03	0.00 ± 0.02
0.05	0.5	40	7.63 ± 0.06	6.08 ± 0.06
0.025	0.5	30	4.23 ± 0.04	2.71 ± 0.02
0.1	0.5	30	0.65 ± 0.04	0.55 ± 0.04
0.2	0.5	30	0.40 ± 0.03	0.30 ± 0.05
0.05	0	30	0.02 ± 0.01b)	0.02 ± 0.02b)
0.05	0.1	30	1.00 ± 0.02b)	1.39 ± 0.02b)
0.05	1.0	30	10.5 ± 0.1b)	8.98 ± 0.02b)
0.5	1.0	30	2.89 ± 0.02b)	5.80 ± 0.03b)

a) Result for ^{241}Am solutions except as otherwise noted.
b) Result for ^{243}Am solutions.

4. Penneman, R. A.; Asprey, S. B.; Proc. 1st Int. Conf. on the Peaceful Uses of Atomic Energy, Geneva(1955), 1956, **7**, 355.
5. Yakovlev, G. N.; Kosyakov, V. N.; Proc. 2nd Int. Conf. On the Peaceful Uses of Atomic Energy, Geneva(1957), 1958, **28**, 373.
6. Hara, M.; Bull. Chem. Soc. Japan, 1970, **43**, 89.
7. Hara, M.; Suzuki, S.; Bull. Chem. Soc. Japan, 1974, **47**, 635.
8. Hara, M.; Suzuki, S.; Bull. Chem. Soc. Japan, 1975, **48**, 1431.
9. Moore, F. L.; Anal. Chem., 1963, **35**, 715.
10. Kolthoff, I. M.; Miller, I. K.; J. Am. Chem. Soc., 1951, **73**, 3055.
11. Lundell, G. E. F.; J. Am. Chem. Soc., 1923, **45**, 2600.
12. Johnson, D. A.; Sharpe, A. G.; J. Chem. Soc., 1964, 3490.
13. Conley, H. L.; Thesis, UCRL-9332, 1960.
14. Hara, M.; Suzuki, S.; J. Radioanal. Chem, 1977, **36**, 95.
15. Hara, M.; Suzuki, S.; Bull. Chem. Soc. Japan, 1979, **52**, 1041.
16. Olin, A.; Acta Chem. Scand., 1957, **43**, 1445.
17. Kasha, M., J. Chem. Phys., 1949, **17**, 349.
18. Myasoedov, B. F.; Mikhailov, V. M.; Lebedev, I. A.; Koiro, O. E.; Frankel, V. Ya.; Radiochem. Radioanal. Lett., 1973, **14**, 17.
19. Moskvin, A. I.; Peretrukhin, V. F.; Radiokhimiya, 1964, **6**, 206.
20. Kenotkina, R. G.; Shevelenko, U. B.; Zh. Neorg. Khim., 1967, **12**, 2345.

Radiopolarographic Study of Americium and Curium

Y. Shiokawa and S. Suzuki
The Research Institute for Iron, Steel and Other Metals,
Tohoku University
Katahira 2-1-1, Sendai 980
Japan

ABSTRACT. The radiopolarographic behavior of americium and curium in non-complexing aqueous media was examined as a function of concentration, specific activity and pH. At pH=3.0, a single wave of americium was observed below [Am]=8.0×10^{-7} M having a plateau at a potential more negative than -2.02 V (main wave). An additional pre-wave was observed above [Am]=9.6×10^{-7} M. At lower pH(<2.6), the pre-wave appeared even at a concentration smaller than 8.0×10^{-7} M and the potential at maxmium height was shifted slightly with the concentration. The radiopolarographic behavior of curium was found to be similar to that of americium. The main waves of americium and curium were diffusion-controlled, but the pre-waves were not. It was found that the pre-waves occured independently of the main waves, and that the main waves were due to the reduction to the metallic state. The transfer coefficients α and the cathodic rate constants K_c^0 at E=0 V were obtained as α=0.41 and $K_c^0=8.8\times10^{-40}$ cm/s for Am(III)/Hg(Am); α=0.50 and $K_c^0=9.1\times10^{-48}$ cm/s for Cm(III)/Hg(Cm), respectively.

INTRODUCTION

The dc polarograms of lanthanide(III) ions exhibit two waves beyond the hydrogen ion discharge in a non-complexing aquoues medium, with the exception of Sm(III), Eu(III) and Yb(III). According to Large and Timnick, the first wave can be attributed to the reduction of the hydrogen ions produced by the protolytic dissociation of the hydrated lanthanide(III), while the second wave is due to the adsorption of the hydrolytic products of lanthanide(III)(1). Both waves are not diffusion-controlled. Therefore, the reduction of lanthanide(III) into the metallic state cannot be observed by dc polarography. There is no data available on the kinetic parameters of the electrode reaction of lanthanide(III) in an aqueous solution.

When a radioisotope is reduced to the metallic state at the mercury dropping electrode with the potential E, the radioactivity A collected in a given number of mercury drops is proportional to the number of reduced ions. The activity A - potential E curve is the radiopolarogram.

In radiopolarography, the observed quantity is not the flux of electrons(current), but the amount of amalgamated radioisotopes. This quantity corresponds to the mass transfer accompanied with the amalgamation and not to the electron transfer. Therefore, the reduction of lanthanide(III) into the amalgam can be studied by radiopolarography

N. M. Edelstein et al. (eds.), Americium and Curium Chemistry and Technology, 105–114.

and from the data the kinetic parameters of this reduction can be derived. We have studied the reduction wave for the reduction Gd(III)/Hg(Gd) by radiopolarography(2).

The reduction of Am(III) and Cm(III) into the amalgam in an aqueous solution can be observed by radiopolarography, because their chemical behavior is quite similar to lanthanide(III). The radiopolarographic behavior of Am(III) and Cm(III) has been investigated by Samhoun and David(3), but detailed behavior and the kinetic parameters of amalgamation are still unknown. In order to determine these parameters, we have investigated the radiopolarographic behavior of Am(III) and Cm(III).

Materials and Methods

Radiopolarographic Cell

In a conventional radiopolarographic cell, the solution examined is in direct contact with the insulating phase(CCl_4), in which the radioactive amalgam is accumulated. In such a cell the radiopolarographic reduction, in an organic solution or in an aqueous solution containing some soluble substance in the carbon tetrachloride phase, is difficult because of the direct contact with the organic phase(CCl_4). Also some experimental troubles can result from the direct contact.

In this work we used a new radiopolarographic cell, shown in Fig. 1, in which the solution examined is in direct contact with an argon phase, not with the carbon tetrachloride. The inner surface of the part of capillary E(inner diameter, 3 mm) below the stopcock F as is shown in Fig. 1, was treated with dimethyldichlorosilane vapor. The increased surface tension between the solution and the inner surface of the capillary E held the solution above the argon phase, but the mercury drops passed through the solution and the argon phase, and were accumulated into the carbon tetrachloride phase.

Procedures

The ^{241}Am, ^{243}Am and $^{243,244}Cm$ used in this work were purified by the standard ion exchange technique(4,5). An aliquot of the americium or the curium solution(ca. 1M HCl) was evaporated to dryness by an infrared lamp. The residue was dissolved in a small amount of distilled water and the solution was evaporated to dryness. The residue was dissolved in 0.1 ml of 0.1 M HCl and 1 ml of 1 M LiCl and the solution was diluted to 10 ml. The pH was adjusted with HCl and/or LiOH and then solution was heated. After cooling the radiopolarographic reduction was performed. The final concentration of supporting electrolyte LiCl was 0.1 M.

The radiopolarographic procedures were as follows. The deoxidized carbon tetrachloride and the solution to be examined were introduced into the cell, and the tube connected to the needle valve I was filled with argon. A constant potential was imposed at the dropping mercury electrode(DME) for 10 min, while the mercury drops were accumulated in the compartment on the stopcock G. Then the stopcock F was closed and the needle valves I and J, and the stopcock G were opened. By the flow of CCl_4 (J⟶G⟶I), the amalgam was rinsed in order to remove the radioactive solution and dropped into the compartment on the stopcock H. After the stopcock G was closed, the amalgam was taken out from the cell with a small amount of carbon tetrachloride. The dried amalgam was weighed, and then the amalgamated metal was extracted with 1 ml of 0.5 M HNO_3 for 10 min. An aliquot of the extraction solution was determined by means of α-ray or γ-ray spectrometry.

The pH value did not differ by more than 0.1 pH unit before and after experiment. The radiopolarographic cell was thermostated at 25±1°C in a constant-temperature box. The polarograph and the DME drop knocker which controlled the constant lifetime of the mercury drop, were designed in our laboratory.

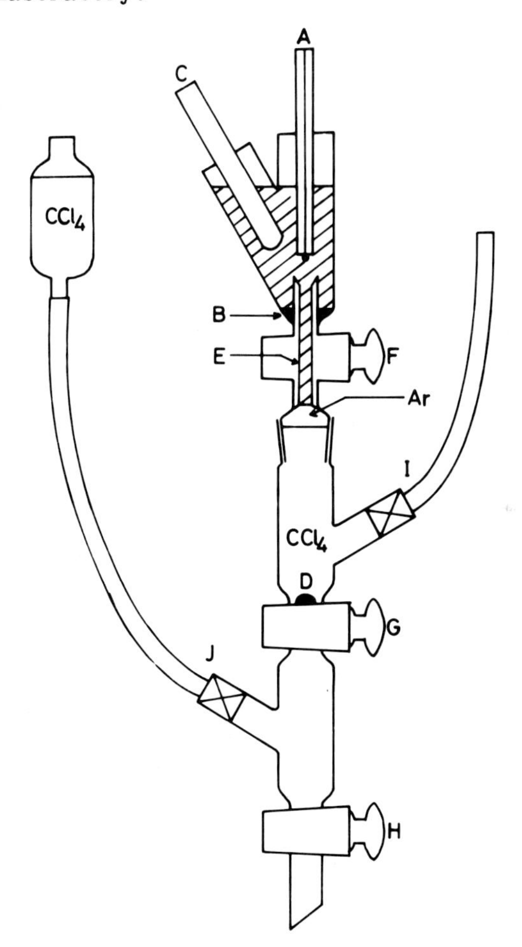

Figure 1. Radiopolarographic cell.

A:DME, B:Hg anode, C:SCE, D:amalgam, E:capillary
F,G and H:stopcocks, I and J:needle valves.

Results and Discussion

Radiopolarographic Behavior

The radiopolarographic behavior of Am(III) and Cm(III) in 0.1 M LiCl was studied as a function of concentration, specific activity and pH of the solution.

At pH=3.0, the concentration and the specific alpha activity of americium were varied from 3.3×10^{-6} M to 1.1×10^{-8} M and from 0.009 μCi/ml to 0.65 μCi/ml, respectively. The radiopolarograms are shown in Fig. 2. In solutions with the same concentrations and different specific activities, no significant changes were found between the radiopolarograms. Near the electrode potential of -1.92 V(6) the concentration of americium had a large effect on the radiopolarographic behavior.

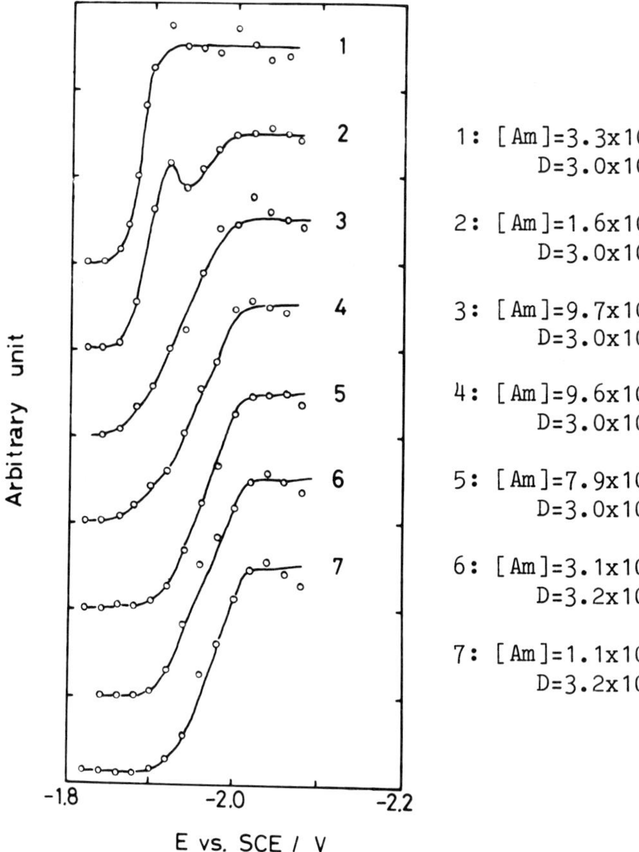

Figure 2. Radiopolarograms of Am(III) at pH=3.0 in 0.1 M LiCl.

A single wave with a half-wave potential $E_{1/2}$ of approximately -1.88 V was observed at 3.3×10^{-6} M and the limiting activity, coresponding to the limiting current, was found at a potential more negative than -1.92 V. But at a concentration of 1.6×10^{-6} M, a pre-wave with a maximum height at approximately -1.92 V and a main wave with a plateau at a potential more negative than -2.02 V, were observed. The pre-wave was observable at a concentration as low as 9.6×10^{-7} M. Below 8.0×10^{-7} M the pre-wave disappeared and only a main wave was observed having a half-wave potential of approximately -1.96 V; the concentration of americium had no effect on the radiopolarogram.

The diffusion coefficients calculated from the limiting activities, shown in Fig. 2, were found to be almost constant and did not depend on the concentration of americium. This shows that the limiting activity and not the sum of the pre-wave height and the limiting activity, is proportional to the concentration. It suggests that the main wave having a plateau more negative than -2.02 V does not vary with the concentration, and that the pre-wave is concentration dependant. Therefore, the changes of the radiopolarogram, as is shown in Fig. 2, can be attributed only to the pre-wave. The diffusion coefficients obtained are lower than the values published in the literature(7). This may be due to deamalgamation in the carbon tetrachloride phase(3).

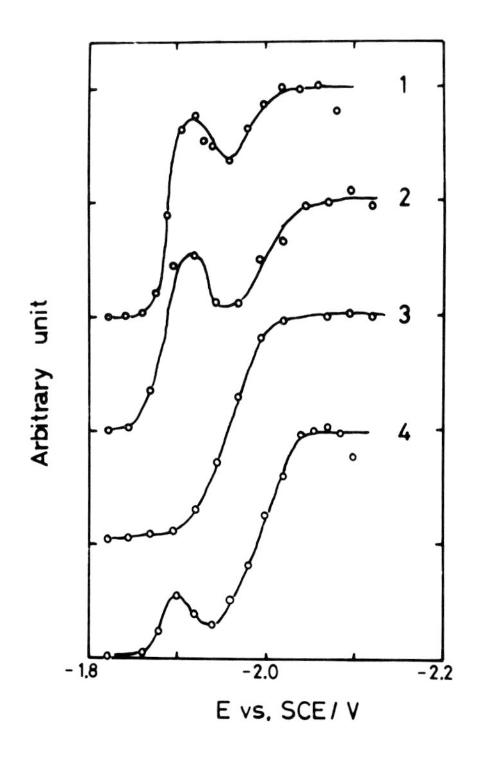

1: [Am]=3.2×10^{-6}M, pH=2.4
D=$4.0 \times 10^{-6} cm^2/s$

2: [Am]=6.7×10^{-7}M, pH=2.2
D=$4.4 \times 10^{-6} cm^2/s$

3: [Am]=6.7×10^{-7}M, pH=2.8
D=$3.0 \times 10^{-6} cm^2/s$

4: [Am]=7.9×10^{-8}M, pH=2.5
D=$4.0 \times 10^{-6} cm^2/s$

Figure 3. Radiopolarograms of Am(III) at various pH in 0.1 M LiCl.

Figure 3 shows the variations of the radiopolarograms with pH. At pH=2.4 and [Am]=3.2×10^{-6} M, a pre-wave and a main wave were observed. The pre-wave was not observable at pH=3.0 and [Am]<8×10^{-7} M, but at a lower pH($\leq$2.5) the pre-waves appeared even at [Am]<8×10^{-7} M, as it is shown in Fig. 3. The potential of the pre-wave was shifted slightly with concentration of americium. Though the half-wave potential of the main waves was independent of pH, the diffusion coefficients at lower pH(<3.0) are slightly larger than the ones at pH=3.0. However, the pre-waves apparently varied with pH.

Figure 4 shows the variation of the radiopolarograms of Cm at pH=3.0. We could not examine a wide range of the concentration because we used ^{243}Cm($t_{1/2}$=28.5y) and ^{244}Cm($t_{1/2}$=18.11y). The pre-waves were not observed at the concentration investigated. The half-wave potential of Cm was found to be approximately -2.01 V. The diffusion coefficients calculated are found to be almost constant, which shows that the wave height is proportional to the concentration.

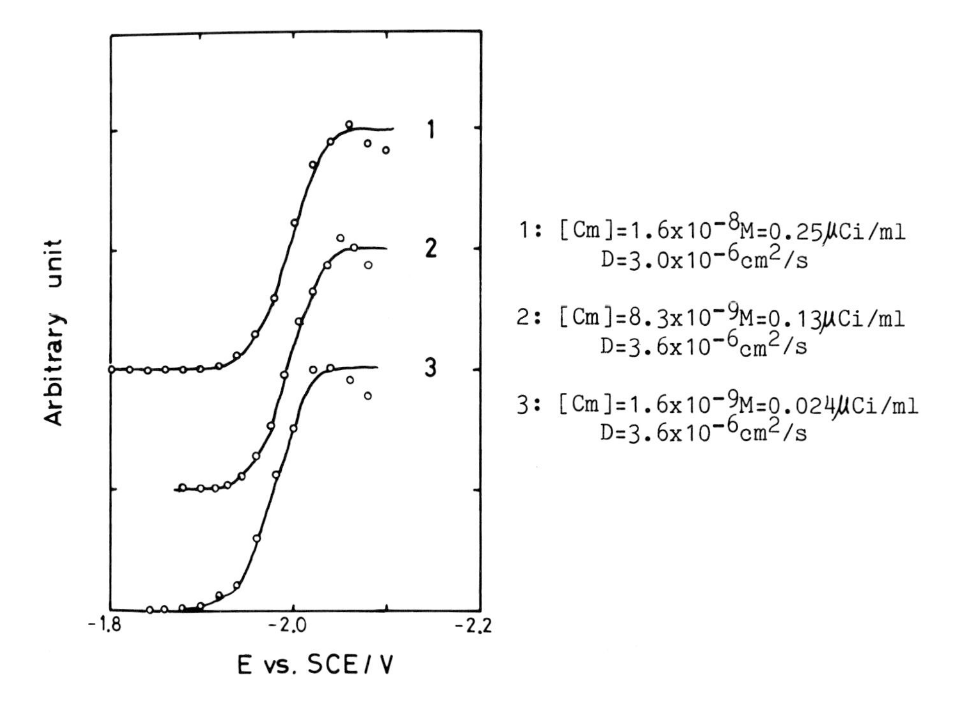

Figure 4. Radiopolarograms of Cm(III) at pH=3.0 in 0.1 M LiCl.

Figure 5 shows the variations of the radiopolarograms of Cm with pH. Similar to Am, the pre-waves of Cm were observed at lower pH, and they disappeared at pH= 2.8. Also, the half-wave potential of main waves did not vary with pH. The potential of the pre-waves of Cm is almost equal to that of Am.

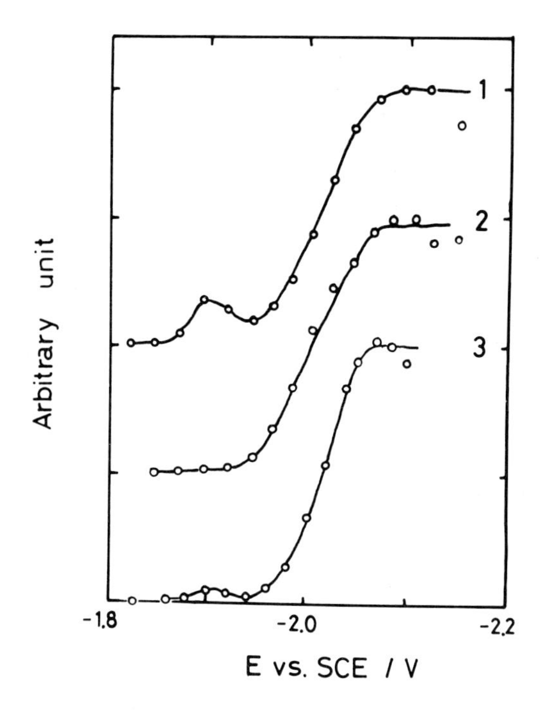

1: [Cm]=2.8×10^{-8}M, pH=2.3
D=4.0×10^{-6}cm^2/s

2: [Cm]=3.5×10^{-8}M, pH=2.8
D=2.6×10^{-6}cm^2/s

3: [Cm]=8.9×10^{-9}M, pH=2.5
D=3.6×10^{-6}cm^2/s

Figure 5. Radiopolarograms of Cm(III) at various pH in 0.1 M LiCl.

Figure 6 shows the dependence of the wave height on the mercury column height. The abscissa is expressed as $m^{2/3}t^{1/6}$ (m;the flow rate of mercury, t;the lifetime of a mercury drop). The radiopolarograms corresponding to the conditions of No.1 for Am and No.4 for Cm of Fig. 6, exhibit single waves, and their limiting activities are found to be diffusion-controlled. The radiopolarograms corresponding to the conditions of Nos.2 and 3 for Am, and Nos.4 and 5 for Cm, exhibit the pre-waves and main waves. The main waves are diffusion-controlled as shown in Nos.2 and 5 of Fig. 6, but the pre-waves are not as shown in Nos.3 and 6. These results show that the pre-waves occur independent of the main waves. Because if this would not be the case, the main waves would not be diffusion-controlled. This consideration is also supported by the fact that the constant diffusion coefficients are obtained independently of existing the pre-waves, as is shown in Fig. 2.

Thus, the pre-waves are not diffusion-controlled, and their height and their potential of the maximum height are dependent on the pH and the concentration of Am and Cm. The pre-waves may be the result of the adsorption. This effect has been reported previously in dc polarographic studies of lanthanide(III) ions(1). At present, we cannot explain explicitly why the pre-waves occur.

On the other hand, the main waves are diffusion-controlled and their heights are proprotional to the concentrations. Also the half-wave potentials are constant within the pH range of our investigation. The corresponding electrode reactions are considered to be totally irreversible.

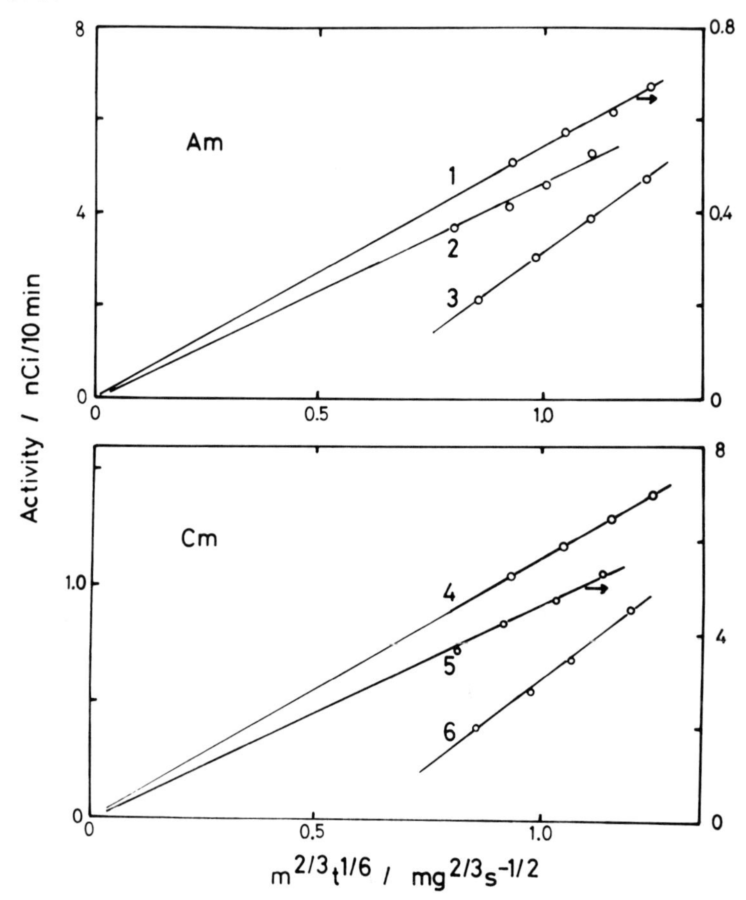

Figure 6. Dependence of the wave height on Hg column height in 0.1 M LiCl.

1: [Am]=7.4×10^{-8}M, pH=3.0 and E=-2.050V
2: [Am]=7.0×10^{-7}M, pH=2.2 and E=-2.075V
3: [Am]=7.1×10^{-7}M, pH=2.2 and E=-1.905V
4: [Cm]=8.5×10^{-9}M, pH=3.0 and E=-2.050V
5: [Cm]=3.5×10^{-8}M, pH=2.2 and E=-2.106V
6: [Cm]=3.5×10^{-8}M, pH=2.2 and E=-1.926V

Kinetic parameters of the electrode reactions

For a totally irreversible electrode reaction, the following relations in dc polarography are derived(8):

$$(RT/\alpha nF) \ln [\ X(5.5-X)/5(1-X)\ ] = E_{1/2} - E \qquad (1)$$

with $X = I/I_d$ and the half-wave potential $E_{1/2}$:

$$E_{1/2} = (RT/\alpha nF) \ln [\ 0.89\ K_c^0\ (t/D)^{1/2}\ ], \qquad (2)$$

where I is the current at the potential E; I_d, the limiting current; D, the diffusion coefficient; t, the lifetime of a mercury drop; α, the transfer coefficient; K_c^0, the cathodic rate constant at E=0 V; n, the number of electrons; R, F and T have usual meanings. Equations (1) and (2) are also valid for radiopolarography, if I and I_d are replaced by the activity A and the limiting activity A_d, respectively.

Now, from Eqs.(1) and (2) we can get the next relation:

$$(RT/\alpha nF)\{\ \ln X(5.5-X)/5(1-X) - 0.5 \ln t\ \} =$$

$$(RT/\alpha nF) \ln (\ 0.89\ K_c^0/D^{1/2}\ ) - E. \qquad (3)$$

According to this, a plot of log [X(5.5-X)/5(1-X)] - 0.5 log t vs. E should result in a straight line for a constant lifetime t and should be reduced to a single straight line for the different lifetimes. From this plot one can obtain the transfer coefficient α and the cathodic rate constant K_c^0.

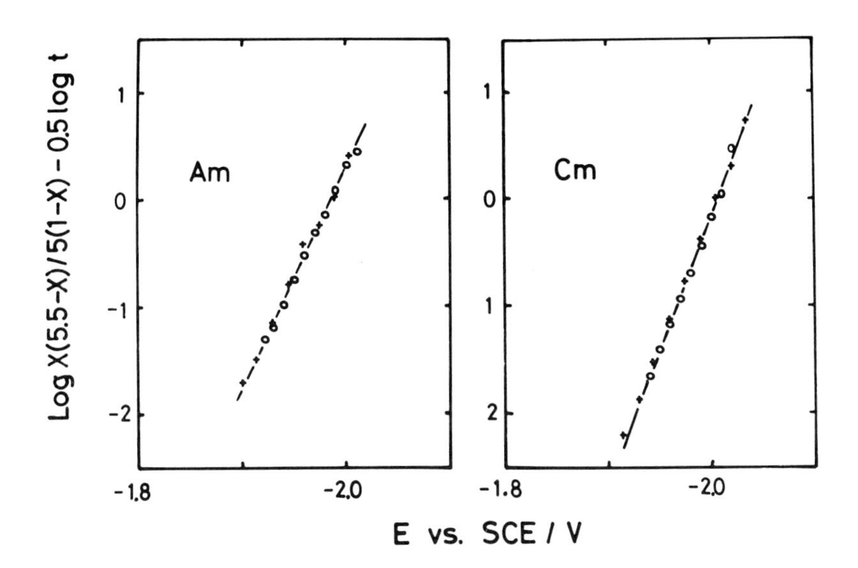

Figure 7. Log plots for waves of Am(III) and Cm(III) at pH=3.0 in 0.1 M LiCl.

o: t=3.2 s, +: t=5.0 s

The radiopolarograms for the different constant lifetimes, t=3.2 s and t=5.0 s, were obtained at pH=3.0 and concentration of [Am]=$6.8x10^{-8}$ M and [Cm]=$8.1x10^{-9}$ M, respectively. They exhibit only main waves. The log plots according to Eq.(3) are shown in Fig. 7. These plots are straight lines which are independent of the lifetime. The results suggest the main waves of Am and Cm are attributed to the reduction to the metallic states and not to a nonfaradic process such as an adsorption.

	α	$K_c^0/D^{1/2}$ $s^{-1/2}$	K_c^0 cm s^{-1}
Am	0.41	$3.6x10^{-37}$	$8.8x10^{-40}$
Cm	0.50	$3.7x10^{-45}$	$9.1x10^{-48}$

Table I Kinetic parameters of electrode reactions for Am(III)/Hg(Am) and Cm(III)/Hg(Cm) in 0.1 M LiCl at pH=3.0 and 25°C.

Table I shows the kinetic parameters of Am(III)/Hg(Am) and Cm(III)/Hg(Cm), where thc diffusion coefficients are assumed to be $6.0x10^{-6}$ cm^2/s for both ions.

We observed the reduction of Am(III) and Cm(III), by radiopolarography, to the metallic state in an aqueous solution and obtained the kinetic parameters of this reaction, which were unknown until now.

Literature Cited

1. Large, R.F,; Timnick, A. Anal.Chem. 1964, **36**, 1258.
2. Shiokawa, Y.; Suzuki, S. Bull.Chem.Soc.Jpn. 1984, **57**, 2910.
3. Samhoun, K.; David, F. J.Inorg.Nucl.Chem. 1979, **41**, 357.
4. Hulet, E.K.; Gutmacher, R.G.; Coops, M.S. J.Inorg.Nucl.Chem. 1961, **17**, 350.
5. Choppin, G.R.; Harvey, B.G.; Thompson,S.G. J.Inorg.Nucl.Chem. 1956, **2**, 66.
6. All electrode potentials in this paper are with respect to the saturated calomel electrode(SCE).
7. Latrous, H; Oliver, J; Chemla, M. Radioanal.Radioanal. Lett. 1982, **53**, 81.
8. Oldham, K.B,; Parry, E.P. Anal.Chem. 1968, **40**, 65.

AMERICIUM TITRATION METHODS

Ph. CAUCHETIER - C. GUICHARD
Commissariat à l'Energie Atomique
IRDI - DERDCA - SEACC
P.O Box 6
F 92260 Fontenay-aux-Roses
France

ABSTRACT. To determine americium, we use either of the following methods : spectrometry at 60 keV, spectrophotometry of Am (III) at 503 nm or isotopic dilution mass spectrometry. These methods need reference materials. Now only certified for activity materials exist. So we devellopped titrimetric methods. The first is an original redox method. After quantitative oxidation to Am (VI) at once by persulfate and argentic oxide, we mix rapidly equal parts of the solution with different knowned quantities of Fe (II) and dilute with molar perchloric acid until fixed volume. Then we measure the Am (III) absorbance and infer its molar extinction coefficient and the americium concentration. The second method is complexometry with EDTA : americium takes the place of mercury in the HgY^{2-} complex and the quantity of free mercuric ions is determined by pulse coulometry. We work in hexamethylene-tetramine medium at pH 5,5 or in ammonia medium at pH 9,0. The results agree in a 2% range. We give obtained molar extinction coefficients.

1. INTRODUCTION

The methods most commonly used to determine americium rely on a previous calibration against a know americium solution, but the only reference materials available are certified not in americium content but in activity only. This implies a proper knowledge of the half-life for which, until very recently, the literature has contained two groups of values different by 5%.

2. USED METHODS

Three principal methods are used in our laboratory to determine americium :
- γ spectrometry on a solid source after evaporation of the solution. The reproducibility obtained is better than 1%. Self-absorption effects are much slighter than α spectrometry, which allows this method to be used in many cases without previous separation. Thus a few ppm of ^{241}Am

N. M. Edelstein et al. (eds.), Americium and Curium Chemistry and Technology, 115–119.

can be measured in plutonium or mixed oxides, as can americium 241 contained in spent fuels, after simple evaporation of the corresponding solutions. The method depends on the existence of reference materials, which means that at present a mass can only be determined through the half-life value, and is also specific to ^{241}Am. ^{243}Am may be measured as well with less sensitivity, but in this case the branching ratios of the two isotopes must also be known.

- Direct absorption spectrophotometry using the characteristic Am (III) peak at 503 nm. The advantages of this method lie in its sensitivity (high molar extinction coefficient), specificity (sharpness of peak) and as the case may be non-destructiveness, since americium is usually in the oxidation state (III) only. We use it regularly, especially for the analysis of ^{243}Am solutions used to prepare tracers for mass spectrometry or to milk off ^{239}Np. It is also used for the determination of americium in spent fuels, without separation as long as the Am/Pu ratio is at least 3.10^{-3}, as well as in fission products. However the technique demands either the use of reference materials of certified americium content or a sufficiently accurate knowledge of the molar extinction coefficients under the measurement conditions, and this can only be obtained experimentally from known solutions.

- Mass spectrometry, which gives access to the isotopic composition and through the isotopic dilution method to the concentration. It is obviously necessary in both cases to separate elements possessing isotopes of identical mass, essentially plutonium and curium. This operation takes place in three stages. Plutonium and uranium are fixed on anionic resin in 12 M HCl solution. The eluate containing rare earths and trivalent actinides is itself separated into groups by passage over anionic resin in nitric-methanol solution. Finally the fraction containing americium and curium is separated over fine-particle cationic resin at 80° C with α-hydroxy-isobutyric acid / 1 /. This method requires tracer solutions (^{241}Am or ^{243}Am according to the isotopic composition of the americium to be measured) of known concentration.

3. CALIBRATION METHODS

We have attempted to calibrate these methods and offset the absence of americium reference materials by the use of titration methods involving quantitative oxidation-reduction or complexation reactions. Coulometric techniques have been proposed / 2,3,4 /, also an EDTA complexometric method / 5 /.
Owing to the instability of higher americium valencies it is difficult to apply coulometric methods and we prefered to use the fast reaction reducing Am (VI) to Am (III) by Fe (II). The solution in question ($\sim 10^{-2}$ M in M $HClO_4$, prepared from pure AmO_2) is oxidised hot by a mixture of argentic oxide and persulphate. After rapid cooling and volume adjustment equal fractions are mixed with exactly known quantities of Fe (II). Each of the solutions obtained is diluted to known volume in M $HClO_4$ and its absorption spectrum measured from 530 to 490 nm.

Finally the peak heights are plotted against the quantities of Fe (II) added (fig.1) : the curve is linear as long as the amount of Fe (II) is sub-stoichiometric and the ordinate at the origin is zero.From this curve we derive both the concentration of the americium solution and the molar extinction coefficient of Am^{3+} in perchloric medium. The quantitativeness of the oxidation and the stability of Am (VI) were checked: the amount of Am (V) present 10 minutes after oxidation, the moment at which Fe (II) is added, is less than 1.5 % and Am (III) is undetectable (0.9 % present after 4 hours).

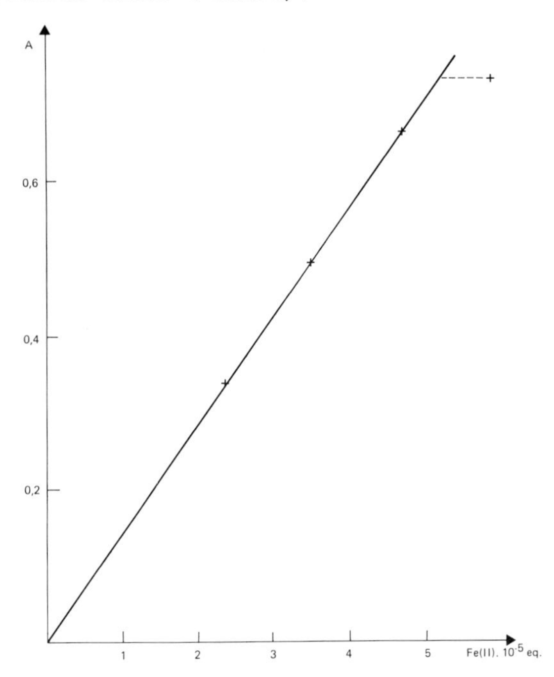

FIGURE 1 : Reduction of Am(VI) by Fe(II): Am(III) absorbance at 503 nm versus added Fe(II) amount.

The accuracy of the results obtained by this method, probably subject to certain interferences (Pu, Np, Cr, Mn..), was checked by another method using the displacement by Am^{3+} of mercuric ions complexed with EDTA (Hg Y^{2-}) and the coulometric reduction of liberated Hg^{2+}. The method is similar to that published by Bergey / 6 / but the medium chosen was hexamethylene-tetramine at pH 5.5, non-complexing for Hg^{2+} and hence more suitable than acetic solution for detection of the equivalent point. This is accomplished by amperometry with two silver electrodes amalgamated by dipping, between which is applied a potential difference of 20 mV. A TACUSSEL TT700 "coulometric burette" combined with a TACUSSEL Titrimax apparatus is used for the pulsed current generation and measurement. An individual relative standard deviation of 1% is obtained for a sample of about 300 µg, and 0.6% in the 2 mg region. It should be noted that neptunium interferes here but not quantitatively, and that about 1600 ppm is formed per year from ^{241}Am.

The same methode has also been used in ammoniacal medium at pH 9 and the results are not significantly different.

4. RESULTS

For purposes of comparison the molar extinction coefficient values for the Am^{3+} ion at 503 nm deduced from these calibration methods are given in table I, and cross-check satisfactorily. They differ from already published values / 7, 8 /. This ties in with the fact that the americium contents found in reputedly pure oxide are lower by about 4 % than expected from the stoichiometry of $Am\ O_2$.

Table I : Molar extinction coefficient values of Am^{3+} in different media, deduced from various calibration methods.
* Obtained by comparaison of spectra (fig. n° 2)
** Obtained by extrapolation of other values

	Calibration method		
Medium	Fe (II)	EDTA pH 5,5	EDTA pH 9,0
$HClO_4$ - 0,9 M	422 **	430 ± 3	426 ± 4
$HClO_4$ - 0,9 M H_2SO_4 - 0,16 M	420 ± 6		
$HClO_4$ 0,08 M H_2SO_4 0,92 M	411 *	419 *	415 *
$HClO_4$ 0,08 M HNO_3 3,68 M	327 *	333 *	330 *
HNO_3 4 M	319 **	322 ± 3	319 ± 3

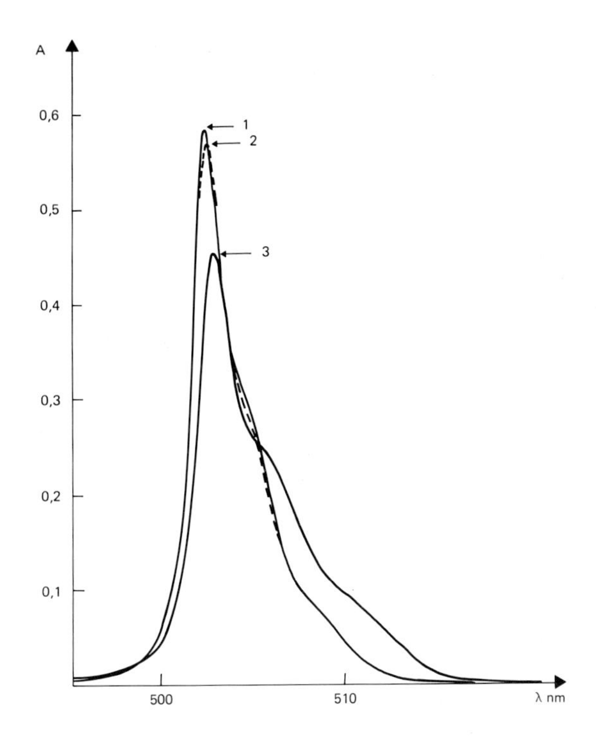

FIGURE 2 : Absorption spectra of solutions with the same Am concentration (395 mg of AmO_2 per liter) in medium :
1 – $HClO_4$ M
2 – $HClO_4$ 0,08 M – H_2SO_4 0,92 M
3 – $HClO_4$ 0,08 M – HNO_3 3,68 M

References :

/ 1 / A. BILLON
J. Radioanal. Chem. 51 (1979), 297

/ 2 / G. KOEHLY
Anal. Chim. Acta 33 (1965), 418

/ 3 / J.R. STOKELY, W.D. SHULTZ
Anal. Chim. Acta 45 (1969), 417

/ 4 / R.C. PROPST
Anal. Chem. 41 (1969), 910

/ 5 / S.S. YAMAMURA
Anal. Chem. 40 (1968), 1898

/ 6 / C. BERGEY, L. FOUCHARD
Talanta 26 (1979), 445

/ 7 / B.F. MYASOEDOV and al.
Analytical Chemistry of transplutonium elements (1972)
Israel Program for scientific translations

/ 8 / A.W. HARMON and al.
LA-UR 81 2885 - Conf. 811025-5 (1981)

Part III

Electronic Structure and Thermodynamics

ELECTRONIC STRUCTURE OF NEUTRAL AND SINGLY IONIZED CURIUM*

Earl F. Worden
Lawrence Livermore National Laboratory
P.O. Box 808, L-464
Livermore, CA 94550

John G. Conway
Lawrence Berkeley Laboratory
Materials and Molecular Research Division
Berkeley, CA 94720

Jean Blaise
Laboratory Aime Cotton
CNRS II
Batiment 505
91405, Orsay, France

ABSTRACT. Extensive observations and analyses of the emission spectra of neutral and singly ionized curium, Cm I and Cm II, have resulted in the determination of 785 Cm I and 598 Cm II energy levels. The levels found combine to classify 9145 of the more than 14,250 lines of ^{244}Cm observed between 240 and 2650 nm. Most of the levels have Lande g-values from Zeeman effect data and isotope shifts from measurements of spectra from sources with various enrichments of ^{244}Cm, ^{245}Cm, ^{246}Cm and ^{248}Cm. These data allowed us to assign many levels to specific electronic configurations. The ground configurations of Cm I and Cm II are [Rn] $5f^7 6d7s^2$ and [Rn] $5f^7 7s^2$, respectively. The relative energies of other electronic configurations of Cm are given and compared with analogous configurations in other actinides and in Gd its lanthanide analogue.

1. INTRODUCTION

There is now considerable knowledge of the electronic structure of neutral and singly ionized curium, Cm I and Cm II. This has resulted from our extensive observations and analyses of the curium emission spectrum. The analyses have yielded over 390 odd and 395 even levels of Cm I (most of these are published in Ref. 1) and over 434 odd and 164 even (nearly all unpublished) levels of Cm II. These odd and even levels of the two spectra combine to classify over 9145 or 64 percent of the more than 14,250 atomic emission spectrum lines of ^{244}Cm

N. M. Edelstein et al. (eds.), Americium and Curium Chemistry and Technology, 123–134.

observed between 240 and 2650 nm. Most levels have Lande g-values from Zeeman effect data and almost all of the levels have known isotope shifts obtained from measurements of the spectra taken with sources enriched in various percentages of the isotopes ^{242}Cm, ^{244}Cm, ^{245}Cm, ^{246}Cm and ^{248}Cm (see Table I and Ref. 2). These data, especially the extensive isotope shift data, have permitted assigning a number of levels to specific electronic configurations and the g-values allow term assignments of the levels.

Table 1. Isotopic Composition of Curium Samples Used for Electrodeless Lamp Preparation.

Sample	Percent of Isotope (>1%)				Use
	244	245	246	248	
1	95	1.5	3		Wavelengths, Zeeman effect, spectrum assignment and reversal. Isotope shift on strong lines.
2	27		16	55	Isotope shift
3	55	23	21		Isotope shift

The analyses of these spectra have progressed to the point where all levels expected below 15,000 cm^{-1} have been identified in each spectrum. As is the case in all very complex spectra, many levels still remain to be found at high energies. However, the analyses are quite complete and give a very good description of the electronic structure of Cm I and Cm II.

2. RESULTS AND DISCUSSION

Levels belonging to the most important electronic configurations in the first, Cm I, and second, Cm II, spectra of curium have been determined. These configurations and the lowest identified level in each are given in Table II for Cm I and in Table III for Cm II. The configurations for each spectrum are grouped as odd or even according to spectroscopic rules. For the configurations considered here, a configuration is odd if the total number of f plus p electrons is odd and it is even if that total number is even. The energy of the lowest level for each configuration is given. In our discussion we will use the energy difference between the lowest levels as the energy difference between the electronic configurations. These differences will be representative of the promotion energy between configurations. Tables IV and V contain similar sets of data for Gd I and Gd II. In the tables the term designation, the Lande g-value and, in the case of Cm, the isotope shift ^{246}Cm - ^{244}Cm are given for each level.

Table II. Lowest levels of identified configurations in Cm I.

Configuration	Lowest Level (cm^{-1})	Desig.	Lande g-value	Isotope Shift (10^{-3} cm^{-1})
Odd				
$5f^7 6d\ 7s^2$	0.00	$^9D_2$	2.563	000
$5f^7 6d^2 7s$	10144.93	$^{11}F_2$	2.873	-339
$5f^8 7s\ 7p$	17656.63	$^9D_6$	1.621	-463
$5f^7 6d^3$	30443.92	$^{11}F_2$	2.53	-619
$5f^7 6d 7s 8s$	34255.16	$^{11}D_3$	2.064	-220
Even				
$5f^8 7s^2$	1214.18	$^7F_6$	1.452	-279
$5f^7 7s^2 7p$	9263.37	$^9P_3$	2.112	+118
$5f^7 6d 7s 7p$	15252.70	$^{11}F_2$	2.835	-269
$5f^8 6d 7s$	16932.72	$^9G_7$	1.466	-586
$5f^7 6d^2 7p$	32581.06 [a]	$^{11}G_2$	2.043	-527
$5f^8 7s 8s$	33013.09	$^9F_7$	1.54	-460

[a]Lowest level of this configuration has not been found. It should be $^{11}G_1$ near 32000 cm^{-1}.

Table III. Lowest levels of identified configurations in Cm II.

Configuration	Lowest Level (cm^{-1})	Desig.	Lande g-value	Isotope Shift (10^{-3} cm^{-1})
Odd				
$5f^7\ 7s^2$	0.00	$^8S_{7/2}$	1.935	000
$5f^7 6d 7s$	4010.66	$^{10}D_{5/2}$	2.492	-496
$5f^7 6d^2$	14830.15	$^{10}F_{3/2}$	3.009	-962
$5f^8 7p$	27065.11	$^8F_{11/2}$	1.51	-972
Even				
$5f^8 7s$	2093.88	$^8F_{13/2}$	1.500	-738
$5f^8 6d$	17150.79	$^8G_{13/2}$	1.415	-1177
$5f^7 7s 7p$	24046.39	$^{10}P_{7/2}$	2.098	-403
$5f^7 6d 7p$	32034.44	$^{10}F_{3/2}$	2.933	-923

Table IV. Lowest levels of identified configurations in Gd I. [a]

Configuration	Lowest Level (cm^{-1})	Desig.	Lande g-value
Odd			
$4f^7 5d\ 6s^2$	0.00	$^9D_2$	2.651
$4f^7 5d^2 6s$	6378.15	$^{11}F_2$	2.980
$4f^7 5d^3$	22429.16	$^{11}F_2$	
$4f^8 6s6p$	25658.06	$^9D_6$	1.580
$4f^7 5d6s7s$	31907.02	$^{11}D_3$	2.496
$4f^7 6s^2 7s$	34719.13	$^9S_4$	1.980
Even			
$4f^8\ 6s^2$	10947.21	$^7F_6$	1.480
$4f^7 6s^2 6p$	13433.85	$^9P_3$	2.225
$4f^7 5d6s6p$	14036.03	$^{11}F_2$	2.955
$4f^8 5d6s$	24255.10	$^9G_7$	1.515
$4f^7 5d^2 6p$	25069.18	$^{11}G_1$	3.764
$4f^8 6s7s$	40439.51	$^9F_7$	1.550

[a] J. Blaise, J.-F. Wyart and Th. A.M. Van Kleef, C.R. Acad. Sc. Paris, 1970, 70, 261-263, and J. Blaise, J. Chevillard, J. Verges, J. F. Wyart and Th. A.M. Van Kleef, Spectrochim. Acta, 1971, 26B, 1-34.

Table V. Lowest level of identified configurations in Gd II. [a]

Configuration	Lowest Level (cm^{-1})	Desig.	Lande g-value
Odd			
$4f^7 5d\ 6s$	0.00	$^{10}D_{5/2}$	2.557
$4f^7 6s^2$	3444.24	$^8S_{7/2}$	1.920
$4f^7 5d^2$	4027.16	$^{10}F_{3/2}$	3.163
$4f^8 6p$	32595.35	$^8D_{11/2}$	1.565
Even			
$4f^8 6s$	7992.27	$^8F_{13/2}$	1.537
$4f^8 5d$	18366.85	$^8G_{15/2}$	1.465
$4f^7 6s6p$	25668.69	$^{10}P_{7/2}$	2.190
$4f^7 5d6p$	25960.07	$^{10}F_{3/2}$	3.095

[a] J. Blaise, J. F. Wyart and T. A.M. Van Kleef, J. Phys. (Paris), 1971, 32, 617, and J. Blaise, J. Chevillard, J. Verges, J. F. Wyart and Th. A.M. Van Kleef, Spectrochim. Acta, 1971, 26B, 1-34.

Comparison of Tables II and IV and III and V show some similarities in the energies of like configurations for the two spectra of the elements but there are also some notable differences. The general ordering of the configurations considered here for the first spectra (Cm I and Gd I, Tables II and IV) is the same except for f^7d^3. The 5d electrons are relatively more tightly bound in Gd than the 6d orbital electrons are in Cm. This is true in both the neutral and singly ionized configurations involving f^7. More striking is the relative positions of configurations involving f^7 and f^8. The even configuration $5f^87s^2$ in Cm I is only 1214 cm^{-1} above the $5f^76d7s^2$ ground configuration in Cm I while in Gd I $4f^86s^2$ is 10947 cm^{-1} above the equivalent ground configuration $4f^75d6s^2$. All equivalent f^8 configurations in Cm I are lower in energy by 7000 to 9000 cm^{-1} than the corresponding configurations in Gd.

Changes in the relative ordering of the electronic configurations because of differences in binding energies of orbital electrons in the two elements is more evident in the singly ionized or second spectra, Tables III and V. Here, the ground state of Cm II is $5f^77s^2$ some 4010 cm^{-1} below $5f^76d7s$ the configuration equivalent to the ground configuration in Gd II, $4f^75d6s$. The promotion energy f^7s^2 to f^8s is only 2094 cm^{-1} in Cm II while it is 4550 cm^{-1} in Gd II. As a result, the configuration ordering in Cm II is f^7s^2, f^8s, f^7ds, f^7d^2 while in Gd II it is f^7ds, f^7s^2, f^7d^2, f^8s. The p orbital electrons do show some differences but not the marked differences exhibited by f and d orbital electrons.

Fig. 1 shows the relative positions of three electronic configurations each in neutral actinides and lanthanides. In the figure, the lowest level of the $f^{N-1}ds^2$ configuration is set at zero energy and the positions of the lowest levels of the f^Ns^2 and f^{N-1} dsp configurations are shown. The promotion energy for s to p is about the same in the actinides as in the lanthanides and nearly constant across the series. This relationship and other trends in these and similar plots with more configurations are useful in the analyses of actinide and lanthanide spectra since estimates can be made of the position of unknown lowest levels of various configurations (dotted lines in the figure).

Fig. 1 shows that the f to d promotion energy changes considerably across the series. The general trend is for the f^Ns^2 configuration to become more stable as the series are ascended. The $5f^N7s^2$ configuration is less stable than $4f^N6s^2$ relative to $f^{N-1}ds^2$ for N less than or equal 7 but the opposite is true when N is equal to or greater than 8. Thus the two curves cross between N equal 7 and 8. The 5f orbital electron is more stable relative to the 6d orbital electron in the heavier actinides than is the 4f orbital electron relative to the 5d orbital electron in the heavier lanthanides.

The increased stability for the 5f orbital electron versus the 6d orbital electron is more clearly shown in Fig. 2 where the f^Ns^2 configuration is set at zero energy. This increased stability of the

5f orbital electron is related to the tendency toward divalent character for the higher members of the series found by nuclear chemists before the positions of the electronic configurations had been established by spectroscopists. Again, one can see that the promotion energy for 7s to 7p orbital electron is nearly constant while the 7s to 6d promotion energy increases across the series.

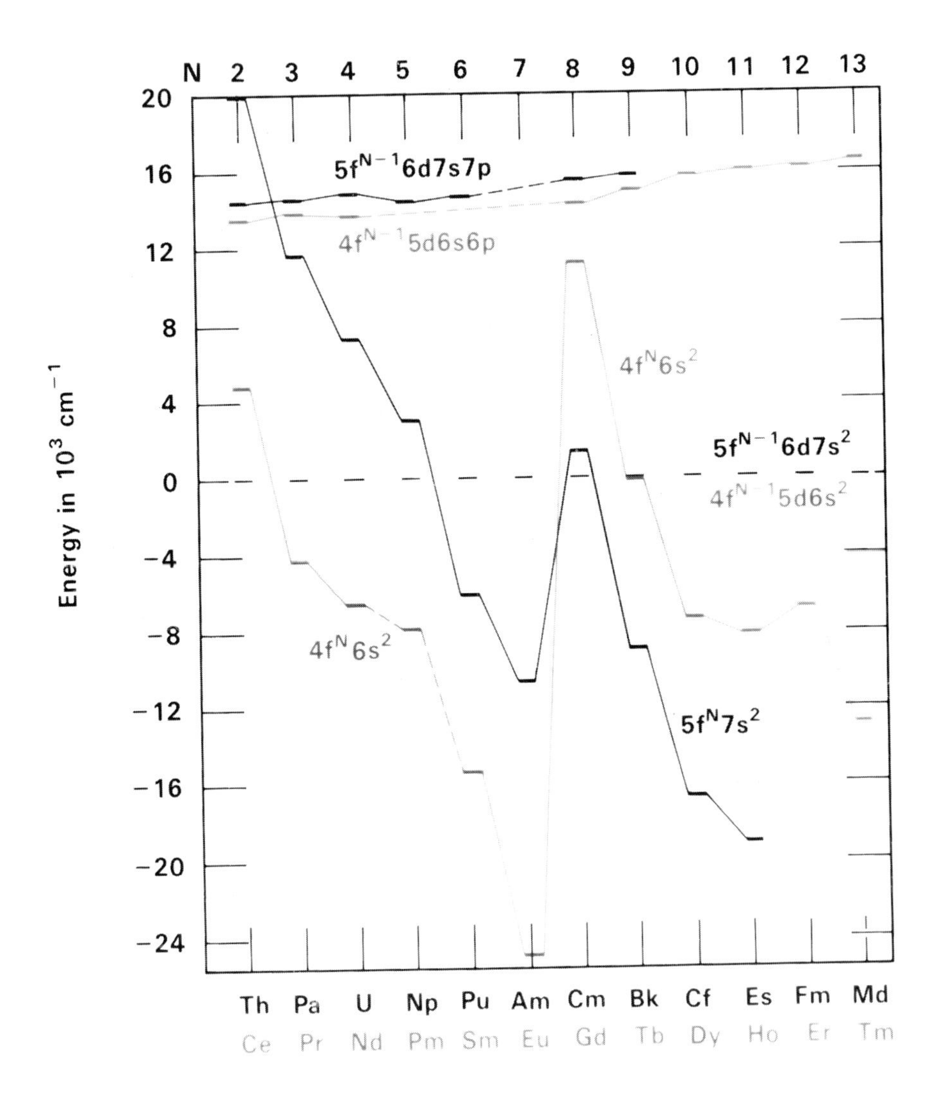

Fig. 1. Energy differences of the lowest levels of the f^Ns^2 and f^{N-1} dsp configurations relative to $f^{N-1}ds^2$ in the neutral or first spectra of some actinides (5f configurations--dark) and lanthanides (4f configurations--light).

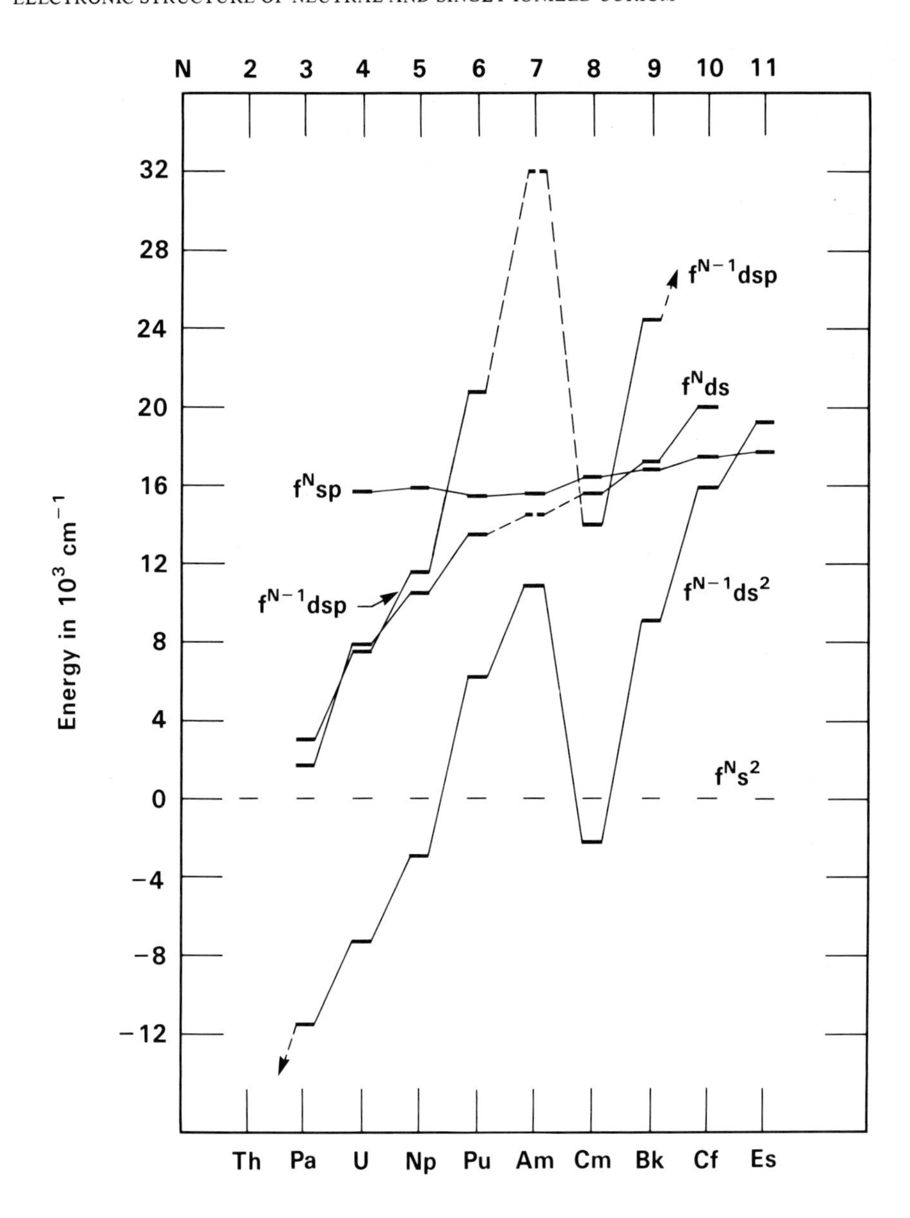

Fig. 2. Energy of the lowest levels of several configurations relative to the lowest level of f^Ns^2 in neutral actinides. In the configurations shown, the electrons are 5f, 6d, 7s and 7p. The increase in stability of the f^Ns^2 configuration across the series is evident.

Only the lowest levels of various configurations have been shown thus far. A large number of complete terms have been found for many of the configurations of Cm I and Cm II. Table VI shows levels of the $5f^8 7s$ configuration of Cm II. This table clearly illustrates the value of isotope shift in making assignments of levels to configurations. The term assignments are made by use of the experimental Lande g-values. The constancy of isotope shift and the agreement of the observed g-values with values from theory or with the LS g-values substantiate the validity of the assignments.

Table VI. Low levels of the $5f^8 7s$ configuration of Cm II.

Level (cm^{-1})	J	g_{obs}	g_{LS}	g_{calc}[a]	IS[b] (mK)	Desig.
2093.88	13/2	1.500	1.540	1.499	-738	^8F
3941.46	11/2	1.424	1.455	1.432	-765	^6F
5919.28	9/2	1.527	1.577	1.529	-736	^8F
6347.92	11/2	1.500	1.554	1.495	-744	^8F
7067.14	7/2	1.485	1.620	1.492	-748	^8F
8144.32	9/2	1.400	1.435	1.404	-765	^6F
8436.11	5/2	1.656	1.716	1.655	-733	^8F
9073.60	7/2	1.444	1.398	1.439	-753	^6F
9127.87	3/2	1.834	2.002	1.847	-733	^8F
9801.32	1/2	3.740	4.007	3.739	-729	^8F
10433.80	5/2	1.300	1.315	1.305	-763	^6F
11250.911	3/2	1.167	1.067	1.155	-759	^6F
11978.47	1/2	-0.420	-0.670	-0.425	-760	^6F

[a]J. Blaise, J. F. Wyart, J. G. Conway and E. F. Worden, Phys. Scripta, 1980, 22, 224.
[b]Isotope shifts are relative to the $5f^7 7s^2$ $^8S_{7/2}$ ground state. They are ^{246}Cm-^{244}Cm in mK (1 mK = 10^{-3} cm^{-1}).

The level structure of the lowest terms in Cm II are shown in Fig. 3. The electronic configurations and LS term designations are indicated. The structure of the levels of the $5f^8 7s$ configuration given in Table VI is more clearly illustrated in the figure. The closeness in binding energy of the 5f, 6d, and 7s orbital electrons is evident.

Fig. 4 shows the range of isotope shifts observed for assigned levels in the Cm I and Cm II spectra. It should be noted that there is overlap of observed shift for a number of configurations. In these cases, theory, Lande g-values, transition intensities and other factors are important for assigning the level to the appropriate configuration. As seen in the figure, we report the shifts relative to the ground state as x. For the shifts shown in Table II, III and VI, x is zero. The shifts are the frequency of ^{246}Cm minus the frequency of ^{244}Cm. As is well known, the isotope shift is caused by interaction with the nucleus of the penetrating s orbital electrons and to a lesser extent by the $p_{1/2}$ electrons. Since f electrons act to shield or reduce the penetrations of the s and p electrons, the configurations with smallest isotope shift of those shown should be f^7d^3 and f^7d^2p in Cm I and f^7d^2 and f^8d in Cm II. Thus the value of x should be about + 0.7 cm^{-1} for Cm I and about +1.2 cm^{-1} for Cm II.

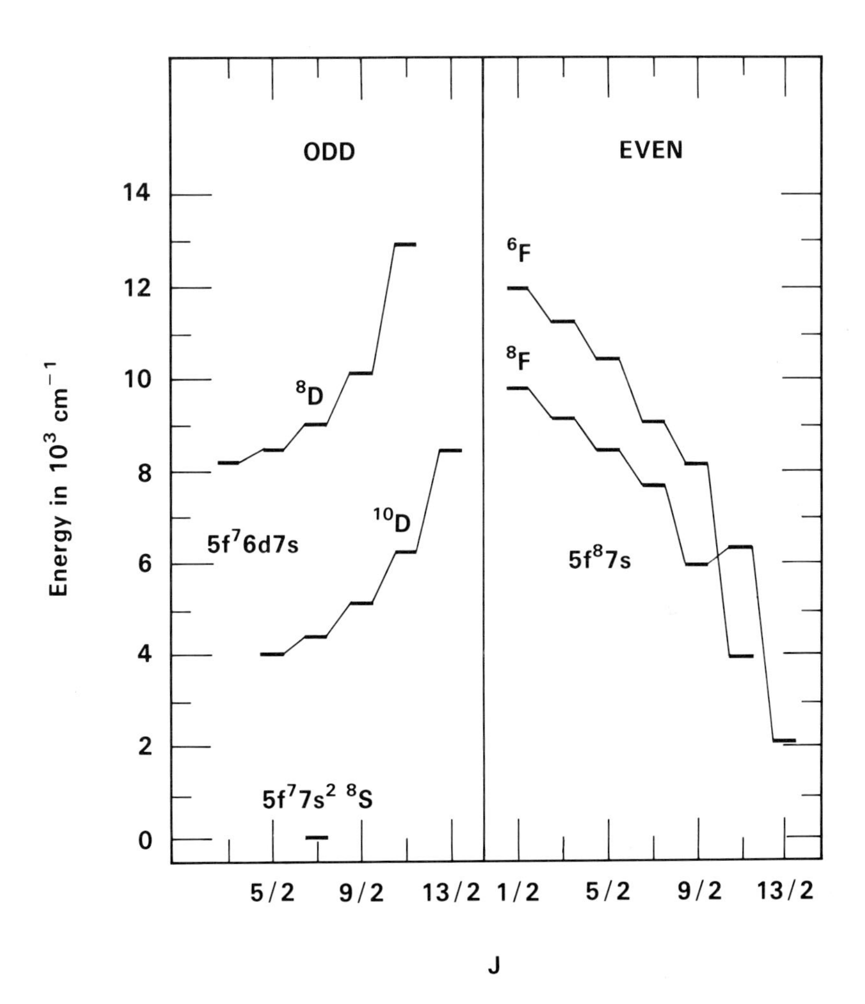

Fig. 3. Low lying energy levels of singly ionized curium. The electronic configurations and LS term designations are given.

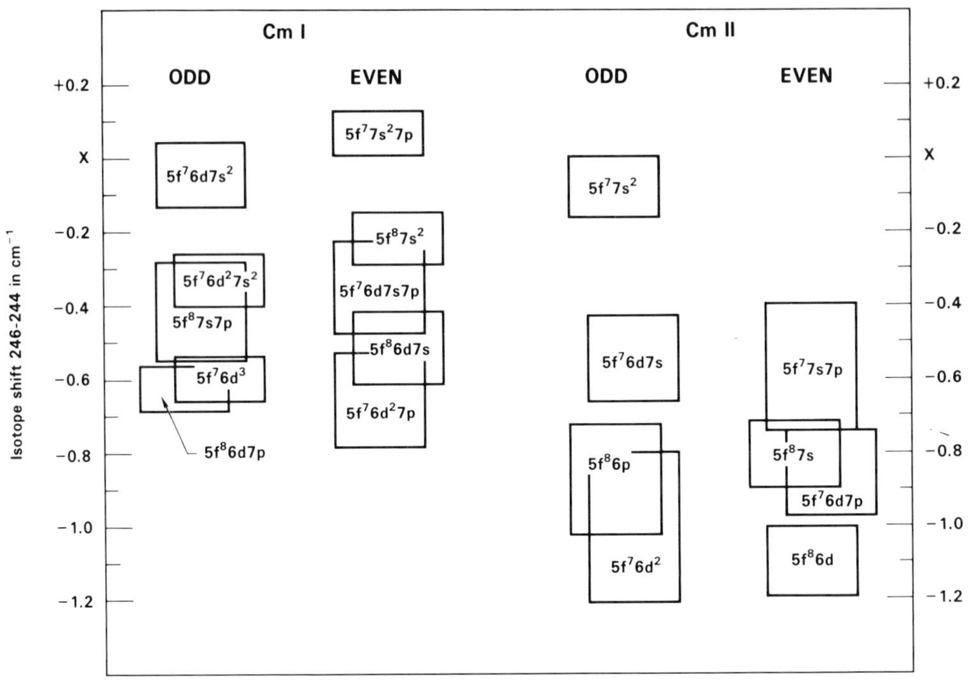

Fig. 4. Range of observed isotope shifts for several electronic configurations of curium I and II. The value of x is about +0.7 cm^{-1} for Cm I and about +1.2 cm^{-1} for Cm II.

Investigation of the isotope shifts with sources 2 and 3 listed in Table I were very important in obtaining reliable isotope shift values for the levels. The use of exposures from all three of these samples on the same plate made resolution of many confusing shifts on individual lines possible. By use of lines with large isotope shift and with negligible or no ^{245}Cm hyperfine structure, we were able to determine the relative isotope shift values shown in Fig. 5. The relative isotope shift for ^{242}Cm was obtained from data of Conway and McLaughlin.[2] For lines with small isotope shift more accurate ^{246}Cm - ^{244}Cm shifts could be calculated from the more accurately measured ^{248}Cm - ^{244}Cm shift and the relative isotope shift values. The odd-even staggering common in the actinides is shown by the position of ^{245}Cm in Fig. 5.

We show in Table VII a listing of papers dealing with observations of the atomic emission spectra and energy level analyses of Cm I and Cm II. The major contribution of each paper is noted in the Table. The early investigations 1 and 2 were limited by the small quantities of material, sources and spectrographs. The production of 2 mg of highly purified ^{244}Cm by E. K. Hulet, the use of electrodeless lamps as sources, the use of the 9.15 m grating spectrograph at Argonne National Laboratory with the associated high quality 2.4 T magnet, and finally the use of the Fourier Transform Spectrometer at Laboratory Aime Cotton all contributed to the successful acquisition of the very high quality data needed for the analyses of these complex spectra.

Table VII. Papers and reports on the atomic emission spectrum of curium, Cm I and Cm II.

	Paper or Report	Major Contribution
1.	J. G. Conway, M. F. Moore and W. W. T. Crane, J. Am. Chem. Soc., 1951, **73**, 1308.	About 200 emission lines from a spark source 252 to 500 nm.
2.	J. G. Conway and R. D. McLaughlin, J. Opt. Soc. Am., 1956, **46**, 91.	Isotope shifts ^{244}Cm-^{242}Cm of 183 lines on large spectrograph 305 to 525 nm.
3.	Leo Brewer, J. Opt. Soc. Am., 1971, **61**, 1101 and J. Opt. Soc. Am., 1971, **61**, 1666.	Energies of electronic configurations of neutral, singly, doubly and triply ionized actinides.
4.	R. H. Gaddy, Applied Spectroscopy, 1972, **26**, 49.	Analytical methods for impurities in Cm and Cm in Am.
5.	E. F. Worden and J. G. Conway, J. Opt. Soc. Am., 1976, **66**, 109.	Energy levels of Cm I, 335 odd and 348 even levels, isotope shift, Zeeman effect, relative energies and isotopic shifts of configurations.
6.	J. G. Conway, J. Blaise and J. Verges, Spectrochim Acta, 1976, **B31**, 31.	Atomic emission lines Cm I and II 850 to 2640 nm, energy levels, and energy level classification of lines.
7.	E. F. Worden, E. K. Hulet, R. G. Gutmacher, and J. G. Conway, Atom. Data and Nucl. Data Tables, 1976, **18**, 459.	Wavelength, isotope shifts, g-value, intensities, spectrum assignment and energy level assignments for 2034 of the strongest ^{244}cm lines of Cm I and II in the 240 - 1120 nm region.
8.	E. A. Lobikov, N. K. Odintsova and A. R. Striganov, "Emission Spectrum of Curium, Report IAE - 3210," I. V. Kruchatov Institute of Atomic Energy, Moscow, 1979.	Emission spectrum of Cm I and II 242 to 701 nm. Wavelengths, intensities self-reversal. 6771 lines reported, no levels, some spectrum assignment. List contains over 800 impurity lines of Pu and lanthanides not found and eliminated by the authors of the paper.
9.	E. A. Lobikov, A. R. Striganov, V. P. Labozin, N. K. Odintsova and V. F. Pomytkin, Opt. and Spectrosc., 1979, **46**, 596.	Curium isotope shifts ^{246}Cm-^{244}Cm and ^{248}Cm-^{246}Cm for 97 lines 404 to 693 nm. Wavelengths and intensities (Classifications taken from papers 5 and 7).

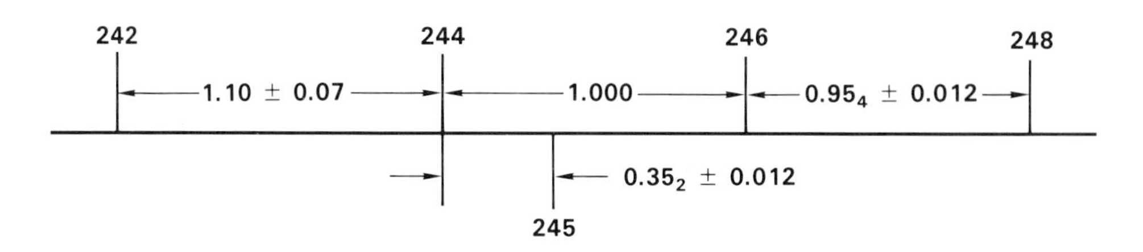

Fig. 5. Relative isotope shift for the atomic emission spectrum lines of five isotopes of curium.

III. ACKNOWLEDGEMENTS

This work was supported in part by the U. S. Department of Energy under Contract No. W-7405-ENG-48 at the Lawrence Livermore National Laboratory, by the Director, Office of Energy Research, Office of Basic Energy Sciences, Chemical Sciences Division of the U. S. Department of Energy under Contract No. DE-AC03-76SF at the Lawrence Berkeley Laboratory and in part by Centre National de la Recherche Scientifique, France.

References

1. E. F. Worden and J. G. Conway, "Energy Levels of the First Spectrum of Curium, Cm I," J. Opt. Soc. Am., 1976, 66, 109-121.
2. J. G. Conway and R. D. McLaughlin, J. Opt. Soc. Am., 1971, 46, 91.

SHIFTED HOMOLOGOUS RELATIONSHIPS BETWEEN THE TRANSPLUTONIUM AND EARLY RARE-EARTH METALS*

John W. Ward
Materials Science and Technology Division
Los Alamos National Laboratory, Los Alamos, NM 87545

ABSTRACT. The physico-chemical properties of the late actinide metals americium through einsteinium are compared with their rare-earth counterparts. Localization of the 5f electrons beginning at americium signals the appearance of true rare-earth-like properties, but the homologous relationship is shifted to place americium below praseodymium, einsteinium is then below europium. The comparison of crystal structures, phase transitions, vapor pressures and heats of vaporization reveals remarkable similarities, especially for Sm-Cf and Eu-Es, where the stability of the divalent metal becomes established and divalent chemistry then follows.

There is of course a major perturbation caused by the half-filled shell at curium, although it may be argued that americium is the anomaly in the so-called second rare-earth series. However, the response of americium, berkelium and californium under pressure reveals the true perturbation to be a thermodynamic one, occurring at curium.

The onset of 5f electron localization in the actinides is probably first seen in the higher temperature phases of plutonium, particularly the fcc delta phase, with a metallic radius of 1.64Å and a rare-earth-like X-ray adsorption-emission response.[1] Nevertheless, no magnetism is observed, and the radius is still somewhat small compared to the rare-earth metals. The return of 5f bonding is shown dramatically in the epsilon and liquid phases, producing (as with neptunium) a very low melting point and a liquid with extremely high viscosity and a high boiling point.

The complexities continue with americium, where a major localization and f-band narrowing occurs but magnetism is again quenched by by the $5f^6$ electronic configuration; this unusual non-magnetic ground-state also allows americium to become superconducting, like lanthanum.[2] Chemically americium is clearly a trivalent metal. The divalent state is only stabilized by a large ligand such as I^-; tetravalency occurs with strong oxidizing agents (O_2,F_2) because the 5f electrons, though localized, are very close (<2eV) to the Fermi

*Work performed under the asupices of the U.S. Department of Energy.

N. M. Edelstein et al. (eds.), Americium and Curium Chemistry and Technology, 135–145.

level. Crystal structures are similar to those of the early rare-earths - dhcp at room temperature and fcc at higher temperatures. High-pressures[3] transform the lattice first to an fcc structure (as do the rare-earths), then to an exotic double body-centered monoclinic phase, and then finally to an α-uranium-type type structure in the range 15-20 GPa, again showing the proximity of the 5f electrons to the Fermi level.

The dhcp room temperature and fcc high temperature phases are also found for curium and the elements berkelium and californium, in marked contrast to the supposed homologs directly above these elements in the periodic table. A suggestion first noted by Johansson[4] shifts the homologous relationship so as to place americium below praseodymium, which then allows comparison of the two series as chemical and structural homologs. This is illustrated in Table 1. The f-bonded early actinides are grouped for convenience under plutonium, which becomes the homolog for (the f-bonded collapsed α-phase of) cerium. It should be noted that cerium is often used as a stand-in for plutonium, where radioactive contamination must be avoided.

Americium then becomes the homolog for praseodymium, as noted above. Both metals can be oxidized to the dioxide, due to the proximity of the f electron energies to the Fermi level. The heat of vaporization for Pr is 85.0 kcal/mol which may be compared with 67.9 kcal/mol for Am, as measured by Ward and coworkers (5,6) on both the ^{241}Am and ^{243}Am isotopes; these data are shown in Fig. 1. A long-standing controversy between the measured and theoretical values for the heat of vaporization has recently been resolved by a new spectroscopic value for the lowest level of the f^6ds^2 Am I configuration, thus finally bringing the theoretical calculations into harmony with experiment. Various theoretical treatments[4,7-10] have had considerable success in predicting the cohesive energies (∿heat of vaporization) for the rare-earths, failing for the early actinides because of no viable model for the itinerant and bonding 5f-electrons. These correlations are based on the spectroscopic values for the gas (free atom) and the heat of solution of the metal; it might therefore be expected that reasonable estimates would be made for the elements americium and beyond, where f-bonding no longer is in the picture. This indeed turns out to be true, with the resolution of the americium calculation just discussed.

It can be seen immediately that a major perturbation in the homologous relationship in Table I occurs at curium, because of the half-filled shell effect. Is curium like neodymium or gadolinium? Comparing crystal structures the former is true; in terms of a heat of vaporization (Cm = 92.2 kcal/mol, Gd = 95.0 kcal/mol, Nd = 73.3 kcal/mol) the latter seems to be true. However, the "problem" is not really a problem at all; the argument is simply a thermodynamic one, since the half-filled shell stabilizes the trivalent Cm(Gd) gas. In terms of solid-state properties, the homologous relationship remains Cm-Nd.

For the elements beyond plutonium, measurements of the heat capacity are difficult-to-impossible, due to the intense radioactivity

TABLE I

COMPARATIVE PHYSICOCHEMISTRY OF LANTHANIDES AND ACTINIDES

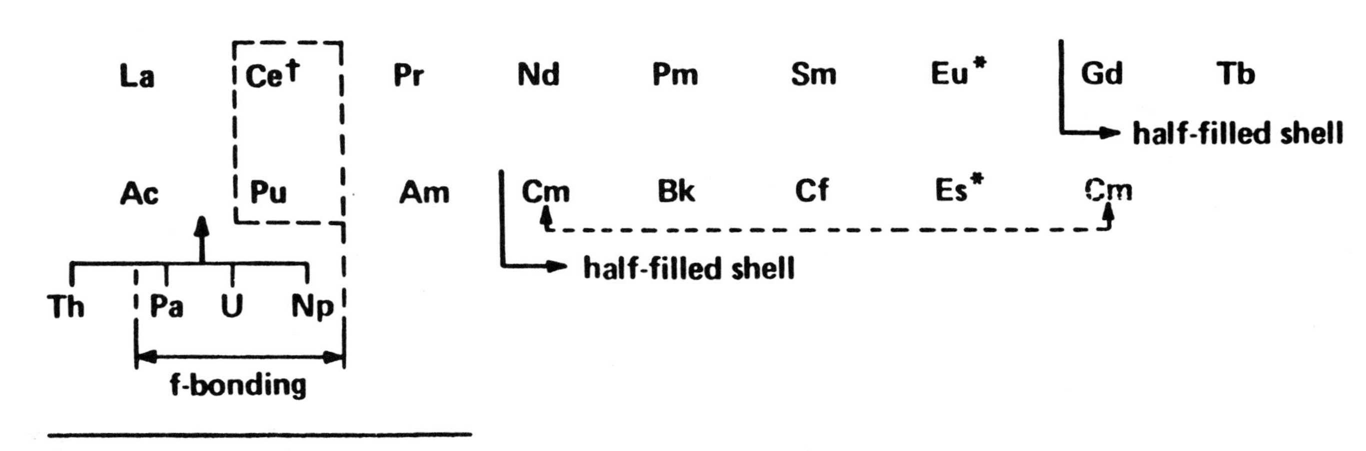

†f-bonding in the collapsed phase

*divalent

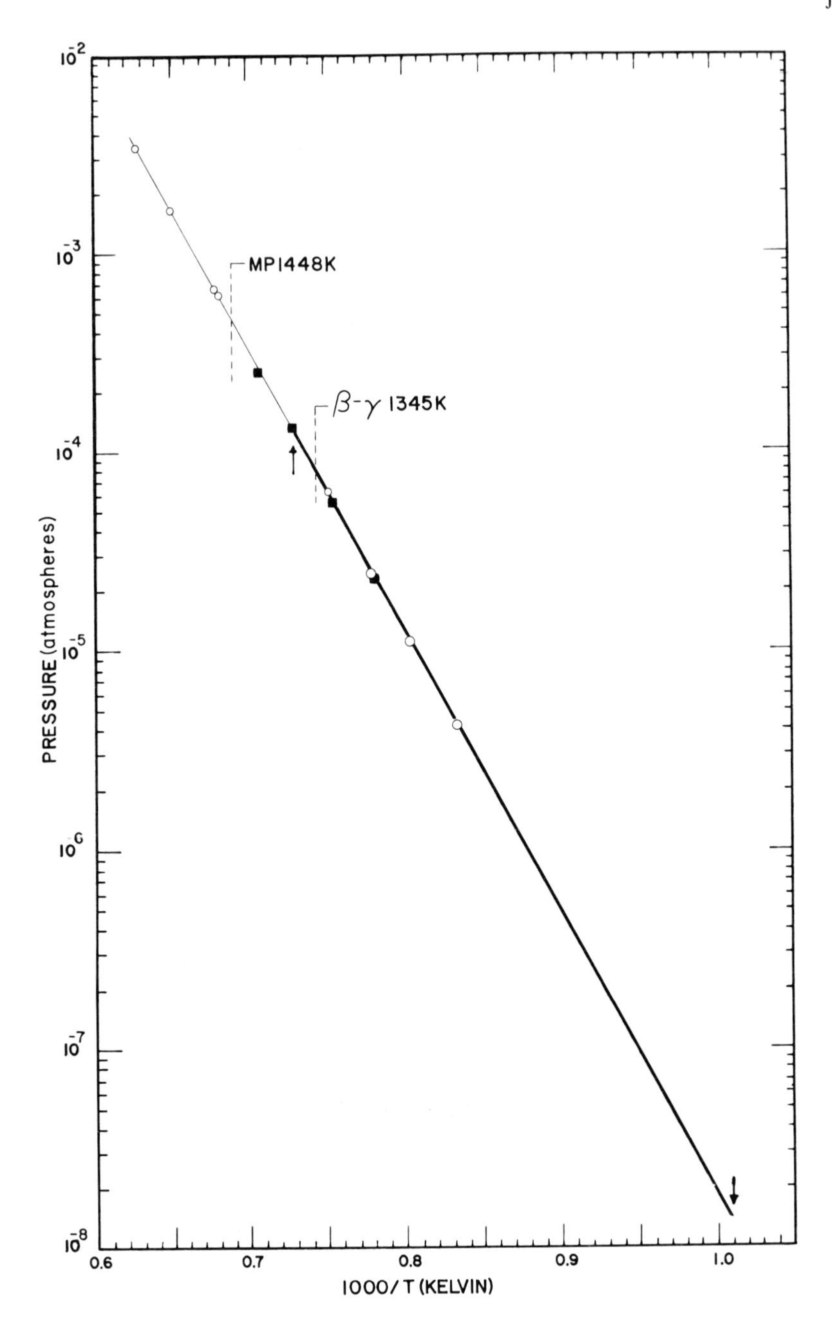

Figure 1.
Vapor pressure data for americium. The heavier line represents 37 data points taken with the 241 isotope (Ref. 5); extent of range is indicated by small arrows. Data points and the lighter line represent values with the 243 isotope, extending into the liquid range: O Los Alamos data; ■ Karlsruhe data (Ref. 6).

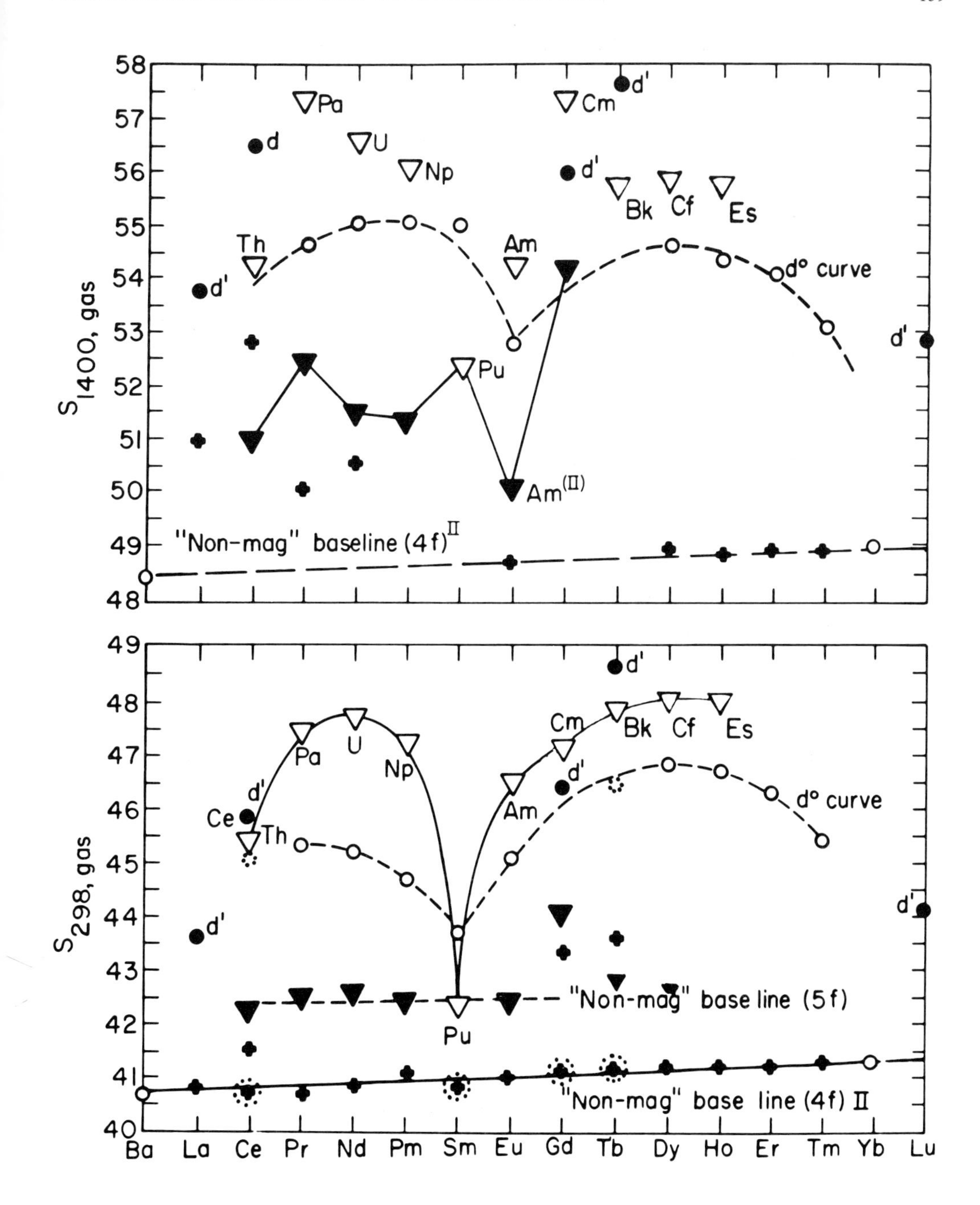

Figure 2.

Comparison of gaseous entropies for the lanthanides and actinides at 298K and 1400K. o: divalent (d°) curve for the lanthanides; ● trivalent (d^1) gases; ▽ fully magnetic actinide gas; ▼ nonmagnetic portion of entropy for each element.

TABLE II

Comparative Values for the Actinides Am-Es and Lanthanides Pr-Eu

Metal Pair(s)	Crystal Phases	M.P. (K)	B. P. (298K)	C^{o}_{p} (Cal/K/Mol)	S^{o}_{298} (Cal/K/mol)	ΔH^{o}_{298} (kcal/Mol)	Chemistry
Am	dhcp, fcc	1449	~2220	5.97	13.20	67.9	III [II, IV]
Pr	dhcp, fcc	1204	3785	6.56	17.67	85.0	III [IV]
Cm	dhcp, fcc	1618	3383	6.62[a]	17.20[a]	92.2	III [IV]
Nd	dhcp, fcc	1289	3341	6.55	16.99	78.3	III [IV]
(Gd)	(hcp, bcc)	(1585)	(3539)	(8.86)	(16.24)	(95.0)	(III)
Bk	dhcp, fcc	1323	2900	6.24[a]	18.68[a]	74.1	III
Pm	dhcp, fcc(?)	~1350	2730(?)	-	-	?	III
Cf	dhcp, fcc	1173	1745	6.70[a]	19.25[a]	46.9	III, II
Sm	-Sm (dhcp under pressure)	1345	2064	7.06	16.61	49.4	III, II
Es	fcc	1133	1241	6.39[a]	21.38[a]	31.8	II
Eu	bcc	1090	1870	6.48	19.31	41.9	II

[ ] = other possible valence states

[a]from the Entropy correlation, Ward & Hill (11)

and/or vanishingly small quantities for the experiment. Ward and Hill[11] have established a correlation relating the crystal entropy S_{298} to metallic radius, atomic weight, magnetic properties and electronic structure. This correlation permits estimates of reliable entropy values for metals for which no data are available, based on comparison with a closely-similar metal for which there are measured data. The early actinides cannot be assessed by this correlation, because there is no model (as for the theoretical calculation) to account for the complex 5f-electron bonding. However, success is expected and realized for the metals americium and beyond. The estimated value for the correlation for americium of 13.2 cal./mol/K has been precisely confirmed recently by experiments on the ^{243}Am isotope.[12]

Noteworthy is the huge change of entropy (4.13 cal/K/mol) upon vaporization of Am from the trivalent non-magnetic solid to the divalent magnetic gas. It should be noted that all of the actinide gases are fully magnetic, like the rare-earths, as shown in Fig. 2. Perhaps this should be obvious, but it is interesting that simple subtraction of the magnetic entropies results in a good nonmagnetic baseline, except for curium, where there are many low-lying excited excited states.

The importance of the crystal entropy lies in the fact that the S_{298} value is the basis point for the free-energy functions for both the solid and gas. From a reasonably accurate estimate for the solid entropy, the gaseous spectroscopic data, and precise vapor pressure measurements, it is possible to calculate all the thermodynamic functions for the metal up to the highest temperatures of measurement. These values have been recently tabulated in the Metals Chapter of the "Handbook for the Physics and Chemistry of the Actinides".[13]

Unfortunately there exist as yet no data for the rare-earth element promethium, so a comparison with berkelium cannot easily be made. However, it is expected on the basis of the regularity seen elsewhere that the comparison will be excellent. In the measurement of the vapor pressure of pure berkelium-249 metal[14] neodymium was used as the comparative element for the entropy correlation, (correcting for the proper magnetic entropies) and the fit was excellent. The measured heat of vaporization for berkelium of 74.1 kcal/mol may be compared with that for neodymium of 78.3 kcal/mol. The expectation is that the the value for promethium, when finally measured, will be lower than for neodymium.

The comparison at samarium/californium is especially revealing. Incipient stabilization of the divalent state is seen in both metals, with divalent chemical compounds easily formed. The surface of samarium metal has been shown to be divalent[15], and there was some question in the early stages of californium metal studies whether the metal was divalent. However, the thermodynamic data[16] clearly establish californium to be a trivalent metal but with a heat of vaporization of only 46.9 kcal/mol (as compared to 49.4 kcal/mol for samarium) and with a vapor pressure midway between that for samarium and europium, as shown in Fig. 3.

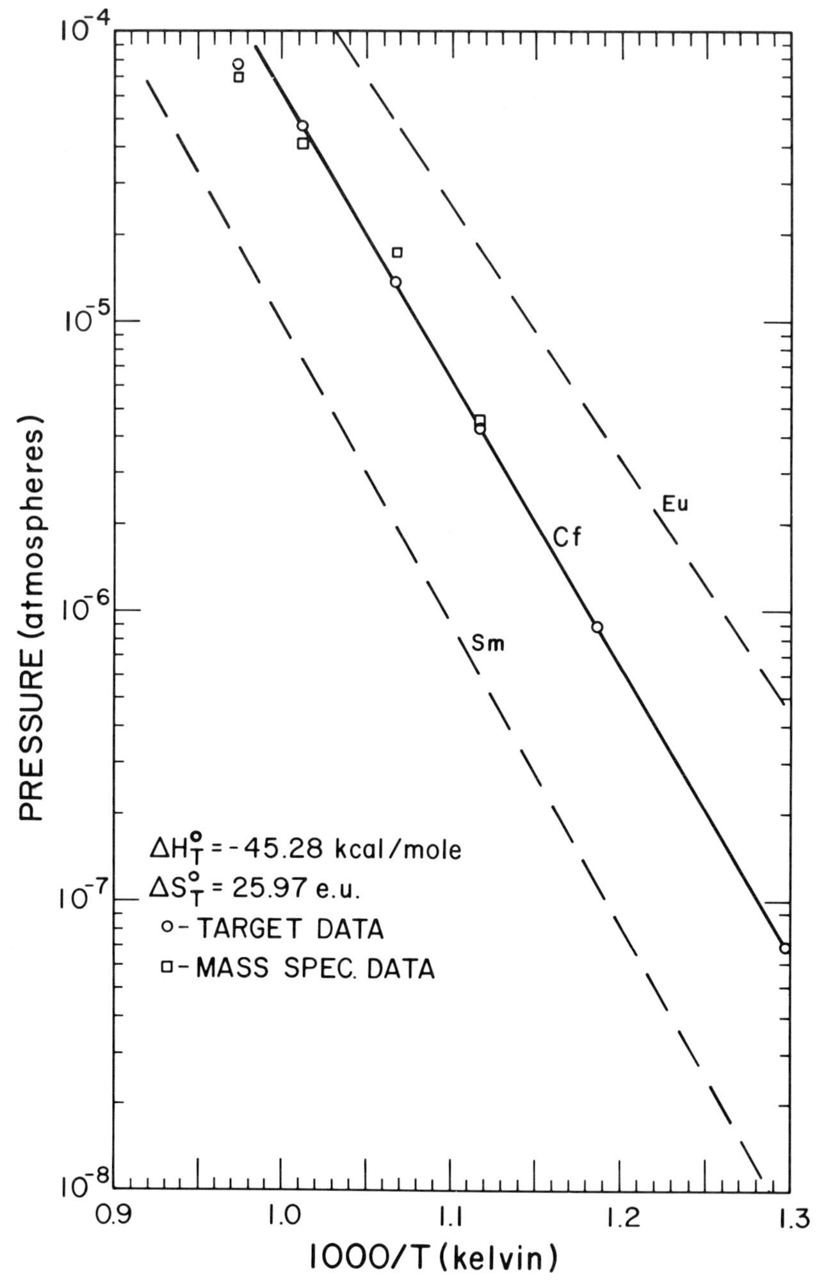

Figure 3.

Vapor pressure of solid californium, compared to data for samarium and europium (Ref. 16). Target data (o) and mass spectrometric data (□) were taken, starting with a 2 mg sample of pure ^{249}Cf (tailoff for the highest points was due to incipient sample depletion).

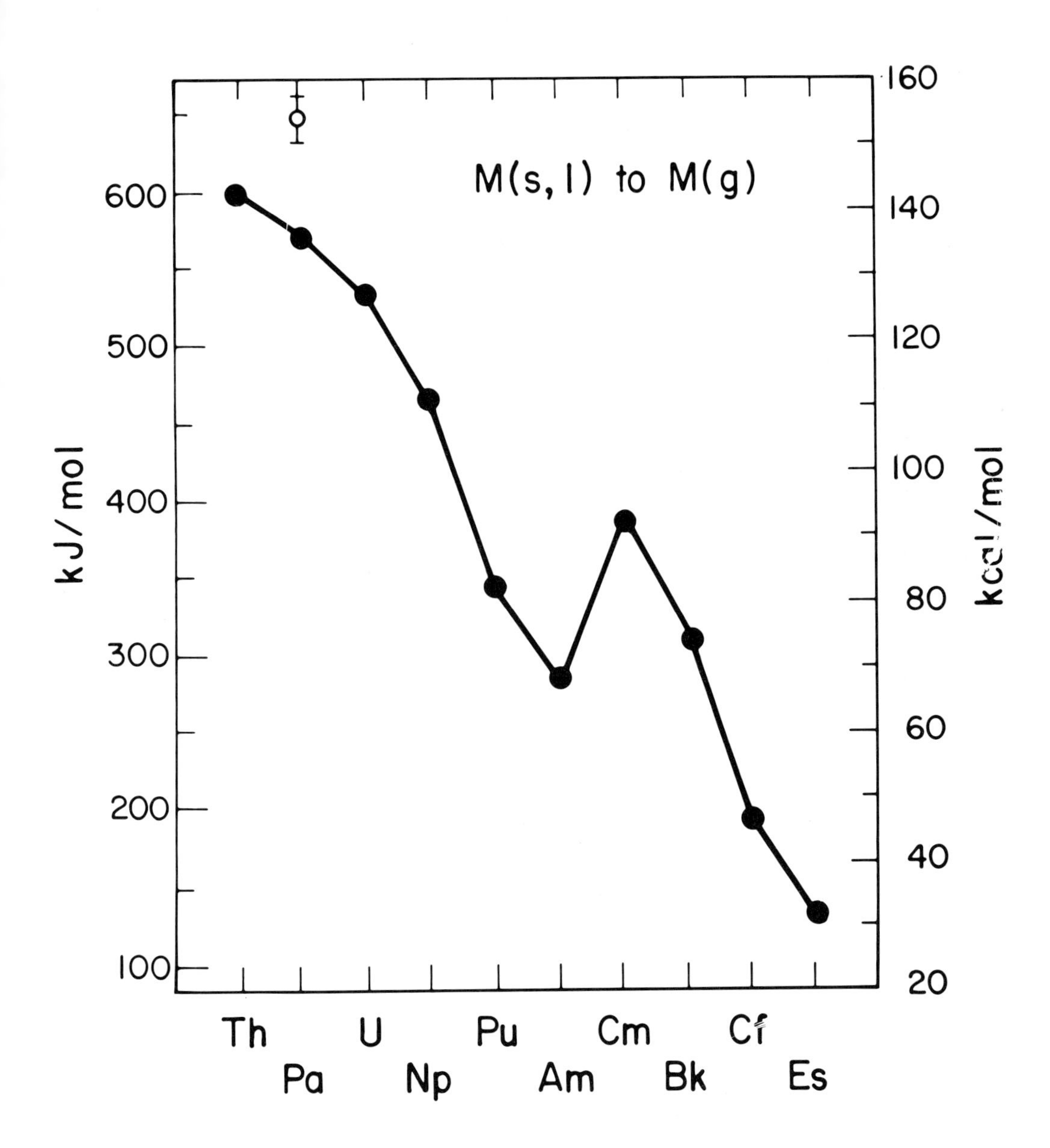

Figure 4.

Comparative plot of the cohesive energies of the actinide metals, in kJ/mol (left ordinate) or kcal/mol (right ordinate). Open data point at Pa represents a still-unresolved uncertainty for this element.

Recent Henry's-Law vaporization studies of einsteinium-253 dissolved in ytterbium show[9,17] the metal to be clearly divalent, with a heat of vaporization of only 31.8 kcal/mol (similar to strontium), thus thus completing the homologous miniseries by analogy to divalent europium.

Some comparative properties of the early lanthanides and late actinides are summarized in Table II. Trends in cohesive energies are illustrated by the measured heats of vaporization of the actinide metals, plotted in Fig. 4.

Beyond einsteinium there is no corresponding half-filled shell in the actinide series, and the divalent state will continue to be the preferred electronic configuration for the metals fermium and beyond. At the moment, half-lives are too short and quantities too small for meaningful bulk measurements beyond einsteinium.

Thus in summary, a shifted homologous relationship exists for the transplutonium actinides, resulting in the miniseries Am-Es corresponding to the series Pr-Eu. The half-filled shell at curium introduces a thermodynamic complication which is however not reflected in the condensed phase properties of this element.

References

1. Bonnelle, C., Structure and Bonding (Berlin) 31 (1976) 23.

2. Smith, J. L., Haire, R. G., Science 200 (1978) 535.

3. Roof, R. B., Zeit. fur Kristall. 158 (1982) 307.

4. Johansson, B., Phys. Rev. B 11 (1975) 2836.

5. Ward, J. W., Müller, W., Kramer, G. F., Transplutonium 1975, Eds W. Muller and R. Lindner, North-Holland, Amsterdam/American Elsevier, New York (1976) 161.

6. Ward, J. W., Kleinschmidt, P. D., J. Chem. Phys. 71 (1979) 3920.

7. Nugent, L., Burnett, J. L., Morss, L. A., J. Chem. Thermo, 5 (1973) 665.

8. Samhoun, K., Ph.D. Thesis, Université de Paris-Sud, Centre d'Orsay, Ser. A, No. 1727, 1976.

9. Kleinschmidt, P. D., Ward, J. W., Matlack, G. M., Haire, P. G., High Temperature Science (1985) - to be published.

10. Johansson, B., J. Less-Commons Metals 100 (1984) 49.

11. Ward, J. W., Hill, H. H., Heavy Element Properties, Eds. W. Muller & H. Blank, North-Holland, Amsterdam/American Elsevier, New York (1976) 65.

12. Müller W., Schenkel, L., Schmidt, H. E., Spirlet, J. C., McElroy, D. L., Hall, R. O. A., Mortimer, M. J., J. Low-Temp. Phys. 30, 561.

13. Ward, J. W., Kleinschmidt, P. D., Peterson, D. E., Chapter "Thermochemical Properties of the Actinide Elements and Selected Actinide-Noble Metal Intermetallics," in the Handbook of the Physics and Chemistry of the Actinides, Eds. A. J. Freeman, G. Lander and C. Keller, North-Holland, Amsterdam (1984) - to be published.

14. Ward, J. W., Kleinschmidt, P. D., Haire, R. G., J. Chem. Phys. 77 (1982) 1464.

15. Allen, J. W., Johansson, L., Bauer, R. S., Lindau, I., Hagstrom, S. B. A., Phys. Rev. Let. 41 (1978) 1499.

16. Ward, J. W., Kleinschmidt, P. D., Haire, R. G., J. de Physique (Paris) 40 (1979) C4-233.

17. Kleinschmidt, P. D., Ward, J. W., Matlack, G. M., Haire, R. G., J. Chem. Phys. 81 (1984) 473.

THERMODYNAMIC SYSTEMATICS OF OXIDES OF AMERICIUM, CURIUM, AND NEIGHBORING ELEMENTS

L. R. Morss
Chemistry Division, Argonne National Laboratory
9700 South Cass Avenue
Argonne, Illinois 60439 USA

ABSTRACT. Recently-obtained calorimetric data on the sesquioxides and dioxides of americium and curium are summarized. These data are combined with other properties of the actinide elements to elucidate the stability relationships among these oxides and to predict the behavior of neighboring actinide oxides.

1. INTRODUCTION

The solid-state chemistry of metallic elements is centered on their oxides. For the transuranium elements, the oxides were the first well-characterized substances to be prepared and have been the most thoroughly studied class of compounds. For the elements beyond plutonium, however, inaccessibility of adequate amounts of long-lived isotopes inhibited the study of their solid-state properties until recently.

Thermodynamic measurements on transplutonium oxides before 1980 were limited to thermogravimetric and manometric studies (1-8) and one solution-calorimetry investigation (9). This paper summarizes more recent results and places these results in the perspective of other relevant thermodynamic data for these and adjacent elements.

Table I shows the known binary actinide oxides, arranged by increasing atomic numbers and oxidation states. Measured enthalpies of formation and standard entropies at 298 K are shown to indicate those oxides which have been thermodynamically well characterized. (Some MO_{2-x} phases have been omitted for clarity.)

Although this paper focuses only on oxides $MO_{1.5}$ and MO_2, it should be noted that there is a rich chemistry of higher actinide oxidation states in complex oxides of protactinium through americium (10). These complex oxides fill many of the blank regions in the lower right region of Table I, and allow the study of spectroscopic, magnetic, electronic, and thermodynamic properties of these oxidation states.

N. M. Edelstein et al. (eds.), Americium and Curium Chemistry and Technology, 147–158.

TABLE I.

Well-Characterized Binary Actinide Oxides. Measured enthalpies of formation (kJ mol^{-1}) and standard entropies (J $mol^{-1} K^{-1}$) are shown beneath corresponding formulas.

An(III)		An(IV)		An(V)		An(VI)
$AcO_{1.5}$						
		ThO_2 -1226.4 ± 3.5 65.23 ± 0.20				
		PaO_2		$PaO_{2.5}$		
		UO_2 -1085.0 ± 1.0 77.03 ± 0.20	$UO_{2.25}$ -1127.6 ± 1.2 83.53 ± 0.18 $UO_{2.33}$ -1142.4 ± 0.9 83.51		$UO_{2.67}$ -1191.6 ± 0.8 94.18 ± 0.17	$UO_3(\gamma)$ -1223.8 ± 2.0 96.11 ± 0.40
		NpO_2 -1074.0 ± 2.5 80.3 ± 0.4		$NpO_{2.5}$		
$PuO_{1.5}$ (-828) 81.51 ± 0.32	$PuO_{1.61}$ -884.5 ± 16.7	PuO_2 -1056.2 ± 0.7 66.13 ± 0.26				
$AmO_{1.5}$ -845 ± 8	$AmO_{1.6}$(?)	AmO_2 -932.2 ± 2.7				
$CmO_{1.5}$ -841.5 ± 6.0	$CmO_{1.714}$	CmO_2 -911 ± 6				
$BkO_{1.5}$	$BkO_{1.8}$	BkO_2				
$CfO_{1.5}$ -826 ± 5	$CfO_{1.714}$	CfO_2				

References for Table I: References 10,11, and Morss, L. R.; Fuger, J. to be published ($CfO_{1.5}$).

2. AMERICIUM OXIDES

The enthalpy of formation of hexagonal Am_2O_3 was determined recently by solution microcalorimetry (11). Although this measurement was not the first thermodynamic property of an actinide sesquioxide to be reported (12-14), Am_2O_3 represents the lightest actinide sesquioxide suitable for solution calorimetry. (Although a large sample of plutonium sesquioxide has been prepared (13) and still exists, its high temperature of synthesis renders it so refractory that it is probably unsuitable either for solution or combustion calorimetry. Lighter actinide sesquioxides such as U_2O_3 and Np_2O_3 are unknown.) Since Am_2O_3 readily oxidizes in air, it had to be handled in an oxygen-free dry box. Fortunately, its reactivity facilitated solution calorimetry in 6M hydrochloric acid, in which it readily dissolves.

Unsuccessful efforts were made by this author to prepare multi-milligram samples of monoclinic Am_2O_3, which may (3) have a narrow stability range (figure 1) or may be observed only when stabilized by a lanthanide impurity (15). The hexagonal sesquioxide was observed upon reduction of the dioxide in hydrogen at temperatures above 625°C, with rapid or slow cooling; lower reduction temperatures yielded only the body-centered cubic sesquioxide. Because of concern that the bcc sesquioxide might not be stoichiometric (3), solution-calorimetry measurements were not undertaken on the bcc phase.

A few years earlier, the enthalpy of formation of AmO_2 was redetermined (16). Although Am sesquioxide, as well as those of Cm, Cf, and

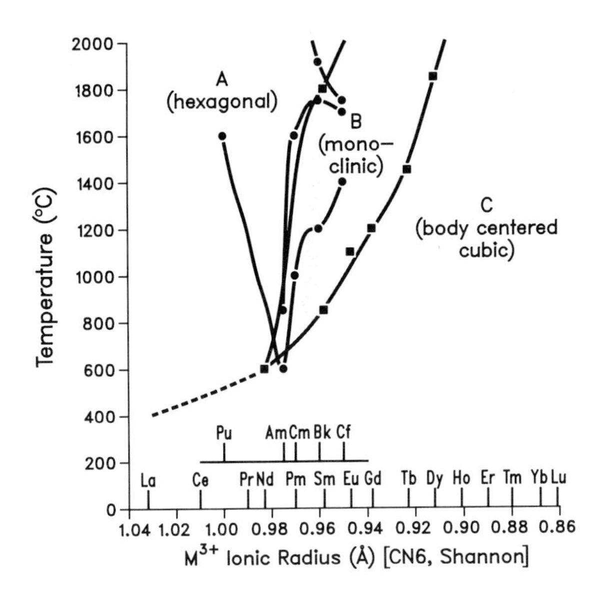

Figure 1. Phase diagram for ■ lanthanide and ● actinide sesquioxides (region above 2000°C omitted).

the lanthanides, dissolves readily in dilute hydrochloric acid (11) (the most convenient calorimetric solvent), the dioxide requires stronger acid to dissolve it. AmO_2 also has been observed to disproportionate partially and nonreproducibly when dissolved in acid (16). Therefore, it was necessary to carry out the solution calorimetry in a complexing acid (H_2SO_4) with a reductant (I^-) to achieve adequate dissolution rates and a well-defined reaction. An important result obtained from this study was a thermochemical estimate of the standard aqueous electrode potential $E°(Am^{4+}/Am^{3+}) = 2.62 \pm 0.09$ V (16).

Ackermann and Chandrasekharaiah (17) correlated the behavior of these oxides with other sesquioxides and dioxides and with the aqueous ions. More recent investigators have refined such correlations in terms of the functional dependence of enthalpy of solution upon ionic size and crystal structure (11,16,18) and in terms of the energetics of gaseous atoms in different electron configurations (18). Even more recently Brewer's reassignment of the energy levels of Am I (neutral Am) (19) has led to a much more consistent comparison (11) of the energetics of Am metal, vapor, and trivalent species with other f elements (by means of the correlation function P(M)) than had earlier been calculated (20,21). A refined treatment of the systematics of lanthanide atomic energy levels (22) should further improve actinide metal/oxide/aquo-ion systematics when it is applied to the 5f elements.

3. CURIUM OXIDES

Curium sesquioxide was the first actinide sesquioxide to be studied by solution calorimetry (14). The determination of $\Delta_f H_m^\circ(Cm_2O_3)$ was facilitated by the availability of multimilligram amounts of ^{248}Cm, a long-lived (half-life 3 x 10^5 y) isotope purified from its radiochemical parent ^{252}Cf in the transplutonium production program at Oak Ridge National Laboratory, by the ease of reduction of CmO_{2-x} to $CmO_{1.5}$, by the relative stability of stoichiometric monoclinic sesquioxide (figure 1) above 700°C, and by its rapid dissolution in hydrochloric acid.

Curium dioxide, on the other hand, has proved to be much more intractable than AmO_2. Early studies with ^{244}Cm (half-life 18.1 y) showed significant substoichiometry (4,5,23), especially as a function of time, due to alpha-particle lattice damage. Despite the more recent availability of ^{248}Cm, which appeared from thermogravimetric measurements and X-ray powder diffraction lattice parameter to yield stoichiometric $CmO_{2.00}$ in oxygen at 1 atm pressure, recent magnetic susceptibility determinations infer significant hypostoichiometry (x near 0.2 in CmO_{2-x}) (24-27). Very careful annealing in pure oxygen for several days at 1 atm pressure and even at 100 atm pressure (25,26) failed to eliminate Curie-Weiss paramagnetism attributable to Cm(III). Since annealed dioxide dissolves very slowly in acidic media, calorimetric solvents required complexing (HSO_4^-, BF_4^-) and reducing (I^-) agents to accelerate dissolution of CmO_2. Typical data (24) (table II) show significant scatter.

TABLE II.

CmO_2 THERMOCHEMISTRY
Solution calorimetry in 0.5 M H_2SO_4 - 0.1 M KI:
$CmO_2(cr) + 4\ H^+\ (aq) + 3/2\ I^- \rightarrow Cm^{3+} + 1/2\ I_3^- + 2\ H_2O(aq)$
Sample A($CmO_{2.00}$): $\Delta H = -223.6 \pm 2.1$ kJ mol^{-1} (5 experiments, 95% confidence)
Sample B($CmO_{1.93}$): $\Delta H = -225 \pm 6$ kJ mol^{-1} (3 experiments, 95% confidence, corrected to $CmO_{2.00}$)
Solution calorimetry in 6M HNO_3 - 0.1 M $NaBF_4$:
$CmO_2(cr) + 3\ H^+(aq) \rightarrow Cm^{3+} + 3/2\ H_2O(aq) + 1/4\ O_2(g)$
Sample C($CmO_{1.986}$): $\Delta H = -160.5$ kJ mol^{-1} (1 experiment)
Sample D($CmO_{1.982}$): $\Delta H = -153.0$ kJ mol^{-1} (2 experiments)
These experimental data, combined with currently accepted auxiliary data and small corrections to standard states, yield $\Delta_f H°(CmO_2,cr) = -911 \pm 6$ kJ mol^{-1} (95% confidence).

The thermochemical measurements on curium oxides have been utilized systematically in a manner parallel to those on americium oxides: correlation of enthalpies of solution with ionic size (18,22), correlation (P(M)) with energetics of gaseous atoms (18), and thermochemical estimation (24) of the aqueous $E°(Cm^{4+}/Cm^{3+}) = 3.0 \pm 0.1$ V. The steps in the latter calculation are shown in table III. Similar thermodynamic behavior is found in the lanthanides praseodymium (aqueous $E°(Pr^{4+}/Pr^{3+}) = 3.2 \pm 0.2$ V) and terbium (aqueous $E°(Tb^{4+}/Tb^{3+}) = 3.1 \pm 0.5$ V (28) or 3.3-3.4 V, see below). These tetravalent lanthanides have been stabilized in complexed form in carbonate solutions (29), but, unexpectedly, Cm(IV) has not yet been stabilized in this medium (30).

TABLE III.

Estimation of $E°(Cm^{4+}/Cm^{3+})$
For the reaction $CmO_2(cr) + 4H^+(aq) \rightarrow Cm^{4+}(aq) + 2H_2O(\ell)$ we predict $\Delta H° = -40 \pm 7$ kJ mol^{-1} from data on other MO_2 and M^{4+} (see fig. 3). Since we have calculated $\Delta_f H°(CmO_2,cr) = -911 \pm 6$ kJ mol^{-1}, we may calculate $\Delta_f H°(Cm^{4+},aq)$ from the above data: $\Delta_f H°(Cm^{4+},aq) = \Delta_f H°(CmO_2,cr) - 2\ \Delta_f H°(H_2O) + (-40 \pm 7)$ $= -911 \pm 6 - 2(-285.83) - 40 \pm 7 = -379 \pm 9$ kJ mol^{-1}. Thus for the reaction $Cm^{4+}(aq) + {}^1\!/_2\,H_2(g) \rightarrow Cm^{3+}(aq) + H^+(aq)$ $\Delta H° = -615 \pm 5 - (-379 \pm 9) = -236 \pm 10$ kJ mol^{-1}. From estimated entropies $\Delta S° = 170 \pm 21$ J K^{-1} mol^{-1} Thus $\Delta G° = \Delta H° - T\Delta S° = -287 \pm 12$ kJ mol^{-1}. Hence $E°(Cm^{4+}/Cm^{3+}) = -\Delta G°/nF = +2.96 \pm 0.12$ V (compared with literature estimates of 3.1 and 3.2 V).

4. RELATIONSHIPS TO OTHER THERMODYNAMIC MEASUREMENTS

There are high-temperature calorimetric measurements on the incremental oxidation of cerium (31) and plutonium (12) oxides. For cerium (32), praseodymium (33), terbium (34), and all actinides thorium through californium (2,5-8,35) there are oxygen-pressure dissociation data as a function of temperature and composition. Each of these methods leads to $\bar{H}(O_2)$, the partial molal enthalpy of oxygen in these systems. High-temperature solid-state EMF measurements also yield thermodynamic information on these oxides (36).

Chikalla and Turcotte (37) compared log $P(O_2)$ data at 1173 K for the composition $MO_{1.9}$ with the aqueous $E°(M^{4+}/M^{3+})$, since both of these parameters are proportional to the free energy change of M(IV)-M(III) species. They predicted $P(O_2)$ for other oxides $MO_{1.9}$. The $E°(Am^{4+}/Am^{3+})$ has been reassessed as 2.62 V (16) and a data point for $CeO_{1.9}$ is available; the revised plot (figure 2) still predicts only slightly more stability for $CmO_{1.9}$ than for $CfO_{1.9}$. Isobars for $^{248}CmO_{2-x}$ show $P(O_2)$ = 0.72 atm for $CmO_{1.9}$ at 723 K (6); measurements are lacking for the CfO_x system above x = 1.72. Nevertheless, CmO_x and CfO_x isobars (6,8) ($1.50 < x < 1.72$) show CmO_x to lose oxygen at lower temperatures than CfO_x. These contradictory interpretations need to be reconciled.

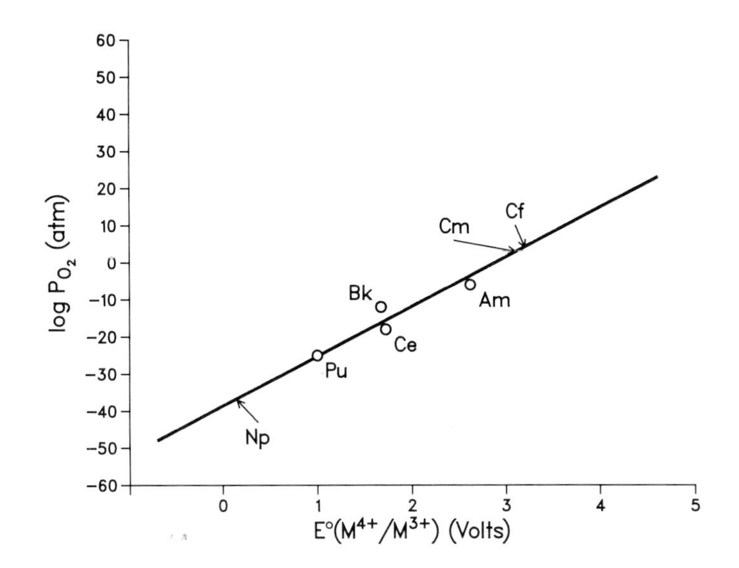

Figure 2. Correlation of log $P(O_2)$ for $MO_{1.90}$ at 1173 K (refs. 32,35, and 37) with aqueous $E°(M^{4+}/M^{3+})$ (ref. 28).

Ackermann and Rauh (35) tabulated $\bar{G}(O_2)$ for $MO_{1.96}$ (M = Th-Am) at 1875 K. There are insufficient data to extrapolate $\bar{G}(O_2)$ for $CmO_{1.90}$ at 1173 K for comparison with Chikalla and Turcotte's plot (37). The stoichiometry of CmO_2 must still be established by new thermogravimetric or analytical measurements.

Enthalpy-of-solution correlations of sesquioxides and dioxides as a function of ionic size lead to interesting conclusions. Such a plot for lanthanide and actinide dioxides shows a single straight line with the exception of terbium (figure 3). This plot permits the prediction of the enthalpies of formation of BkO_2 (-1020 kJ/mol) and of CfO_2 (-858 kJ/mol) using accepted $E°(An^{4+}/An^{3+})$, estimated entropies, and new thermochemical data on Cf^{3+}(aq) (38). Since there is no reason to doubt the reliably determined enthalpy of formation of TbO_2, we must suspect that the estimated (28) enthalpy of formation of Tb^{4+} is too exothermic by ca. 20-30 kJ/mol. Consequently, the estimated aqueous potential $E°(Tb^{4+}/Tb^{3+})$ must be revised from 3.1 (28) to 3.3-3.4 V.

A more practical conclusion that may be reached from figure 3 is that the tetravalent lanthanides and actinides in eightfold oxide coordination are similarly stable with respect to the tetravalent aquo-ions. Thus Ce(IV) is a good model for either Pu(IV) or Np(IV) in nuclear waste isolation studies.

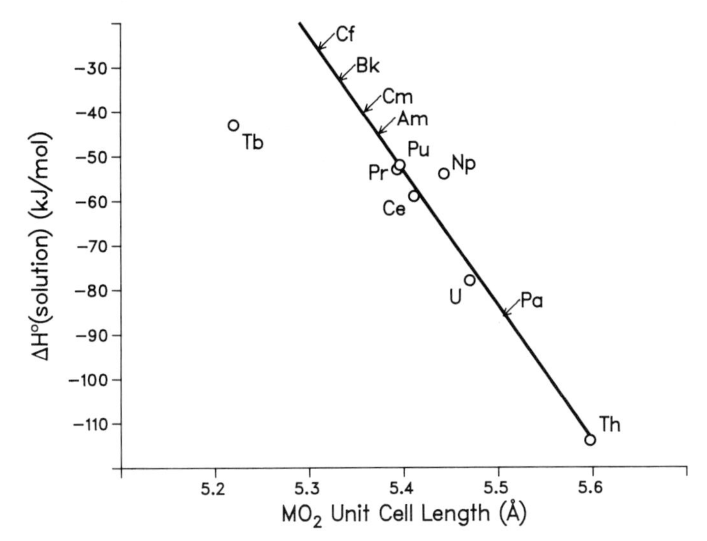

Figure 3. Enthalpy of solution of MO_2 as function of unit cell size.

When a similar plot (figure 4) is examined for the sesquioxides, which have three structure types, it is noted that the enthalpies of solution of actinide sesquioxides of each structure type are ca. 20 kJ/mol less exothermic than those of structurally similar lanthanide sesquioxides. Similar differences are seen for structurally similar hexagonal trichlorides. These systematic trends permit one to predict enthalpies of formation for unstable sesquioxides (for U and Np) as well as for Pu_2O_3 and Bk_2O_3 (table IV). Carrying the comparison one step farther, we have calculated free energies of solution from enthalpy differences and from measured entropies for a lanthanide and actinide ion of similar size in table IV. Not only do figure 4 and table IV show that lanthanides may be used to model trivalent actinide behavior in nuclear waste matrices, but that the lanthanide models provide a conservative safety margin -- the actinide(III) ions in sixfold oxide

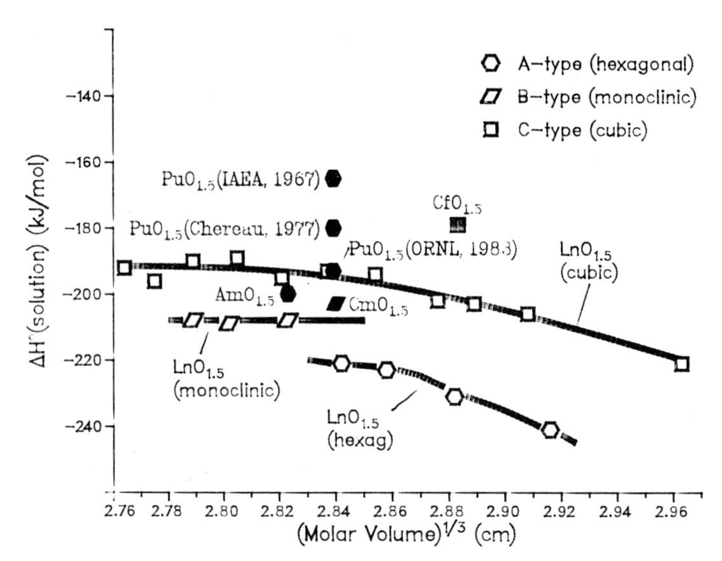

Figure 4. Enthalpy of solution of lanthanide (open symbols) and actinide (filled symbols) sesquioxides as function of interatomic distance. Data from refs. 11 and 12; IAEA Tech. Report 79 ($PuO_{1.5}$, 1967); Besmann, T. M., Lindemer, T. B., J. Am. Ceram. Soc. 1983, **66**, 782-5 ($PuO_{1.5}$, ORNL); and Morss, L. R., et al., to be published ($CfO_{1.5}$).

TABLE IV.

Differences in Enthalpies of Formation of Trivalent Actinide Species (kJ mol^{-1}).

M	$M^{3+}(aq)$[a]	$MO_{1.5}$-M^{3+} Diff.	$MO_{1.5}(s)$[a]	MCl_3-M^{3+} Diff.	MCl_3(cr,hexag)[a]
U	-489	(-239)	(-728)[hexag]	-377	-866
Np	-527	(-234)	(-761)[hexag]	-371	-898
Pu	-592	(-236)	(-828)[hexag]	-368	-960
Am	-617	-228	-845[hexag]	-361	-978
Cm	-615	-226	-841[monocl]	-355	(-970)
Bk	-601	(-246)	(-847)[cubic]	(-350)	(-951)
Cf	-577	-249	-826[cubic]	(-345)	(-922)

[a]From solution calorimetry; entries in parentheses are estimates.

Lanthanide-Actinide Comparison (kJ mol^{-1})

	ΔH(soln)	ΔG(soln)
$NdO_{1.5}(s) + 3H^+(aq) \rightarrow Nd^{3+}(aq) + 3/2\ H_2O(l)$	-221	-169
$PuO_{1.5}(s) + 3H^+(aq) \rightarrow Pu^{3+}(aq) + 3/2\ H_2O(l)$	-193	-145

coordination are ca. 20 kJ/mol more stable than structurally similar lanthanides in comparison with aqueous ions.

Enthalpies of solution represent differences between lattice enthalpies and hydration enthalpies. We hypothesize that actinide sesquioxides have enhanced covalent stabilzation compared to lanthanide sesquioxides because of their more extended 5f orbitals, whereas the actinide dioxides do not manifest a significant stabilization. This conclusion can be conveniently tested (39) for the trivalent ions by comparing enthalpies of solution of similar hydrated salts; fortunately, a suitable pair is available in the trichoride hexahydrates. The enthalpy of solution of $NdCl_3 \cdot 6H_2O$ (corrected to infinite dilution) is -38 kJ/mol (28), and the enthalpy of solution of $PuCl_3 \cdot 6H_2O$ is -35 kJ/mol (40). These nearly identical values reflect the similar coordination in the hydrated salts and hydrated aquo-ions: In the trichloride hexahydrates, the trivalent ion is coordinated to six water molecules and two chloride ions (41) and in the aquo ions the coordination is believed to be ninefold (42).

TABLE V.

Americium and Curium Oxides[a]	
Trivalent	
$AmO_{1.5}$	$CmO_{1.5}$
$LiAmO_2$	
$SrAm_2O_4$, $BaAm_2O_4$	
$Ba_2Am(Nb,Ta,Pa)O_6$	
$AmVO_4$	
$AmNbO_4$	$CmNbO_4$
$AmTaO_4$	$CmTaO_4$
$AmPaO_4$	$CmPaO_4$
$AmAlO_3$	$CmAlO_3$
Tetravalent	
AmO_2	CmO_{2-x}
Na_2AmO_3	
$SrAmO_3$	
$BaAmO_3$	$BaCmO_3$

[a]See also paper by F. Weigel, this volume.

5. COMPLEX OXIDES OF AMERICIUM AND CURIUM

There are known complex oxides of Am(III), Am(IV), Am(V), and Am(VI) (table V) (10). As is the case with the lighter actinides U-Pu, the higher oxidation states are stabilized in complex oxides. There are no quantitative values for this stabilization, even in the case of $BaAmO_3$ (or even for $BaPuO_3$) (43). As of the time of writing, only a few complex oxides of Cm(III) and one of Cm(IV) are known (table V). [Note added in proof: see paper by F. Weigel, this symposium.] Although the perovskites $BaMO_3$ are significantly more stable than the binary oxides (43), the perovskite $BaCmO_3$ still showed significant Cm(III) from magnetic measurements (27). Quantitative results on the enthalpy of stabilization of numerous uranium, neptunium, and most recently plutonium complex oxides should guide efforts to synthesize new complex oxides of Am and Cm (44,45).

6. ACKNOWLEDGEMENT

This work was carried out under the auspices of the U. S. Department of Energy, Office of Basic Energy Sciences, Division of Chemical Sciences, under contract number W-31-109-ENG-38.

7. REFERENCES

1. Asprey, L. B.; Cunningham, B. B. USAEC Report UCRL-329, Berkeley, CA, 1949.
2. Chikalla, T. D.; Eyring, L. J. Inorg. Nucl. Chem. 1967, **29**, 2281-2293.
3. Chikalla, T. D.; Eyring, L. J. Inorg. Nucl. Chem. 1968, **30**, 133-145.
4. Mosley, W.C. J. Inorg. Nucl. Chem. 1972, **34**, 539-555.
5. Chikalla, T.D.; Eyring, L. J. Inorg. Nucl. Chem. 1969, **31**, 85-93.
6. Turcotte, R. P.; Chikalla, T.D.; Eyring, L. J. Inorg. Nucl. Chem. 1973, **35**, 809-816.
7. Turcotte, R. P.; Chikalla, T. D.; Eyring, L. J. Inorg. Nucl. Chem. 1971, **33**, 3749-3763; Turcotte, R. P.; Chikalla, T. D.; Haire, R. G.; Fahey, J. A. J. Inorg. Nucl. Chem. 1980, **42**, 1729-1733.
8. Turcotte, R. P.; Haire, R. G. in Transplutonium Elements, ed. W. Muller and R. Lindner, North-Holland, 1976, pp. 267-276.
9. Eyring, L.; Lohr, H. R.; Cunningham, B. B. J. Am. Chem. Soc. 1952, **74**, 1186-1190.
10. Morss, L. R. in Actinides in Perspective, ed. Edelstein, N., Pergamon, Oxford, 1982, pp. 381-407.
11. Morss, L. R.; Sonnenberger, D. C. J. Nucl. Mater., in press.
12. Chereau, P.; Dean, G.; DeFranco, M.; Gerdanian, P. J. Chem. Thermodyn. 1977, **9**, 211-219.
13. Flotow, H. E.; Tetenbaum, M. J. Chem. Phys. 1981, **74**, 5269-5277.
14. Morss, L. R.; Fuger, J.; Goffart, J.; Haire, R. G. Inorg. Chem. 1983, **22**, 1993-1996.

15. Berndt, U.; Tanamas, R.; Maier, D.; Keller, C. Inorg. Nucl. Chem. Lett. 1974, **10**, 315-321.
16. Morss, L. R.; Fuger, J. J. Inorg. Nucl. Chem. 1981, **43**, 2059-2064.
17. Ackermann, R. J.; Chandrasekharaiah, M. S. Proc. Symp. on Thermodyn. of Nucl. Materials, **Vol.** 2, IAEA, Vienna, 1975, pp. 3-26.
18. Morss, L. R. J. Less-Common Metals 1983, **93**, 301-321.
19. Brewer, L. High Temp. Science, 1984, **17**, 1-30.
20. Nugent, L. J.; Burnett, J. L.; Morss, L. R. J. Chem. Thermodyn. 1973, **5**, 665-678.
21. David, F.; Samhoun, K.; Guillaumont, R.; Edelstein, N. J. Inorg. Nucl. Chem. 1978, **40**, 69-74.
22. Johansson, B.; Munck, P. J. Less-Common Metals 1984, **100**, 49-70.
23. Kanellakopulos, B.; Muller, W. Report AED-Conf-78-074-002; AED: Berlin, 1978; INIS Atomindex 1979, **10**, 425659.
24. Morss, L. R.; Fuger, J.; Goffart, J.; Haire, R. G. Abstracts, "Actinides-1981," Rept. LBL-12441, 1982, p. 263.
25. Goffart, J.; Edelstein, N.; Morss, L. R. Unpublished measurements on CmO_{2-x}, 1982.
26. Kanellakopulos, B.; Peterson, J. R. Unpublished measurements on CmO_{2-x}, 1983.
27. Nave, S. E.; Huray, P.; Haire, R. G. Phys. Rev. 1983, **B28**, 2317-2327.
28. Morss, L. R., "Yttrium, Lanthanum, and the Lanthanide Elements;" Martinot, L.; Fuger, J. "The Actinide Elements," in Oxidation-Reduction Potentials in Aqueous Solutions, ed. Bard, A. J.; Jordan, J.; Parsons, R., in press, Marcel Dekker (1985).
29. Hobart, D. E.; Samhoun, K.; Young, J. P.; Norvell, V. E.; Mamantov, G.; Peterson, J. R. Inorg. Nucl. Chem. Lett. 1980, **16**, 321-328.
30. Hobart, D. E.; Varlashkin, P. G.; Samhoun, K.; Haire, R. G.; Peterson, J. R. Rev. Chim. Minerale, 1983, **20**, 817-827.
31. Campserveux, J.; Gerdanian, P. J. Chem. Thermodyn. 1974, **6**, 795-800.
32. Bevan, D. J. M.; Kordis, J. J. Inorg. Nucl. Chem. 1964, **26**, 1509-1523.
33. Hyde, B. G.; Bevan, D. J. M.; Eyring, L. Trans. Roy. Soc. London 1966, **A259**, 583-614.
34. Lowe, A. T.; Lau, K. H.; Eyring, L. J. Sol. St. Chem. 1975, **15**, 9-17; Hyde, B. G.; Eyring, L. Rare Earth Research III, Gordon & Breach, New York, 1965, pp. 623-624.
35. Ackermann, R. J.; Rauh, E. G. Rev. Internat. Hautes Temps. Refract. 1978, **15**, 259-280.
36. Lott, U.; Rickert, H.; Keller, C. J. Inorg. Nucl. Chem. 1969, **31**, 3427-3436.
37. Chikalla, T. D.; Turcotte, R. P. Proc. 5th Mat. Res. Symp., U. S. Nat. Bur. Stand. Spec. Pub. 364, 1972, pp. 319-330.
38. Fuger, J.; Haire, R. G.; Peterson, J. R. J. Less-Common Metals 1984, **98**, 315-321.
39. The author is indebted to Professor R. Connick (University of California, Berkeley) for suggesting this comparison.

40. Fuger, J.; Parker, V. B.; Hubbard, W. N.; Oetting, F. L.; The Chemical Thermodynamics of Actinide Elements and Compounds, Part 8: The Actinide Halides, International Atomic Energy Agency, Vienna, 1983.
41. Habenschuss, A.; Spedding, F. H. Cryst. Struct. Commun. 1980, **9**, 71-76.
42. David. F. "Redox and Thermodynamic Properties of Cm and Transcurium Elements," to be published.
43. Williams, C. W.; Morss, L. R.; Choi, I-K. in "Geochemical Behavior of Disposed Radioactive Waste," ed. Barney, G. S.; Navratil, J. D.; Schultz, W. W. American Chemical Society Symposium Series 246, Washington, D. C., 1984, pp. 323-334.
44. Gens, R.; Fuger, J.; Morss, L. R.; Williams, C. W. J. Chem. Thermodynamics, in press.
45. Fuger, J. J. Nucl. Mater., in press.

PREPARATION AND PROPERTIES OF SOME NEW CURIUM COMPOUNDS

F.WEIGEL AND R.KOHL
Radiochemische Abteilung des Instituts für Anorganische
Chemie der Universität München
Meiserstr. 1
8000 München 2
F.R.G.

ABSTRACT. The following compounds of element 96, Curium, have been prepared and studied using ultramicrochemical techniques: $CmPO_4 \cdot H_2O$, $Cm(ReO_4)_3 \cdot xH_2O$, $Cm(ReO_4)_3$, $Cm_2(MoO_4)_3$, $CmVO_4$, $CmCrO_4$, $CmAsO_4$, $CmScO_3$, $CmVO_3$, $CmCrO_3$, $CmFeO_3$, $Cm_2(C_2O_4)_3 \cdot 10H_2O$, $Cm(HCOO)_3$, CmOF, CmOCl, CmOBr. A new compound of tetravalent curium, Li_2CmO_3, and, what we believe to be the first representative of pentavalent curium, Na_3CmO_4, have also been prepared. A new synthesis for CmF_3 has been developed, and some thermodynamic properties of $CmCl_3$ have been studied.

1. INTRODUCTION

Even though element 96, Curium, was discovered no less than 40 years ago, in July 1944 (1)(2), and was isolated as the isotope ^{242}Cm in 1948 (3)(4)(5), only relatively few solid compounds have been characterized to date. In the earliest time, when only short-lived ^{242}Cm ($T_{1/2}$ = 162.5 days, S = 3320 Curies/g = 7.4×10^9 dpm/µg) was available in sufficient quantities, it was possible only in a few exceptional cases to obtain X-ray data of curium compounds. Asprey and Ellinger (6) were the first ones to obtain lattice constants of CmF_3 and Cm_2O_3 using ^{242}Cm in the preparation of these compounds.

The situation became somewhat better, when in August 1953, weighable quantities of ^{244}Cm became available by irradiation of "napkin rings" in the MTR (7). ^{244}Cm ($T_{1/2}$ = 17.9 years, S = 82 Curies/g = 1.82×10^8 dpm/µg) was much better suited for chemical work than ^{242}Cm, and a number of compounds were prepared and characterized by X-ray crystallography.

Besides the fluoride, CmF_3, which had been known from the earlier preparation (6), Asprey, Keenan, and Kruse (8) prepared the halides $CmCl_3$, $CmBr_3$, and CmI_3. The only oxyhalide known as early as 1955 was CmOCl, which was also obtained by Weigel et al (9) in a study of the vapor phase hydrolysis of $CmCl_3$. Asprey, Ellinger, Fried and Zachariasen (10) prepared CmF_4. Keenan reported the preparation of complex fluorides of

N. M. Edelstein et al. (eds.), Americium and Curium Chemistry and Technology, 159–191.

tetravalent curium: $LiCmF_5$ (11), $Na_7Cm_6F_{31}$ (12), $K_7Cm_6F_{31}$ (13), and Rb_2CmF_6 (14). Damien (15)(16) prepared the chalcogenides, Charvillat et al (17) the pnictides, Weigel et al (18) the silicides.

Quite a few papers deal with the curium oxygen system. The phase diagram was established by Chikalla and Eyring (19)(20). Some phase transformations of this system were studied by Mosley (21), Haug (22), and by Sudakov and Kapshukov (23). The melting point of Cm_2O_3 was determined by Smith (24), the radiation-induced structural change of C-Cm_2O_3 into A-Cm_2O_3 was reported by Wallmann (25), and by Noé (26).

Relatively few ternary compounds of Cm have been studied, using ^{244}Cm. These include $CmPO_4$(27)(28)(29), $CmNbO_4$(28), $CmTaO_4$(28), and $Cm_{0.5}Pa_{0.5}O_2$ (28).

A few papers have also been published using the rare isotope ^{248}Cm, which is the daughter of ^{252}Cf and is particularly well-suited for chemical work because of its long half life ($T_{1/2} = 3.5\times10^5$ years, S = 4.24×10^{-3} Curies/g = 9.42×10^3 dpm/μg). Compounds of ^{248}Cm, which have been studied, include $CmCl_3$ (30), CmOCl (31), CmO_2 (32)(33), CmO_{2-x} (34) CmF_4 (35), $BaCmO_3$ (35), and Cm_2O_3 (36).

Because we have been supplied with a total of 40 mg of ^{244}Cm by courtesy of the US Department of Energy, Isotopes Distribution Office, Oak Ridge, TN, we decided to contribute to the solid state chemistry of curium by synthesis and characterization of additional, hitherto unknown compounds, and by restudying some of the known compounds using more advanced techniques. This work is reported in this paper.

2. EXPERIMENTAL

2.1. Safety Precautions.

All work with ^{244}Cm was carried out in a Berkeley standard glove box, which was shielded with 2 mm lead on its front side. Even though ^{244}Cm, despite its short half life, has no serious radiation hazard due to penetrating gamma radiation, and its fission yield is so small so that even several milligrams pose no neutron hazard, we found it suitable to use the well established techniques of Cunningham (37)(38), and those of Fried and Davidson (39) because of their elegance, and because they allow the preparation of individual compounds inside the X-ray capillary or in its vicinity without difficult manipulation. The amount of ^{244}Cm used in an individual preparation was limited to less than 100 μg, because such an amount still gives a good pattern without any film-fogging due to radioactive background radiation.

2.2. Recovery and Purification of Curium.

The curium, which was used in this work had the isotope composition shown in Table 1. It was supplied in two batches by the Isotope distri-

Table 1. Isotopic Composition of Curium used in this work [a]

Isotope	Abundance
^{244}Cm	92.42 %
^{245}Cm	1.162 %
^{246}Cm	6.26 %
^{247}Cm	0.101 %
^{248}Cm	0.057 %

(a) Data supplied by ORNL

Table 2. Conditions for X-ray diffraction work

X-ray generator:	Seifert Iso-Debyeflex 2002
Cameras:	Seifert Debye-Scherrer Cameras with 57.5 mm and 114.6 mm diameter
X-ray tube:	Cu 2000 W 1×10 mm^2 F-60-10, AEG
High voltage:	50 kV
Current:	35 mA
Radiation:	Cu K_α
Filter:	0.02 mm Ni foil in front of film, unfiltered primary beam
Collimator:	0.6 mm ID
Exposure times:	57.5 mm camera: 20 min, 40 min, 1 hour
	114.6 mm camera: up to 10 hours
Film:	57.5 mm camera: OSRAY M3
	114.6 mm camera: DuPONT MRF 31

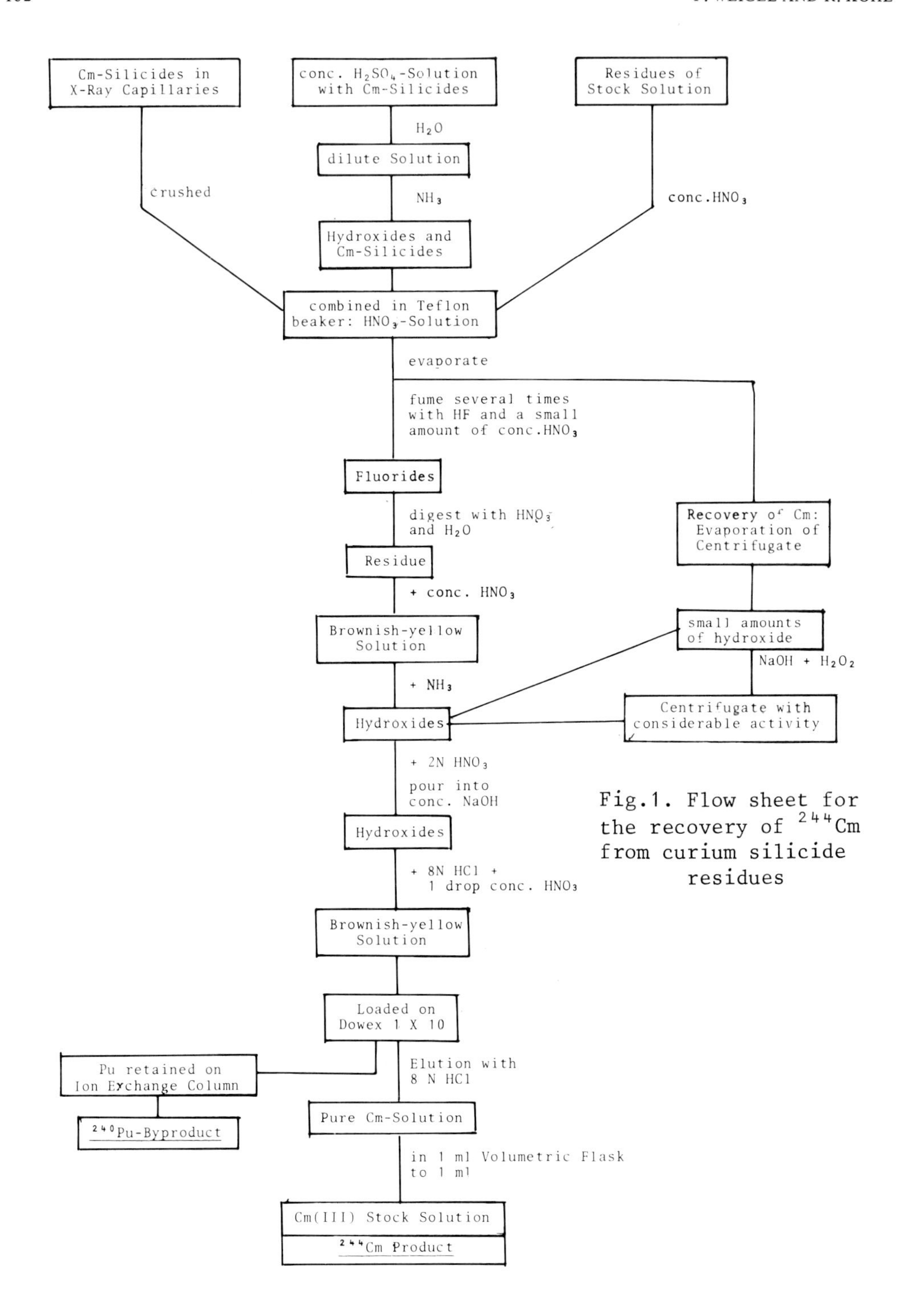

Fig.1. Flow sheet for the recovery of ^{244}Cm from curium silicide residues

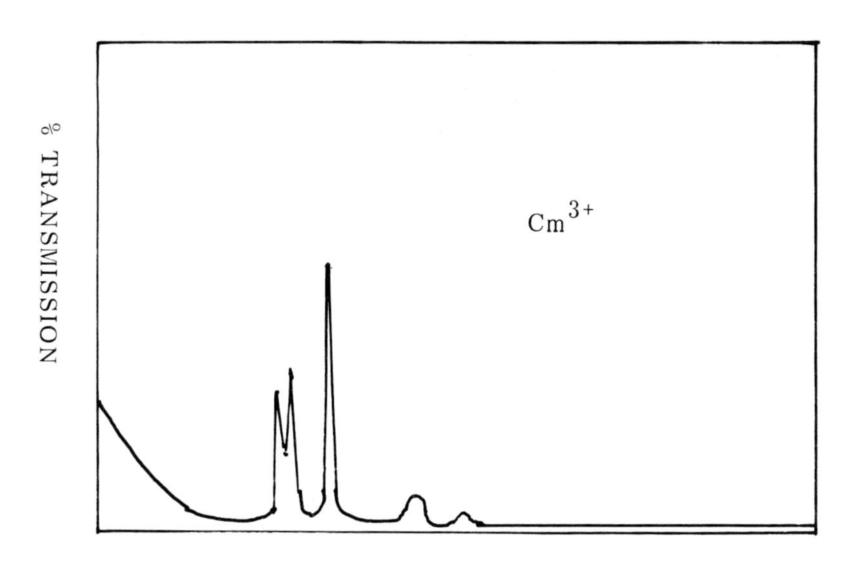

Fig.2. Absorption spectrum of curium(III) in eluate from Dowex 1 column, 8 M HCl

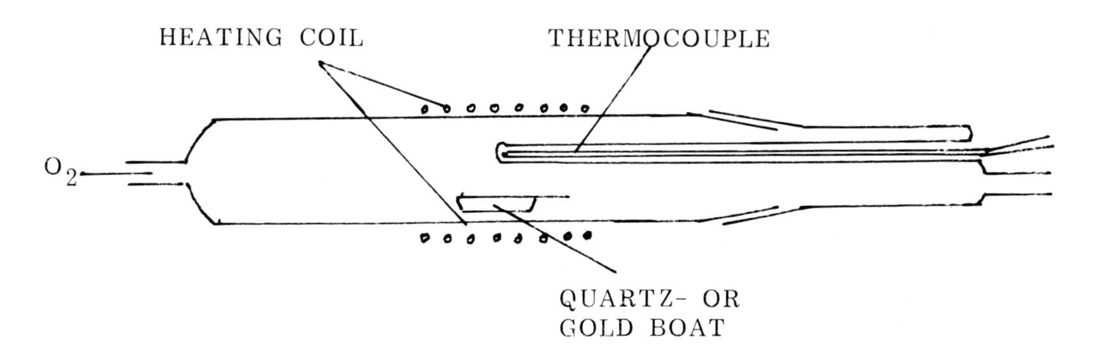

Fig.3. Decomposition apparatus for the pyrolysis of mg quantities of curium(III)-oxalate to curium(IV)-oxide.

bution Office, Oak Ridge National Laboratory, Oak Ridge, TN. The first batch had been used in work on curium silicides (18). The combined residues of this work consisted of

a) mixed curium silicide residues from X-ray capillaries, mixed with debris from the quartz capillaries

b) an assortment of residues in strong sulfuric acid, containing both, small amounts of dissolved curium, and suspended undissolved curium silicides.

c) the dried residue of an old stock solution

The second batch of curium consisted of virgin curium oxide, as supplied by ORNL.

The combined materials from the first batch were worked up in the manner described by the flow sheet shown in Fig.1. The ^{240}Pu, which had been retained on the ion exchange column, was recovered by dissolving the resin in H_2SO_4- H_2O_2 followed by precipitation of the Pu by gaseous ammonia. The purified ^{240}Pu hydroxide was set aside for preparation of a stock solution.

An aliquot of the curium stock solution prepared by the process shown in Fig.1 was counted in a low-geometry counter and was found to contain 2.44 mg ^{244}Cm. This stock was used in the preparation of compounds by precipitation from aqueous solution.

The virgin curium oxide was purified by the following procedure: The CmO_2 was dissolved in a small quantity of 8M HCl. A drop of concentrated HNO_3 was added, and the solution was loaded on the top of a Dowex 1×10 column (100-200 mesh), preconditioned with 8M HCl. The curium was eluted with 8M HCl. Fe and Pu are retained on the column under these conditions. The absorption spectrum of the resulting pale-yellow curium solution showed the absence of other actinides (Fig.2)

From the curium product solution in HCl, the Cm was precipitated as the hydroxide. The $Cm(OH)_3$ was washed several times with distilled water, dissolved in dilute HNO_3 , and precipitated as the oxalate. The oxalate was transferred to the quartz reaction tube of a Fried Davidson apparatus (39), and was ignited to the oxide using the following procedure:

The oxalate was first ignited in air at 750°C for 30 minutes. The temperature of 750°C was maintained for another 15 minutes, then, the oxide product was slowly cooled to 330-350°C in an atmosphere of oxygen. Large quantities of curium (2-3 mg) were treated in the same manner using the apparatus shown in Fig. 3. The black reaction product was characterized by its X-ray powder pattern. It was found to be cubic with a = 5.369 ± 0.003 A, in good agreement with the value reported by Nave et al (35). The curium oxide purified by this method was used either directly in solid state experiments, or a weighed aliquot of it was dissolved in a

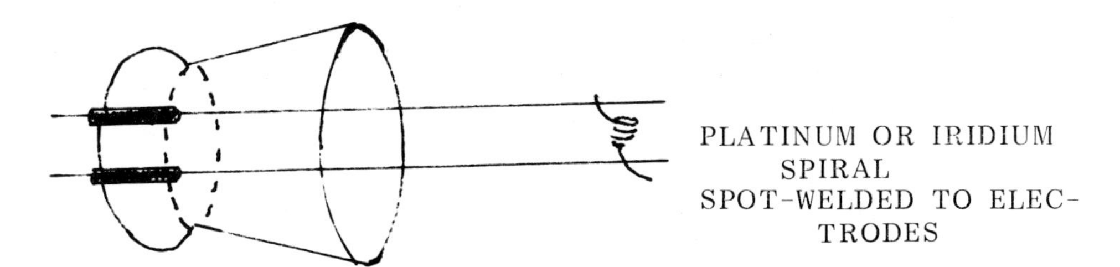

Fig.4. Ground joint cap with tungsten electrodes and platinum (iridium) spiral crucible

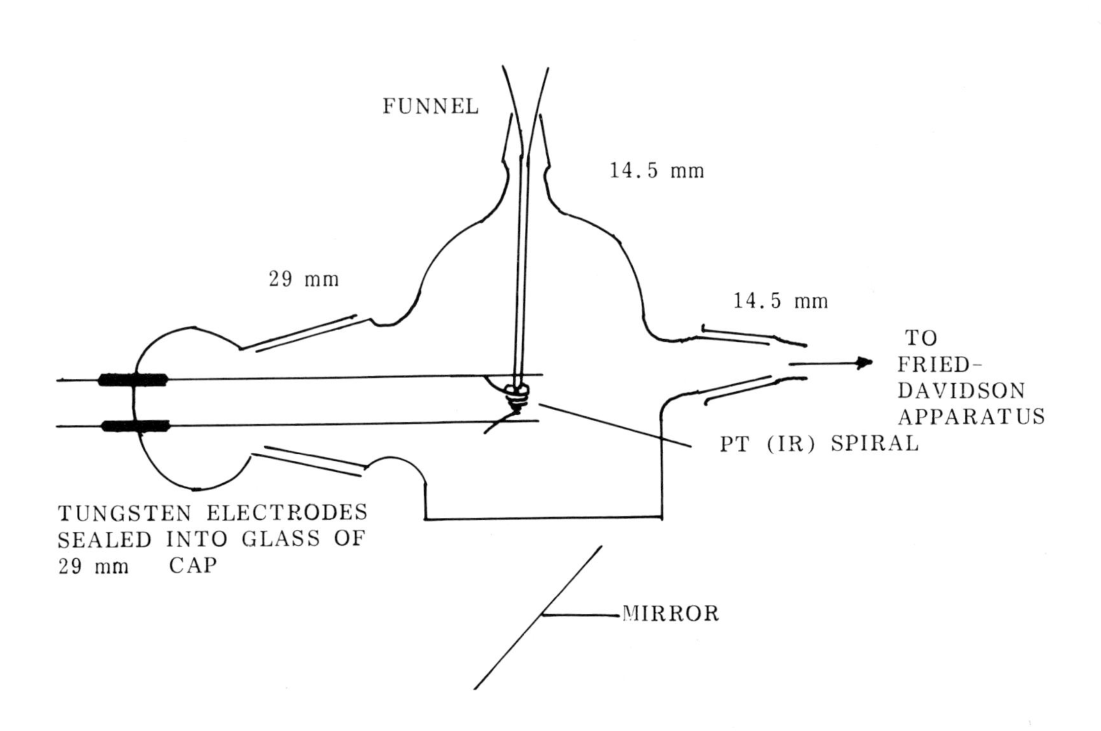

Fig.5. Apparatus for work with platinum and/or iridium spiral crucibles

small amount of concentrated HCl or HNO_3, and HCl, or HNO_3 added to 1 ml.

The glass of the vessels, in which such curium stock solutions were kept for some time, became crazed after a while due to the attack of the α-radiation onto the glass surface. Prior to most preparative experiments, the silica accumulated in the solutions by this effect was removed by evaporating the solution to dryness, and taking up the residue in water.

2.3. Microchemical Preparation Techniques

The microchemical preparation techniques used in this work were essentially the same as those developed by Cunningham (37)(38), but modified to fit our requirements. Dry chemistry work was usually done in a Fried-Davidson apparatus (39). The version used in our work was modified in such a way that the apparatus was set up outside the glove box, but the capillary or reaction tube was located inside, to prevent spread of contamination.

High temperature experiments, such as solid state reactions between oxides were carried out in an iridium spiral furnace as shown in Figs. 4 and 5. The Pt- or Ir-spiral served both as the crucible and as the heating element, and were spot-welded to the tungsten electrodes.

The spiral furnaces were charged by means of a drawn-out glass funnel, which guided the charge material into the interior of the spiral (Fig.5). The spiral temperature was regulated by means of a variac. Temperatures were measured by means of a Haase micropyrometer. The Pt spirals allowed to attain temperatures up to 1300°C, the Ir-spirals up to 1700°C *).

X-ray capillaries and quartz reaction tubes used in microwork were heated by means of small, self-built quartz-kanthal or quartz-platinum microfurnaces, which are in routine use in our laboratory.

Weighings of radioactive material were made with a Sartorius electronic microbalance Type 4401 having a sensitivity of 0.1 μg/scale division on the digital indicator. This balance was calibrated using a Rodder ultramicro quartz fiber torsion balance **). Accuracies of ±1 μg were easily to attain.

Thermal and caloric effects in solid state reactions and phase transformations (melting points, heats of fusion) were measured with a

---

*) Not every iridium wire was found to be suitable for spiral winding. We found that the wire supplied by Alpha Inorganics was the one best suited because of its ductility

**) Sensitivity 85 Nanograms/scale division

Linseis Micro DTA apparatus ***), which allowed the measurements of samples weighing less than 1 mg.

2.4. X-Ray Work

All X-ray data were taken under conditions, and with the equipment specified in Table 2. X-ray capillaries used were in all cases 0.3 mm quartz capillaries, which were drawn out to an OD of 0.1 - 0.2 mm, and were sealed with their tips to little quartz rods (0.2 mm OD, 5 - 7 mm long to facilitate handling. Capillaries were filled and sealed inside the glove box using the techniques described by Haug (40).

To save space, the detailed evaluations of the individual X-ray patterns are not reported here, they will be submitted to the ASTM card catalog for general availability.

2.5. Preparation of Key Compounds

The compounds, which are called key compounds in this context, are the ones, which are used as starting materials for other compounds. They are themselves easily accessible from stock solutions. In this connection, they include curium(III)-oxalate-decahydrate, curium(III)-oxide, curium(III)-formate, and curium(IV)-oxide.

2.5.1. Curium(III)-oxalate-decahydrate, $Cm_2(C_2O_4)_3 \cdot 10H_2O$

For the preparation of a well-defined curium(III)-oxalate-decahydrate, 100 μl of 0.1 M oxalic acid solution were added to 3.6×10^{-3} M curium(III)-solution in HNO_3. (HCl or H_2SO_4 should be avoided, because chloro- or sulfato-oxalates may be formed.) On addition of 7 M NH_4OH solution to neutralize excess acid, pale greenish-yellow curium oxalate decahydrate $Cm_2(C_2O_4)_3 \cdot 10H_2O$ precipitates according to the following equation:

$$2Cm^{3+} + 3C_2O_4^{2-} + 10H_2O \rightarrow Cm_2(C_2O_4)_3 \cdot 10H_2O$$

A small quantity of the compound was sealed in an X-ray capillary and the powder pattern was taken immediately after preparation.

$Cm_2(C_2O_4)_3 \cdot 10H_2O$ is monoclinic, space group $P2_1/c - C^5_{2h}$ (No.14) with the following lattice constants: a = 11.23±0.06 A, b = 9.60±0,05 A, c = 10.22±0.05 A, β = 114.64°.

In the dark, $Cm_2(C_2O_4)_3 \cdot 10H_2O$ shows a strong orange luminescence, which is probably due to radiation-induced transition of the first excited state to the ground state of Cm(IV). Cm(III)-oxalate-decahydrate is isomorphous with the corresponding oxalates of actinium (41), plutonium (42), and americium (43).

***) Manufactured by Linseis Meßgeräte GmbH, Vielitzer Str. 43 8672 Selb, Ofr. W.-Germany.

2.5.2. Curium(III)-formate, $Cm(HCOO)_3$.

Curium(III)-formate was obtained by reacting $Cm(OH)_3$ or aged $Cm_2(C_2O_4)_3 \cdot 10H_2O$ with formic acid. 50 µg $Cm(OH)_3$ or aged $Cm_2(C_2O_4)_3 \cdot 10H_2O$ were suspended in 20 µl distilled water, and 50 µl 98% formic acid were added. In the case of the oxalate, a gas evolution (probably CO_2) was observed, The mixture was heated under an IR lamp to approx. 70°C for 1 hour. It reacts to form $Cm(HCOO)_3$ by the following equation:

$$Cm(OH)_3 + 3\,HCOOH \rightarrow Cm(HCOO)_3 + 3H_2O$$

The resulting white precipitate was centrifuged, the supernatant removed, and the precipitate washed three times with absolute methanol. From the dried, coursely crystalline curium(III)-formate, an X-ray sample was prepared.

$Cm(HCOO)_3$ is hexagonal or rhombohedral, respectively, with the following lattice constants:

hexagonal	$R3m - C_{3v}^5$ (No.160)	rhombohedral
a_H = (10.606 ± 0.017) A		a_R = (6.272 ± 0.011) A
c_H = (4.069 ± 0.007) A		α = 115.46°
v_H = 396.3 A		v_R = 132.1 A
		Z_R = 1
		D_x = 4.76 gcm^{-3}

Curium(III)-formate, on ignition at 800°C for 4 hours yields cubic $C\text{-}Cm_2O_3$.

2.5.3. Curium(III)-nitrate, $Cm(NO_3)_3 \cdot xH_2O$.

30 µg $Cm(OH)_3$ or 30 µg aged $Cm_2(C_2O_4)_3 \cdot 10H_2O$ were placed into a microcone and 15 µl conc. HNO_3 or 8M HNO_3 were added. The solution was evaporated to near-dryness under the IR lamp, and completely dried with a stream of air.

$$Cm^{3+} + 3NO_3^- + xH_2O \rightarrow Cm(NO_3)_3 \cdot xH_2O$$

The pale-yellow, nonhygroscopic nitrate was sealed in a quartz X-ray capillary, and the powder pattern was taken.

The powder diagram of $Cm(NO_3)_3 \cdot xH\ O$ turned out to be very complicated. We have measured 30 lines, but were not able to interpret this diagram. It appears that $Cm(NO_3)_3 \cdot xH\ O$ is either monoclinic or triclinic, and that it is isomorphous with with $Am(NO_3)_3 \cdot 6H_2O$.

The thermal decomposition of $Cm(NO_3)_3$ in oxygen atmosphere at 780°C for 4 hours yields a cubic oxide, possibly $C\text{-}Cm_2O_3$, or, more likely, the σ-phase ($CmO_{1.6}$). Pyrolysis for 6 hours at 330°C yields in CmO_2.

2.6. Ternary Oxides of Cm(III)

There are numerous compounds, which may be described as ternary oxides. They include compounds of the following types: $CmXO_3$, $CmXO_4$, $Cm_2(XO_4)_3$ and $Cm(XO_4)_3$. Most of the compounds we prepared could be characterized by their X-ray powder data.

2.6.1. Curium(III)-perrhenate, $Cm(ReO_4)_3 \cdot xH_2O$.

70 µl of a 3.6×10^{-3} M $Cm(NO_3)_3$-solution was precipitated with gaseous ammonia to form pale-yellow $Cm(OH)_3$. This was washed several times with distilled water, and 80 µl of a 0.01 M $HReO_4$-solution was added. The mixture was continuously heated under an IR lamp, and the resulting pale-yellow solution evaporated under the IR lamp.

$$Cm^{3+} + 3\ ReO_4^- + xH_2O \rightarrow Cm(ReO_4)_3 \cdot xH_2O$$

$Cm(ReO_4)_3 \cdot xH_2O$ is a white, hygroscopic solid. The X-ray powder pattern could be interpreted with an orthorhombic lattice with a = 5.636±0.026 A, b = 7.674±0.036 A, c = 15.401±0.072 A. It is isomorphous with the corresponding Pu-compound, which had been prepared by J.P.Silvestre (44)

2.6.2. Anhydrous Curium(III)-perrhenate, $Cm(ReO_4)_3$

This compound was prepared by two different roads, as shown below:

$$Cm(ReO_4)_3 \cdot xH_2O \xrightarrow[5\ hr]{350^\circ C} Cm(ReO_4)_3$$

$$CmO_2 + 3Re + 5O_2 \xrightarrow[20\ hr]{280-300^\circ C} Cm(ReO_4)_3$$

The dehydration of $Cm(ReO_4)_3 \cdot xH_2O$ is carried out in a quartz tube, which was pre-degassed in vacuo and attached to a Fried-Davidson apparatus. The hydrated compound was slowly heated in air to 350°C, and maintained at this temperature for 5 hours. The white reaction product was identified by its X-ray powder pattern.

In the second method of preparation, 35 µg CmO_2 and 70 µg Re metal *) were intimately mixed **) and placed into the quartz reaction tube of a Fried-Davidson apparatus and heated in air for a period of 20 hours.

---

*) The Re metal was prepared by reduction of Re_2O_7 in dry hydrogen at 600°C. The reaction product was identified as Re metal by its powder pattern.

**) For mixing, a mortar and pestle set (Cat.No. 6230, Sierra Misco Co. 1825 Eastshore Highway, Berkeley, CA. 97410) or a tungsten wire, ground to a small ball, together with an Al_2O_3 microcone were used.

The powder pattern of the white reaction product was the same as that obtained with the reaction product of the dehydration experiment.

An interesting aspect in the preparation of $Cm(ReO_4)_3$ is the reduction of initial Cm(IV) to Cm(III). This reduction is in agreement with the observations of Silvestre in the synthesis of $Am(ReO_4)_3$ and $Pu(ReO_4)_3$ at 400°C.

$Cm(ReO_4)_3$, at 650°C, decomposes to Cm_2O_3 and volatile Re_2O_7; it is thus less stable than $Pu(ReO_4)_3$ (decomposition at 700°C) and $Am(ReO_4)_3$ (decomposition at 850°C).

$Cm(ReO_4)_3$ crystallizes in the hexagonal system, β-$Eu(ReO_4)_3$-type, with a = 10.065±0.059 A, c = 6.294±0.034 A.

2.6.3. Curium(III)-molybdate, $Cm_2(MoO_4)_3$

30 μg CmO_2 and 29 μg $(NH_4)_6Mo_7O_{24}\cdot 4H_2O$ (commercial "ammonium molybdate") are intimately mixed, and reacted with each other in the quartz tube of a Fried-Davidson apparatus by heating at 750°C for 3 hrs; then at 930°C for 6.5 hrs. The brownish-yellow curium molybdate, which is formed according to the reaction

$$14\ CmO_2 + 3\ (NH_4)_6Mo_7O_{24}\cdot 4H_2O \xrightarrow[6\ hrs]{650°C} \xrightarrow[6.5\ hrs]{930°C}$$

$$\rightarrow 7\ Cm_2(MoO_4)_3 + 18NH_3 + 21\ H_2O + 3.5\ O_2$$

was identified by its X-ray powder pattern. It crystallizes in the Scheelite type lattice, tetragonal, space group $I4_1/a - C^6_{4h}$ (No. 88) with a = 5.250±0.007 A, c = 11.496±0.016 A, Z = 4, and by assuming $Cm_{2/3}\square_{1/3})MoO_4$, its density was calculated to be 6.76 gcm^{-3}.

The intensity calculations in $Cm_2(MoO_4)_3$ were done with the SHELX 76 program and comparing with the results of α-$Nd_2(MoO_4)_3$ (45), using isotropic temperature factors, an R-value of 0.072 was obtained. There is good agreement between calculated and observed intensities. By assuming 4 formula 7nits of the type $(Cm_{2/3}\square_{1/3})MoO_4$ in the unit cell, the atomic positions of $Cm_2(MoO_4)_3$ given in Table 3 were calculated. Each Cm atom is surrounded by a distorted oxygen dodecahedron. The $Cm_2(MoO_4)_3$ compound prepared in our work is isomorphous with α-$Nd_2(MoO_4)_3$ and α-$Am_2(MoO_4)_3$. A curium molybdate corresponding to β-$Am_2(MoO_4)_3$ could not be prepared so far.

2.6.4. Curium(III)-tungstate, $Cm_2(WO_4)_3$

80 μg CmO_2 and 88 μg WO_3 are weighed on the microbalance, intimately mixed, and transferred to the quartz tube of the Fried-Davidson apparatus. The mixture was heated at 890°C for 24 hours. It reacted according to

$$2\ CmO_2 + 3\ WO_3 \xrightarrow[24\ hrs]{810°C} Cm_2(WO_4)_3 + 0.5\ H_2O$$

Table 3. Atomic coordinates in curium(III)-molybdate

Atom	Wyckoff notation	Parameters x	y	z
2/3 Cm	4 b	0.0	0.250	0.625
Mo	4 a	0.0	0.250	0.125
O	16 f	0.138	0.997	0.198

Table 4. Atomic coordinates in curium(III)-tungstate

Atom	Wyckoff notation	Parameters x	y	z
Cm	8 f	0.334	0.377	0.407
W_1	8 f	0.145	0.394	0.051
W_2	4 e	0.0	0.132	0.250
O_1	8 f	0.161	0.045	0.221
O_2	8 f	0.132	0.208	0.388
O_3	8 f	0.222	0.320	0.195
O_4	8 f	0.359	0.450	0.038
O_5	8 f	0.053	0.459	0.423
O_6	8 f	0.446	0.241	0.058

The resulting, pale-yellow reaction product was X-rayed immediately after its preparation.

$Cm_2(WO_4)_3$ was found to be monoclinic, α-$Eu_2(WO_4)_3$ type, space group C2/c (No.15) with the following lattice constants: a = 7.714±0.031 A, b = 11.556±0.046 A, c = 11.450±0.045 A, β = 109.14°, V = 964.3 A . With Z = 4 formula units in the cell, the density was calculated to be D_x = 8.48 gcm^{-3}. The intensities were calculated using the LAZY PULVERIX program and the atomic parameters used by Templeton and Zalkin (55). Good agreement between measured and calculated intensities was found. One has to keep in mind that the position of the oxygen atoms cannot be determined from the measured intensities of the powder diagram, because the scattering power of each oxygen atom is only 4% of the scattering power of a curium atom.

2.6.5. Curium(III)-phosphate hydrate, $CmPO_4 \cdot 0...0.5H_2O$

Attempts to prepare and to unequivocally characterize $CmPO_4 \cdot 0...0.5H_2O$ have been made at earlier times by Weigel (29), Kohl (46) and by Walter (47). In all these earlier attempts, no crystalline product was obtained in the precipitation of Cm^{3+} with PO_4^{3-}. We succeeded in preparing crystalline $CmPO_4 \cdot 0...0.5H_2O$ by a hydrothermal synthesis method in a similar manner as described by Weigel and Scherer in the case of promethium(III)-hydroxide (48).

To 50 μg of a 3.6 10^{-3}M $Cm(NO_3)_3$ solution in a microcone, 30 μl of 0.1M $(NH_4)H_2PO_4$ solution were added. In order to precipitate the $CmPO_4$, the acid in the solution was carefully neutralized with ammonia. The yellowis white precipitate, which formed according to

$$Cm^{3+} + H_2PO_4^- + 2NH_3 + xH_2O \rightarrow CmPO_4 \cdot xH_2O + 2\ NH_4^+$$

was washed with distilled water to neutral reaction. The phosphate precipitate was slurried with distilled water into a quartz capillary tube (ID 1.5 mm, length 40 mm). The tube was filled halfways with distilled water and sealed with a hydrogen oxygen torch. The quartz tube, together with some water, was sealed into an outside Duran tube, which was also sealed. The whole assembly was placed into an autoclave, which was maintained for 16 hours at 150°C. After cooling and opening, the pale-yellow curium compound was transferred to a glass centrifuge cone, the supernatant was pipetted off and the $CmPO_4 \cdot 0...0.5H_2O$ was washed with acetone and dried in a current of air. X-ray data were taken immediately after preparation.

$CmPO_4 \cdot 0...0.5H_2O$ was found to be hexagonal, isomorphous with the corresponding phosphates of Nd, Ac, Pu, and Am, space group $P6_222$-D_6^4 (No.180) with a = 7.00±0.08 A, c = 6.39±0.07 A, V = 271.1 A^3. For a unit cell with Z = 3, the density was calculated to be 6.39 gcm^{-3}. The monoclinic $CmPO_4$, which was observed by Weigel and Haug (27)(29) was not observed under the conditions of preparation used here.

2.6.6. Curium(III)-arsenate, $CmAsO_4$.

20 µl 0.1 M $(NH_4)_2HAsO_4$ solution in distilled water were added to 30 µl of 3.6×10^{-3}M Cm^{3+} stock solution in a microcone. On neutralization of the free acid with ammonia, a white, voluminous precipitate was formed according to

$$Cm^{3+} + HAsO_4^{2-} \rightarrow CmAsO_4 + H^+$$

The precipitate was washed with acetone and dried under the IR lamp. It was then transferred to the quartz reaction tube of the Fried-Davidson-apparatus, and slowly heated in air. Within a period of 90 min, the temperature was raised to 950°C, and maintained at that temperature for 4 hours. After cooling, the brownish-yellow reaction product was studied by X-ray diffraction.

$CmAsO_4$, just as $PuAsO_4$ or $AmAsO_4$, was found to be monoclinic, monazite type, space group $P2_1/n - C_{2h}^5$ (No.14) with a = 6.823±0.009 A, b = 7.036 ±0.009 A, c = 6.621±0.009 A, β = 105.19°, V = 306.7 A^3 . With Z = 4 molecules in the unit cell, the density was found to be D_x = 8.29 gcm^{-3}.

This shows an interesting aspect in the structure systematics of the lanthanide and actinide arsenates. In the lanthanide series, $LaAsO_4$ through $NdAsO_4$ exhibit the monoclinic monazite type structure, $PmAsO_4$ through $LuAsO_4$ have the tetragonal zircone type structure. It appears that curium is rather to be considered the chemical homologue of Nd than that of Pm. It van be expected that $BkAsO_4$ probably will have the zircone type structure.

2.6.7. Curium(III)-vanadate(V), $CmVO_4$.

74 µg CmO_2 were intimately mixed with 31 µg NH_4VO_3, and placed into the the quartz reaction tube of the Fried-Davidson apparatus. The mixture was heated to 650°C. After a period of 7 hours, the reaction temperature was raised to 930°C and maintained at this temperature for 8 hours. The brownish to honey-yellow reaction product of the reaction

$$2\ CmO_2 + 2NH_4VO_3 \rightarrow 2CmVO_4 + 2NH_3 + H_2O + 0.5\ O_2$$

was X-rayed immediately after preparation.

It was found to be isomorphous with $AmVO_4$ and with the lanthanide vanadates, $LnVO_4$, in particular, it agrees well with $NdVO_4$ (56). $CmVO_4$ is tetragonal,zircon type structure, space group $I4_1/amd - D_{4h}$ (No. 141) with a = 7.290±0.013 A, c = 6.417±0.012 A, V = 314.0 A^3. For Z = 4 formula units in the unit cell, the density was calculated to be 6.99 gcm^{-3}.

Good agreement between observed and calculated intensities was found with 4 Cm in 4 (a), 4 V in 4(b), and 16 O in 16 (h) with the parameters x = 0.437 and z = 0.212, origin at $\bar{4}$m 2 at 0, 1/4, 1/8 from center (2/m).

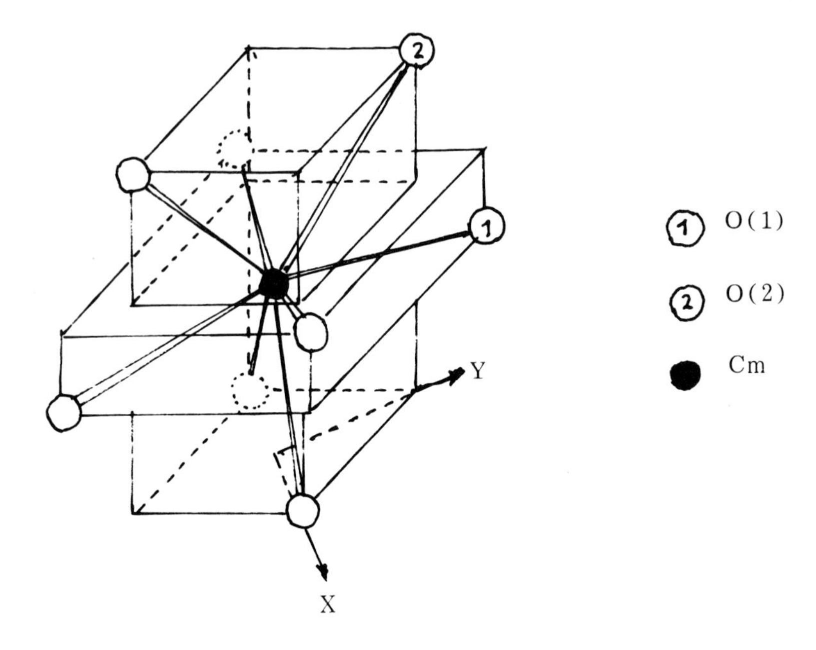

Fig.6. CmO_8-coordination polyhedron in $CmVO_4$, where the distances are Cm-O(1) = 2.347 Å, and Cm-O(2) = 2.560 Å.

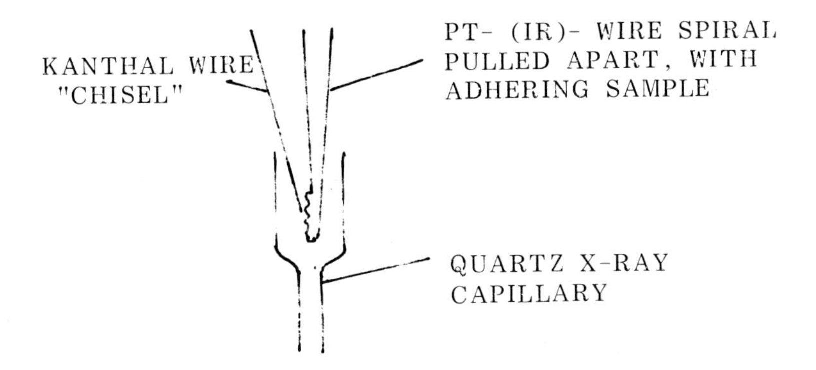

Fig.7. Charging of materials prepared in the Pt- or Ir-spiral furnace into the X-ray capillary

2.6.8. Curium(III)-chromate(V), $CmCrO_4$.

19.5 μl of 0.01 M $Cr(NO_3)_3$-solution were added to 50 μl of 3.6×10^{-3}M Cm^{3+} solution. By addition of an excess of 7M NH_4OH and heating, the hydroxides were precipitated. The light green hydroxide precipitate was washed several times with distilled water and methanol, and dried under the IR lamp. The dried hydroxide mixture was placed into the quartz reaction tube of a Fried-Davidson apparatus, and was heated at 610°C in an atmosphere of oxygen. The dark green reaction product, which was formed according to the reaction

$$2(Cm,Cr)(OH)_3 + O_2 \xrightarrow[15\ hr]{610°C} 2CmCrO_4 + 3H_2O$$

was identified by its X-ray diffraction pattern.

The powder diffraction data of $CmCrO_4$ showed that the compound contained Cm_2O_3 as an impurity. So far, we were not yet able to prepare pure $CmCrO_4$. From the reflexions assigned to $CmCrO_4$, it was found that $CmCrO_4$ is isomorphous with $NdCrO_4$. It has a tetragonal $ZrSiO_4$ lattice, space group $I4_1/amd - D^{19}_{4h}$ (No. 141) with a = 7.242±0.024 A, c = 6.353± 0.021 A, V = 333.2 A^3. With Z = 4 formula units in the cell, the calculated density is D_x = 7.18 gcm^{-3}.

2.6.9. Curium(III)-scandium oxide, $CmScO_3$

300 μl of 1.16×10^{-3} M Cm^{3+} stock solution were carefully evaporated under the IR lamp, and were mixed with 35 μl of 0.01 M $Sc(NO_3)_3$ solution. 35 μl of 7 M NH_4OH were added, and the resulting pale-yellow hydroxide precipitate was washed several times with water, and was then transferred to the Pt-spiral furnace (Figs. 4 and 5). The reaction recipient was evacuated twice, and was flushed with purified argon. The Pt-spiral was heated to 1160°C and maintained at this temperature for 30 minutes. Then, the reaction temperature was raised to 1275°C, maintained at this latter temperature for 15 minutes, and cooled.

After the Pt-spiral had cooled, it was carefully pinched off from the tungsten electrodes of the furnace, and the windings of the spiral were pulled apart under the stereomicroscope inside the wide part of the X-ray capillary. The part, which stuck to the wire was chipped off by means of a micro "chisel" consisting of a kanthal wire (Fig. 7). All these operations required extreme cleanliness to prevent contamination of the samples and great dexterity to prevent losses of material, because the chipped-off material tended to jump away.

The $CmScO_3$was X-rayed immediately after preparation. It was found to have a perowskite type structure,$GdFeO_3$ type, space group Pbnm - D^{16}_{2h} (No. 62) with an orthorhombic lattice with a = 5.532±0.014 A, b = 5.753 ±0.015 A, c = 7.987±0.020 A, V = 254.2 A . With Z = 4 formula units in the unit cell, the denisty is D_x = 8.81 gcm^{-3}

Intensities were calculated using the atomic positions of space group

Pbnm : Cm in 4 (c) with x = 0.015, y = 0.059 , O(1) in 4 (c) with x = 0.109 and y = 0.452, and O(2) in 8 (d) with x = 0.664, y = 0.327, and z = 0.059. These data are based on those published for $EuScO_3$ (52). Experimental and calculated intensities are in good agreement.

2.6.10. Curium(III)-vanadate(III), $CmVO_3$

Starting from 53 μg CmO_2 and 23 μg NH_4VO_3, brownish-yellow $CmVO_4$ was prepared in the manner described on page 174. This was reduced with hydrogen at 890°C in the quartz reaction tube of the Fried-Davidson apparatus for a period of 15 hours. Every 15 minutes, the Fried-Davidson apparatus was evacuated, and the hydrogen was replaced by fresh hydrogen. The resulting dark-green solid, which was obtained according to the reaction

$$CmVO_4 + H_2 \xrightarrow[15\ hrs]{890°C} CmVO_3 + H_2O$$

X-ray powder patterns of $CmVO_3$ were taken immediately after preparation. The powder pattern of $CmVO_3$ shows close resemblance with the patterns of $PuVO_3$ (53) and of $AmVO_3$ (47). $CmVO_3$ has a perowskite type lattice, orthorhombic $GdFeO_3$ type, space group Pbnm - D_{2h}^{16} (No. 62) with a = 5.428±0.019 A, b = 5.571±0.020 A, and c = 7.766±0.028 A, V = 234.8 A^3. With Z = 4, the density was calculated to be D_x = 9.70 gcm^{-3}.

2.6.11. Curium(III)-chromate(III), $CmCrO_3$

$CmCrO_3$ has been prepared by solid state reaction between Cm_2O_3 and Cr_2O_3 and by ignition of mixed hydroxides of Cm(III) and Cr(III), which had been precipitated by ammonia from a mixture of Cm^{3+}- and Cr^{3+} solutions:

$$\left.\begin{array}{l} Cm_2O_3 + Cr_2O_3 \\ (Cm,Cr)(OH)_3 \end{array}\right\} \xrightarrow{850°C-900°C} CmCrO_3$$

a) 135 μg CmO_2 and 38 μg Cr_2O_3 were weighed on a microbalance, were intimately mixed, and charged into the quartz reaction tube of a Fried-Davidson apparatus. The mixture was heated in air for 33 hours at 700°C. After 15 hours, the heating was interrupted and the reaction product was ground again to insure homogeneity. The resulting light-green reaction product was identified by its X-ray powder pattern.

b) 100 μl of a 3.6×10^{-3} M Cm^{3+} solution was mixed with 38.8 μl 0.01 M Cr^{3+} solution. The dark green hydroxide mixture was precipitated from the heated solution with 7 M NH_4OH. The precipitate was washed, dried, and transferred to the quartz reaction tube of the Fried-Davidson apparatus, and ignited at 900°C. The reaction product was X-rayed immediately after preparation.

In both preparations, unreacted Cr_2O_3 was found besides $CmCrO_3$. This also applied to a preparation, which was carried out at 1250°C in the Pt-spiral furnace, and which was also contaminated with Cr_2O_3. This, on the other hand, had the advantage that the Cr_2O_3-lines could be used as an internal

$CmCrO_3$ has a perowskite type structure, orthorhombic $GdFeO_3$ type, space group Pbnm - D_{2h}^{16} (No. 62) with the following lattice constants: a = 5.403±0.021 A, b = 5.508±0.021 A, c = 7.669±0.030 A, V = 228.2 A^3. With Z = 4 formula units per cell, the calculated density was found to be D_x = 10.01 gcm^{-3}.

2.6.12. Curium(III)-Iron(III)-oxide, $CmFeO_3$

The preparation of this compound was done by ignition of the mixed hydroxides in the same manner as in the preparation of $CmScO_3$. 200 µl 1.16×10^{-3} M Cm^{3+} solution were mixed with 34 µl 6.95×10^{-3}M Fe^{3+} solution, and the hydroxides were precipitated with ammonia. The resulting hydroxide mixture was washed, dried, and transferred to the Pt spiral furnace. It was reacted to $CmFeO_3$by ingition at 1160°C and 1275°C, respectively. The dark brown reaction product was identified by its X-ray powder pattern.

$CmFeO_3$ is also a perowskite type compound, $GdFeO_3$ type. It is orthorhombic, space group Pbnm - D_{2h}^{16} (No.62) with the lattice constants a = 5.414±0.010 A , b = 5.579±0.015 A, c = 7.739±0.015 A, V = 233.8 A^3. With Z = 4 formula units in the cell, the caculated density is D_x = 9.88 gcm^{-3}.

2.7. Oxyhalides of Curium(III), CmOX.

Until recently, only very little information was available on the oxyhalides of curium(III), CmOX, where X = F, Cl, Br, or I. In 1971 Peterson (31) prepared ^{248}CmOCl and determined its lattice constants, in 1977, Weigel, Wishnevsky, and Hauske (9) have studied the vapor phase hydrolysis of $CmCl_3$ and determined the heat of formation of CmOCl. We therefore thought it of interest to prepared and characterize some of the oxyhalides.

2.7.1. Curium(III) oxyfluoride, CmOF

In preliminary mockup experiments, we had found that the thermal decomposition of gadolinium monofluoroacetate, $Gd(FCH_2COO)_3$, yielded pure GdOF. It was therefore assumed that curium monofluoroacetate, $Cm(FCH_2COO)_3$, would react in the analogous manner.

An excess of freshly precipitated $Cm(OH)_3$ or aged $Cm_2(C_2O_4)_3$ was added to 50 µl of 0.01 M monofluoro acetic acid in a microcone. After the acid had reacted with the curium hydroxide, or the oxalate, the mixture was centrifuged, and the supernatant separated from the solid residue. The clear supernatant solution was rapidly evaporated to dryness. The remaining pale yellow amorphous residue was transferred to the quartz reaction tube of the Fried-Davidson apparatus. It was then heated slowly to 730°C, and maintained at this temperature for a period of 4 hours. The greyish-yellow reaction product was X-rayed immediately after preparation. It was identified as CmOF, isomorphous with EuOF and PuOF.

The powder pattern of curium oxyfluoride was indexed in terms of a hexagonal-rhombohedral lattice, space group $R\bar{3}m - D_{3d}^{5}$ (No. 166) with the following lattice constants:

hexagonal	rhombohedral
a_H = 3.864±0.007 A	a_R = 6.843±0.011 A
c_H = 19.408±0.034 A	α_R = 32.80°
V_H = 250.5 A^3	V_R = 83.6 A^3
Z_H = 6	Z_R = 2

$$D_x = 11.08\ gcm^{-3}$$

An intensity calculation was carried out using the LAZY PULVERIX program and the parameters determined for YOF by Mann and Bevan (60). With z_{Cm} = 0.241, z_F = 0.372, and z_O = 0.117, good agreement between observed and calculated intensities was obtained. The Cm-O distances were found to be 2.34 A and 2.41 A, the Cm-F distances 2.49 A, and 2.54 A. The observation that Cm-O distances were smaller than Cm-F distances was also observed for the corresponding distances in YOF and EuOF.

2.7.2. Curium(III) oxychloride, CmOCl

For the preparation of CmOCl, the reagent vessel of the Fried-Davidson apparatus was charged with concentrated HCl. Apporx. 50 μg of curium (III)-oxalate (2.5.1, p. 5) was charged into the quartz reaction tube. The oxalate was slowly heated to 500...520°C, and was reacted for 8 hours in a gaseous mixture of HCl and H O vapor. Every 15 minutes,the gas mixture was evacuated, and was replaced with fresh mixture. The resulting pale-yellow reaction product was sealed in an X-ray capillary for identification.

In accordance with the study of Peterson on ^{248}CmOCl (31), the ^{244}CmOCl studied in our laboratory showed the same symmetry, space group and lattice constants within the limits of error. ^{244}CmOCl, like the ^{248}Cm-compound, is tetragonal, PbFCl structure, space group P4/nmm - D_{4h}^{7} (No.129) with the following lattice constants: a = 3.983 0.003 A, c = 6.738 0.005 A, V = 106.9 A^3. With Z = 2 formulas in the cell, the calculated density is D_x = 9.18 gcm^{-3}.

The preparation of ^{244}CmOCl offered an opportunity to study the change of lattice constants as a function of time. If a solid undergoes external or internal irradiation with heavy particles. such as α-particles, a change of the lattice constants is frequently observed. As a rule, they become larger, i.e. the lattice expands.

This effect, so far, has only been studied in cubic compounds, such as CmO_2 (26)(32), or $CmAlO_3$ (57)(58). So far, no studies of compounds with lower symmetry had been made. The availability of tetragonal CmOCl

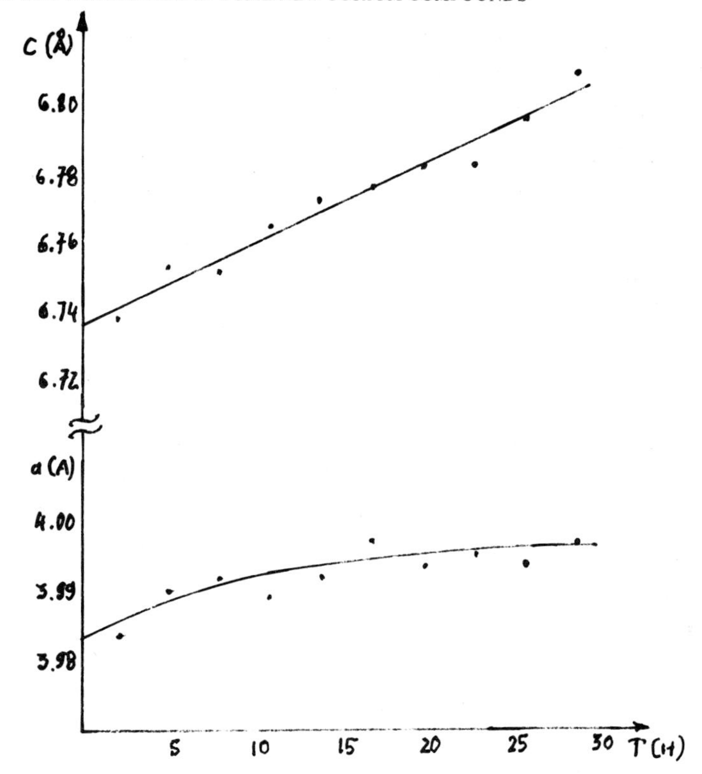

Fig.8. Change of lattice constants in CmOCl with time (^{244}CmOCl)

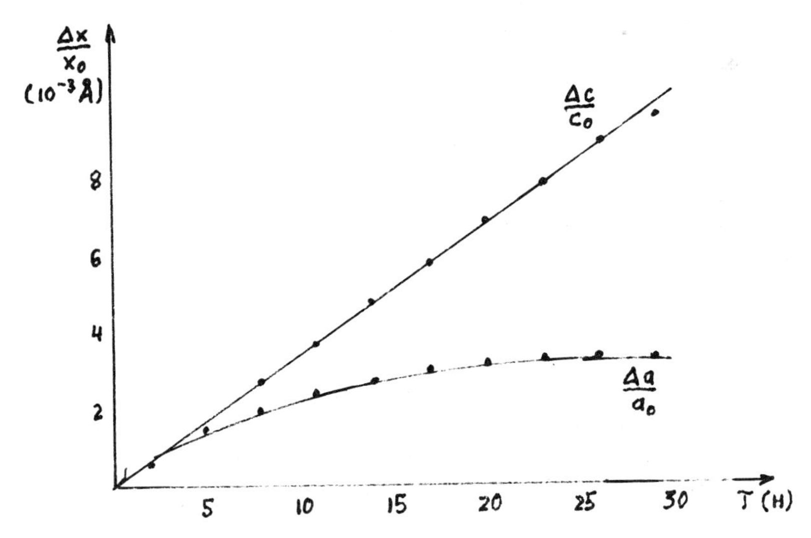

Fig.9. Relative change of lattice constants in CmOCl with time (^{244}CmOCl)

offered us a chance to study the behavior of two different lattice constants if the lattice was subjected to self-irradiation.

We have therefore taken the powder pattern of a CmOCl sample at 3 hour intervals for a total time of 30 hours. The observed change of the CmOCl lattice constants is shown in Fig. 8 . Fig. 9 shows the relative change of the same constants with time. It was not possible to continue the measurements beyond a time period of 40 - 50 hours, because the X-ray patterns become increasingly diffuse and more and more difficult to evaluate.

The interesting observation is, that the a-axis approaches a limiting value of a = 3.995 A after a few hours, whereas the c-axis continues to grow. This behavior may be understood from the layer structure of CmOCl, which had been studied by Peterson and Burns (30) on ^{248}CmOCl. Each curium atom is surrounded by 4 O-atoms, 4 Cl-atoms from one neighbouring layer, and 1 Cl atom from the next layer. The layers stacked in the c-direction in the sequence Cl-Cm-O-Cm-Cl.... have most certainly a smaller bond strength between each other than the Cl-O- or Cm-atoms within one layer. The observed behavior is best explained by the assumption that a Cm atom, in its decay, is hardly ruptured from its position relative to its nearest neighbours, but that the whole layer close to the decaying atom is dislocated.

After a period of 80 - 90 hours, no more X-ray reflexions can be observed in a CmOCl powder pattern. The order of the lattice may be restored, however, by heating the sample in a mixture of HCl and H_2O.

The lattice parameters of ^{244}CmOCl extrapolated to t = 0 hours were compared with the corresponding data of ^{248}CmOCl. It was found that the two sets of lattice constants for the two isotopic compounds were in almost complete agreement with each other, i.e., within the margin of error, it is of no influence, if a ^{244}Cm atom is exchanged for a ^{248}Cm atom.

The dependance of lattice parameters on time is taken into account for all other Cm compounds studied. In all cases, X-ray patterns were taken immediately after preparation, at the utmost, a period of 15 - 30 minutes after preparation is allowed.

2.7.3. Curium(III)-oxybromide, CmOBr

CmOBr was prepared from aged $Cm_2(C_2O_4)_3 \cdot 10H_2O$ in the reaction tube of an available vapor phase hydrolysis apparatus. 35 µg of curium oxalate decahydrate were charged into a micro platinum boat (6×1.5×1.5 mm^3), which was placed onto a quartz slide, and was carefully introduced into the reaction tube (Fig. 10). The sample was heated to 360°C, and a stream of N_2 was passed over it. At the same temperature, the N_2 was replaced by HBr gas, and the curium sample was heated for a period of 40 minutes in the HBr to form $CmBr_3$. The $CmBr_3$ was hydrolyzed for another 40 minutes at 360°C in a stream of $HBr/H_2O/N_2$. In the same gas stream, the temperature was raised to 600°C, and maintained for 1 hour. The pale-yellow reaction product was X-rayed after cooling. The reason for the complicated preparation technique is the low melting point of $CmBr_3$, which is approx. 400°C (46). The melting of the $CmBr_3$ must be avoided to prevent that the hydrolysis stops at a too early time.

CmOBr is isomorphous with AmOBr (50). It crystallizes in the tetragonal symmetry, space group P4/nmm - D^7_{4h} (No. 129) with a = 3.956±0.008 A, and c = 7.936±0.015 A, V = 124.4 A^3. With Z = 2 formula units in the cell, the density is calculated to be D_x = 9.09 gcm^{-3}. Intensities were calculated with 2 Cm in 2 (c), 2 Br in 2 (c), and 2 O in 2 (a), and with the parameters z_1 = 0.192, and z_2 = 0.610, with the origin located in the center of symmetry (2/m).

2.7.4. Curium(III)-fluoride, CmF_3

Curium(III)-fluoride was one of the first two curium compounds, which were prepared in the solid state, and were characterized by their X-ray powder patterns (6). The standard preparation procedure for this compound involves the treatment of CmO_2 with anhydrous hydrogen fluoride at 500°C, or the precipitation of Cm^{3+} with F^- in aqueous solution, followed by dehydration of the precipitate in high vacuo. In both cases, vacuum equipment made of monel or Ni is required.

We have observed that the thermal decomposition of curium(III)-trifluoroacetate, $Cm(CF_3COO)_3$, yields anhydrous CmF_3, probably by a reaction such as

$$Cm(CF_3COO)_3 \rightarrow CmF_3 + 3COF_2 + 3CO$$

The same is true for the preparation of GdF_3 and AmF_3 from the corresponding trifluoroacetates. The procedure for the preparation of CmF_3 is as follows:

To 80 µl of 0.01 M trifluoroacetic acid, excess $Cm(OH)_3$ is added. After a reaction time of 15 minutes, the supernatant is separated from undissolved curium(III)-hydroxide and is cautiously evaporated to dryness under the IR lamp. The isolated pale-yellow residue, which is amorphous to X-rays, is placed into the quartz reaction tube of a Fried-Davidson apparatus and is heated in vacuo, or in argon atmosphere to a temperature of 550°C within 3 hours. The reaction temperature is maintained for another 3 hours, then the product is sealed in an X-Ray capillary, and a powder pattern is taken immediately.

In agreement with earlier results (8), CmF_3 was found to be hexagonal LaF_3 -type, space group $P6_3$/mcm - D^2_{6h} (No.193) with a = 6.992±0.013 A, c = 7.177±0.013 A, V = 303.9 A . With Z = 6 formula units in the cell, the density was calculated to be D_x = 9.87 gcm^{-3}.

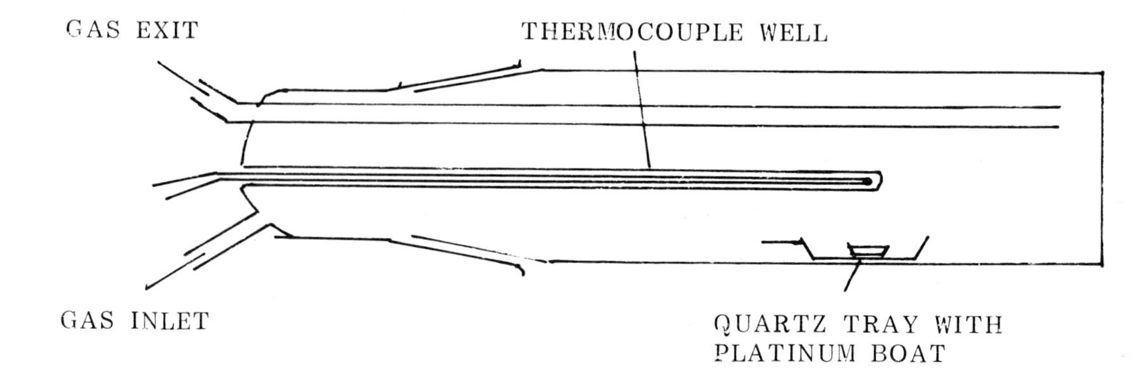

Fig.10. Reaction tube of vapor phase hydrolysis apparatus, modified for preparation of CmOBr

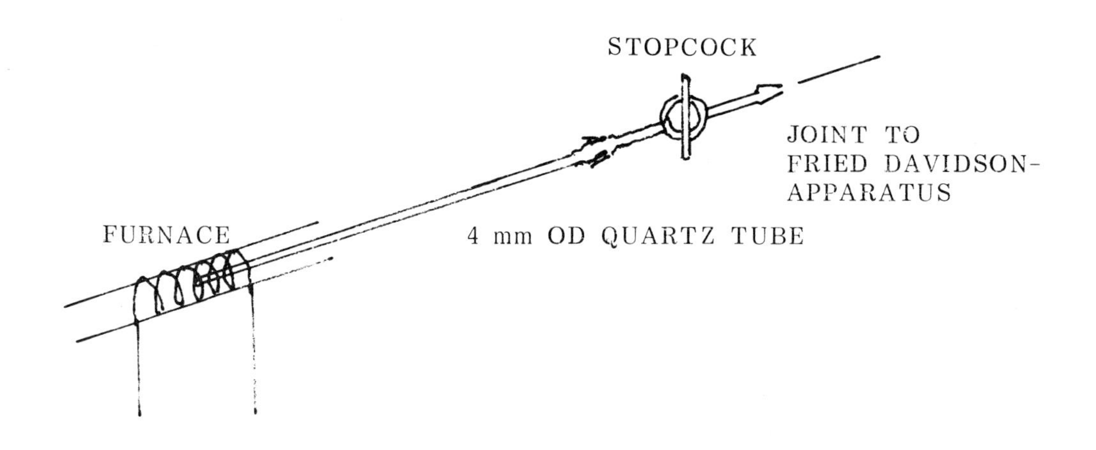

Fig.11. Quartz reaction tube for Fried-Davidson-apparatus for preparation of $AmCl_3$ and $CmCl_3$.

2.7.5. Curium(III)-chloride, $CmCl_3$, and Americium(III)-chloride, $AmCl_3$

Curium(III)- and americium(III)-chlorides, both well-known compounds, were prepared for supporting research in connection with work on the determination of the vapor pressure of americium(III)-chloride.

For preparation of $CmCl_3$, approx. 1.2 mg CmO_2 were placed into the quartz reaction tube of 4 mm OD (Fig.11), which was equipped with a stop cock. This stop cock is necessary to prevent the entry of moisture into the tube, when it is transferred into a different glove box. The reagent vessel of the Fried-Davidson apparatus was charged with CCl_4 dried over P_2O_5. The CmO_2 was slowly heated to 400°C, then, the reaction tube was evacuated, and filled with CCl_4 vapor. In the course of the reaction, the temperature was slowly raised to 800...850°C, and the gas in the reaction tube was changed in 10 minute intervals. The sublimation of the $CmCl_3$ can be easily observed; after a while, it starts to collect outside of the furnace zone as a pale-yellow ring. After a period of 3 hours, the CmO_2 starting material has been almost completely converted to $CmCl_3$. By carefully moving the furnace over the reaction tube, the $CmCl_3$ sublimate could be concentrated in a narrow zone inside the quartz reaction tube. The stopcock was then closed, and the tube transferred to an inert gas glovebox. A small quantity of the $CmCl_3$ was sealed in a quartz X-ray capillary, and was X-rayed immediately after preparation. The major part of the $CmCl_3$ was sealed in a quartz sample tube (2 mm OD) for the DTA apparatus. The $AmCl_3$ sample was prepared in exactly the same manner as the curium compound, except that 2.5 mg AmO_2 were used as the starting material.

The X-ray data of $CmCl_3$ were in agreement with earlier work (8)(30). $CmCl_3$ was found to be hexagonal, space group $P6_3/m - C^2_{6h}$ (No. 176) with $a = 7.366\pm0.020$ A, $c = 4.204\pm0.011$ A, $V = 197.5$ A^3. With Z = 2 formula units in the cell, the density was calculated to be $D_x = 5.89$ gcm^{-3}.

The determination of the melting points, heats and entropies of fusion in the DTA apparatus yielded the values summarized in Table 5.

2.8. Compounds of Curium(IV) and Curium(V)

Only very few examples of curium compounds having another valency than +3 are known, such as CmO_2, CmF_4, complex curium(IV)-fluorides, and the oxo compound $BaCmO_3$ (48)(36). In the course of this work, we decided to attempt the preparation of other. hitherto unknown compounds of curium(IV) and, possibly, of curium (V). We succeeded in preparing dilithium-trioxo-curiate(IV), Li_2CmO_3, and trisodium-tetraoxocuriate(V), Na_3CmO_4.

2.8.1. Dilithium-trioxo-curiate(IV), Li_2CmO_3.

The preparation of Li_2CmO_3 was done in an apparatus similar to that shown in Fig. 3, except that a gold boat was used instead of a quartz

Table 5. Melting points and thermodynamic data for fusion of $AmCl_3$ and $CmCl_3$.

		$AmCl_3$	$CmCl_3$
melting point	°C [a]	718	704
		(715)	(695)
	K	991	997
		(988)	(968)
ΔH_f	(kcal/mol)	11.5±0.1	11.5±0.4
	(kJ/mol)	48.1±0.4	47.9±1.6
ΔS_f	(cal/mol·K)	11.6±0.1	11.7±0.4
	(J/mol·K)	48.5±0.4	49.0±1.6

a) melting point values in parentheses are from Peterson and Burns (30)

or alumina boat. A slow stream of oxygen, which had been purified from CO_2 and H_2O was passed over the gold boat, which had been charged with a mixture of Li_2O and CmO_2 in the ratio 2:1. The gold boat was heated at 550°C for 30 minutes. The black, slightly hygroscopic reaction product was immediately sealed in an X-ray capillary, and its powder pattern was taken.

The compound Li_2CmO_3 was only stable in the presence of excess Li_2O, but did not combine with the latter to form higher oder compounds, In the Li_2CmO_3 pattern, the Li_2O lines were clearly discernible and could be used as internal standard.

The observed powder pattern could be interpreted in terms of a tetragonal lattice, unknown space group, with a = 8.300±0.022 A, and c = 8.611±0.023 A. Li_2CmO_3 is isomorphous with Li_2AmO_3 (47).

The compound is insoluble in CsF- and in 5 M LiOH solution. The insolubility is probable due to the reduction of Cm(IV), because Cm(VI) is most likely been stabilized in concentrated CsF solution. Li_2CmO_3 dissolves in 1 N $HClO_4$. The brief appearance of an olive green color in the solution indicates the transient presence of Cm(IV) in the dissolving solvent.

Despite our attempts in this direction, we were not able to prepare compounds such as Li_8CmO_6 ,Li_7CmO_6, or Li_6CmO_6 by reacting Li_2CmO_3 with corresponding amounts of Li_2O. On the other hand, it is surprising that Li_2CmO_3, in the presence of Li_2O, is stable up to 650°C, as is demonstrated by high temperature X-ray diffraction data.

2.8.2. Trisodium-tetraoxo-curiate(V), Na_3CmO_4.

For the preparation of Na_3CmO_4, Na_2O_2 and CmO_2 were mixed in the ratio 3:1 and were reacted with each other in an atmosphere of oxygen. In analogy to the preparation of Li_2CmO_3, an excess of Na_2O_2 was necessary. The mixture of Na_2O_2 and CmO_2 was charged into the gold boat, and heated at 450°C for 30 minutes in a stream of oxygen, which was free of H_2O, and CO_2. The reaction product was transferred to the inert gas glove box in a desiccator filled into an X-ray capillary, and the powder pattern was taken immediately after sealing the X-ray capillary.

The dark brown, extremely hygroscopic reaction product showed the presence of Na_3CmO_4 besides CmO_2. The reflexions assigned to Na_3CmO_4 could be interpreted in terms of a cubic NaCl structure, space group Fm3m $-O_h^5$ (No. 225). The lattice constant was 4.8207±0.0026 A, V = 112.0 A^3, with Z = 1 formula unit in the cell, the density was calculated to be D_x = 5.59 gcm^{-3}. The compound was found to be isomorphous with Na_3UO_4 (47).

Assuming 1 formula unit in the cell, the curium and sodium atoms are distributed statistically on the position 4 (a), the oxygen atoms occupy the 4 (b) positions. I.e. the Na_3CmO_4 structure consists of two fcc lattices, where the oxygen atoms and the metal atoms form one of these

Table 6. Summary of compounds prepared in the course of this work

Compound	Symm.	a(Å)	b(Å)	c(Å)	α(°)	β(°)	Color
CmO_2	(c)	5.369					black
$CmScO_3$	(o)	5.532	5.753	7.987			pale yellow
$CmVO_3$	(o)	5.428	5.571	7.766			dark green
$CmCrO_3$	(o)	5.403	5.508	7.669			pale green
$CmFeO_3$	(o)	5.414	5.579	7.739			dark brown
$CmPO_4 \cdot 0.5H_2O$	(h)	7.00		6.39			pale yellow
$CmAsO_4$	(m)	6.823	7.036	6.621		105.19	pale yellow
$CmVO_4$	(t)	7.290		6.417			honey yellow
$CmCrO_4$	(t)	7.242		6.353			dark green
$Cm_2(MoO_4)_3$	(t)	5.250		11.496			brown yellow
$Cm_2(WO_4)_3$	(m)	7.714	11.556	11.450		109.14	light yellow
$Cm(ReO_4)_3 \cdot xH_2O$	(o)	5.636	7.674	15.401			white
$Cm(ReO_4)_3$	(h)	10.065		6.294			white
CmOF	(r)	6.843			32.80		greyish-yellow
CmOCl	(t)	3.983		6.738			pale yellow
CmOBr	(t)	3.956		7.936			pale yellow
CmF_3	(h)	6.992		7.177			white
$CmCl_3$	(h)	7.366		4.204			pale yellow
$Cm(HCOO)_3$	(r)	6.272			115.46		pale yellow
$Cm_2(C_2O_4)_3 \cdot 10H_2O$	(m)	11.23	9.60	10.22		114.64	pale green
Li_2CmO_3	(t)	8.300		8.611			dark brown
Na_3CmO_4	(c)	4.821					dark brown

Symmetries: (c) cubic
(t) tetragonal
(h) hexagonal
(r) rhombohedral
(o) orthorhombic
(m) monoclinic

two individual lattices. An intensity calculation was carried out using the atomic positions given above, and the LAZY PULVERIX program. The calculated intensities are in good agreement with the observed values. The lattice constant is compared with the lattice constants of the corresponding lattice constants of Na_3UO_4 (54) and Na_3AmO_4(47). The structures of the compounds Na_3NpO_4 (59) and Na_3PuO_4 (59) are not yet evaluated. From X-ray data, for these latter compounds, a lattice of lower symmetry with NaCl superstructure is suspected. It is still unknown, why Na_3NpO_4 and Na_3PuO_4 have another structure than Na_3UO_4,Na_3AmO_4 and Na_3CmO_4. This behavior warrants a systematic X-ray study of the oxometallates (V) from uranium through curium.

It is remarkable that, even though Na_3CmO_4 has a NaCl type structure, the expected formation of CmO_2^+ , the linear "curyl" configuration in the solid, is not observed. Rather, the 6 oxygen atoms located in the corners of an octahedron around each curium atom have all the same distance to this actinide central atom.

3. DISCUSSION

A number of hitherto unknown curium compounds have been synthesized.Also, some well-known compounds have been prepared in order to study specific physical properties, or novel methods of synthesis.

Most of the compounds prepared involve trivalent curium. In this valency state, the curium behaves in no way different from other trivalent acinides. A new type of tetravalent curium compound, and, what we believe the first compound of pentavalent curium have been identified.

It has been observed that monofluoracetic acid and trifluoroacetic acid form curium(III)salts, which pyrolyze to form CmOF and CmF, respectively. These reactions are the basis for a new synthesis for the trifluoride, and for the oxyfluoride.

The availability of more sophisticated X-ray diffraction programs allowed the improvement of many X-ray data of curium compounds. It was possible to make intensity calculations to determine the atomic positions of the heavy atoms. One has to keep in mind, however, that the atomic positions of the light atoms are uncertain, because of their low contribution (a few percent in most cases) to the intensities. They should therefore be treated with reservation. A summary of all compounds prepared is given in Table

4. ACKNOWLEDGMENTS

We wish to express our thanks to the US Department of Energy, Washington, D.C. for supplying us with the curium used in this work. To Mr.A. Ochs, we wish to express our gratitude for his masterful glass- and quartz blowing work.

To the Bundesministerium für Forschung und Technologie, Bonn Bad Godesberg, we wish to express our thanks for the financial support of this work. We also express our thanks to the Deutsche Forschungsgemeinschaft for financial support (Grant We 324/34).

5. REFERENCES

1. G.T.Seaborg, R.A.James, and A.Ghiorso
   Paper 22.2, p.1554-1571 in
   G.T.Seaborg, J.J.Katz and W.M.Manning
   The Transuranium Elements Book 2
   National Nuclear Energy Series Div. IV Vol.14B
   McGraw Hill Book Co. Inc. New York, Toronto, London 1949

2. G.T.Seaborg, R.A.James, and A.Ghiorso
   Metallurgical Laboratory Chemistry Division Summary Report for August 1944
   'Heavy Isotopes by Bombardment of Pu^{239}'
   CS-2135 (Sept. 13, 1944)

3. L.B.Werner and I.Perlman
   'First Isolation and Properties of Curium'
   UCRL-156 (AECD-2729) (July 28, 1948)
   UCRL-156(Rev) (July 28, 1948)

4. L.B.Werner and I.Perlman
   'First Isolation of Curium'
   UCRL-156 (2nd Rev)(AECD-3155) (Feb.9,1951)

5. L.B.Werner and I.Perlman
   Paper 22.5, p. 1586-1594 in
   G.T.Seaborg, J.J.Katz and W.M.Manning, l.c. Reference (1)

6. L.B.Asprey and F.H.Ellinger
   'The First Crystallographic Data for Curium Compounds'
   AECD-3627 (1953)

7. C.H.Youngquist and P.R.Fields
   Proceedings, 6th Hot Laboratory Conference, Chicago, Il. Mar. 19-21 (1958) Supplement, p.48-51

8. L.B.Asprey T.K.Keenan, and F.H.Kruse, *Inorg. Chem.* 4, 985-986 (1965)

9. F.Weigel, V.Wishnevsky and H.Hauske, *J. less-common Metals* 56, 113 - 123 (1977)

10. L.B.Asprey, F.H.Ellinger, S.Fried and W.H.Zachariasen, *J.Am.Chem. Soc.* 79, 5825 (1957)

11. T.K.Keenan, *Inorg.Nucl.Chem.Letters* 2, 153-156 (1967)

12. T.K.Keenan, *Inorg.Nucl.Chem.Letters* 2, 211-214 (1966)

13. T.K.Keenan, *Inorg.Nucl.Chem.Letters* 3, 391-396 (1967)

14. T.K.Keenan, *Inorg.Nucl.Chem.Letters* 3 453-457 (1967)

15. D.Damien and R.G.Haire, *J.less-common Metals* 68, 159-165 (1979)

16. D.Damien, R.G.Haire, J.R.Peterson, *Journal de Physique, Colloque No.4*, C4-95 - C4-100 (1979)

17. J.D.Charvillat, U.Benedict, D.Damien, C.H.DeNovion, A.Wojakowski, and W.Müller, p. 79-84 in
W.Müller and R.Lindner, *Transplutonium Elements*
North Holland Publishing Co., Amsterdam 1976

18. F.Weigel and R.Marquart, *J.less-common Metals* 90, 283-290 (1983)

19. T.D.Chikalla and L.Eyring
'The Curium Oxygen System'
BNWL-CC-1569 (Feb.13,1968)

20. T.D.Chikalla and L.Eyring, *J.inorg.nucl.Chem.* 31, 85-93 (1969)

21. W.C.Mosley, *J.inorg.nucl.Chem.* 34, 539-555 (1972)

22. H.O.Haug, *J.inorg.nucl.Chem.* 29, 2753-2758 (1967)

23. L.V.Sudakov and I.I.Kapshukov, p.201-204
V.I.Spitsyn and J.J.Katz
Chemistry of the Transuranium Elements - Proceedings of the Moscow Conference 1976
Pergamon Press, Oxford 1976

24. P.K.Smith, *J.inorg.nucl.Chem.* 31, 241 - 245 (1969)

25. J.C.Wallmann, *J.inorg.nucl.Chem.* 26, 2053 - 2057 (1964)

26. M Noé
Contribution a l'Etude du Comportement des Oxydes du Curium-244 et de Plutonium-238. Sous l'Effet de l'Irradiation Interne
Doctoral Dissertation Liége 1972

27. F.Weigel and H.Haug, *Radiochim.Acta* 4, 227 (1925)

28. C.Keller and K.H.Walter, *J.inorg.nucl.Chem.* 27, 1253 - 1260 (1965)

29. F.Weigel
Präparative und röntgenographische Untersuchungen an stark radioaktiven Stoffen
Habilitationsschrift (Instructor's Thesis) Univ. of Munich 1965

30. J.R.Peterson and J.H.Burns, *J.Inorg.Nucl.Chem.* 35, 1525-1530 (1973)

31. J.R.Peterson, *J.Inorg.Nucl.Chem.* 34, 1603-1607 (1972)

32. M.Noé and J.R.Peterson, *Inorg.Nucl.Chem. Letters* 8, 897-902 (1972)

33. J.R.Peterson and J.Fuger, *J.Inorg.Nucl.Chem.* 33, 4111-4117 (1971)

34. R.P.Turcotte, T.D.Chikalla, and L.Eyring, *J.Inorg.Nucl.Chem.* 35, 809-816 (1973)

35. S.E.Nave, R.G.Haire, and P.G.Huray, *Phys.Rev.* B28 , 2317- (1983)

36. L.R.Morss, J.Fuger, J.Goffart, and R.G.Haire, *Inorg.Chem.* 22, 1993-1996 (1983)

37. B.B.Cunningham, *Microchemical Methods used in Nuclear Research* UCRL-424 (AECD-2703) (Aug.30,1949) - *Nucleonics* 5, 62 (Nov.1949)

38. B.B.Cunningham, *Microchem. Journal, Symposium Issue* 1, 69-93 (1961)

39. S.Fried and N.R. Davidson,
Paper 15.5., p.1072-1096 in
G.T.Seaborg, J.J.Katz, and W.M.Manning
The Transuranium Elements Book 2
National Nuclear Energy Series, Div.IV,Vol.14B
McGraw Hill Book Co. Inc. New York, Toronto, London 1949

40. H.Haug *Über Untersuchungen in den Systemen Uranoxid-Europiumoxid und Plutoniumoxid-Europiumoxid*
Doctoral Dissertation, Univ. of Munich, 1963, p. 50

41. F.Weigel and H.Hauske, *J.less-common Metals* 55, 243-247 (1977)

42. I.L.Jenkins, F.L.Moore and M.J.Waterman, *J.Inorg.Nucl.Chem.* 27, 77-87 (1965)

43. F.Weigel and N.ter Meer,*Inorg.Nucl.Chem.Letters* 3, 403-408 (1967)

44. J.P.Silvestre, *Rev.Chim.Minér.* 14, 225 (1977)

45. P.P.Jamieson, S.C.Abrahams, and J.L.Bernstein, *J.Chem.Phys.* 50, 86-94 (1969)

46. R.Kohl, *Ein Beitrag zur präparativen Mikrochemie des Curiums.*
M.S.Thesis (Diplomarbeit), Univ. of Munich 1981

47. K.H.Walter, *Ternäre Oxide des drei- bis sechswertigen Americiums.*
Doctoral Thesis, University of Karlsruhe 1965
KFK-280 (Jan.1965)

48. F.Weigel and V.Scherer, *Radiochimica Acta* 7, 72-74 (1967)

49. D.Brown, *The Halides of the Lanthanides and Actinides*
Wiley Interscience, New York 1968

50. F.Weigel, V.Wishnevsky and M.Wolf *J.less-common Metals* 63, 81 - 86 (1979)

51. R.G.Haire and J.Y.Bourges
p. 19 in B.Johansson and A.Rosengren (Eds.)
Proceedings of the 10ème Journées des Actinides
May 27-28, 1980, Stockholm

52. M.Faucher and P.Caro, *Mat.Res.Bull.* 10, 1 (1975)

53. L.E.Russell, J.D.Harrison, and N.H.Brett, *J.Nucl.Mat.* 2, 310-320 (1960)

54. R.Scholder and H.Gläser, *Z.anorg.allg.Chem.* 327, 15-27 (1964)

55. D.H.Templeton and A.Zalkin, *Acta Cryst.* 16 762- 766 (1963)

56. *ASTM Card Catalog*, Card No. 15-769

57. W.C,Mosley, *J.Am.Ceram.Soc.* 54, 475 (1971)

58. M.Noé and J.Fuger, *Inorg.Nucl.Chem.Letters* 11, 451 (1975)

59. L.Koch, *Über ternäre Oxide des 5- und 6-wertigen Neptuniums und Plutoniums mit Lithium und Natrium.*
Doctoral thesis, Univ. of Karlsruhe 1966
KFK-196 (Feb. 1966)

60. A.W.Mann and D.J.M.Beran, *Acta Cryst.* B26, 2129-2131 (1970)

THE ELECTRONIC AND MAGNETIC PROPERTIES OF Am AND Cm

Norman Edelstein
Materials and Molecular Research Division
Lawrence Berkeley Laboratory
University of California
Berkeley, California 94720 U.S.A.

ABSTRACT. A review of the present status of the analyses of the optical spectra of Am and Cm in various oxidation states will be given. From these analyses, the magnetic properties of the ground states of these ions can be determined. These predicted values will be compared with the various magnetic measurements available.

1. INTRODUCTION

The optical and magnetic properties of Am and Cm have now been studied for over 30 years. The common oxidation state for these elements is the trivalent one, and their lanthanide analogues in this oxidation state have magnetic properties which differ from the rest of the lanthanide series. This fact has made the study of the magnetic properties of the Am^{3+} and Cm^{3+} ions of special interest in order that the differences (or similarities) between the 4f and 5f series could be determined. In this paper the optical properties of the various accessible oxidation states of these two elements will be briefly reviewed, followed by a review of their magnetic properties. The optical properties of the atoms in the gaseous phase or as free ions will not be covered.

2. Optical Spectra

The optical spectra for AmX_3 (X = Cℓ, Br, I) and Am^{3+} and Cm^{3+} in single crystals of $LaCℓ_3$ have been measured and analyzed [1-4]. The free ion

N. M. Edelstein et al. (eds.), Americium and Curium Chemistry and Technology, 193–211.

energy levels of Cm^{3+} (aquo) have also been assigned on the basis of the correlation between observed and calculated band intensities [5]. Other oxidation states of both Am (divalent, tetravalent to hexavalent) and Cm (tetravalent) are known and low resolution spectra of the tetravalent state in solids and solution have been obtained [6-9]. Recently, calculated free ion spectra for Am^{4+} and Am^{2+} have been published [10]. Selective laser excitation experiments on Cm^{3+} in D_2O have also been carried out [11].

3. Magnetic Properties

Magnetic properties of materials determined by bulk magnetic susceptibility or electron paramagnetic resonance (epr) measurements are usually determined by the energy levels of the materials which are populated at the temperatures of the measurement. Most epr spectra of actinide ions are measured at 4.2 K so usually only the magnetic properties of the ground crystal field state are determined. Magnetic susceptibility measurements are performed in a range of temperatures (~2-300 K) so the splittings of lowest J level may sometimes be determined [12,13]. If the ground state is a singlet (non-magnetic), the magnetism of the material is determined by the mixing of the higher lying magnetic states into the ground state by the magnetic field. This type of magnetic behavior is independent of temperature (if the magnetic state lies much higher than kT). Table 1 lists the configurations of various oxidation states of Am and Cm for which magnetic data have been measured. Each of these configurations will be discussed individually.

4. $5f^5$ - Am^{4+}

The ground term for the Am^{4+} ion is a nominally $^6H_{5/2}$. However because of the strong spin-orbit coupling for actinide ions, this state is less than 66% pure. This number comes from calculations on Pu^{3+} [14]. Table 2 shows the eigenvector components for Sm^{3+} and Pu^{3+} free ions. Am^{4+} has a larger spin-orbit coupling constant than Pu^{3+}, thus the Pu^{3+} eigenvector represents a lower limit for an Am^{4+} intermediate-coupled

Table 1. Accessible Oxidation State for Am and Cm

	Am	Cm	Ground State
$5f^5$	Am^{4+}	-	J = 5/2
$5f^6$	Am^{3+}	Cm^{4+}	J = 0
$5f^7$	Am^{2+}	Cm^{3+}	J = 7/2

Table 2. Largest Eigenvector Components for Sm^{3+} and Pu^{3+} Ground J = 5/2 State (Ref. 14)

Sm^{3+}

96.0% ^6H + 2.3% $^4G4$ + 1.4% $^4G1$ + ⋯.

Pu^{3+}

66.0% ^6H + 14.3% $^4G4$ + 9.6% $^4G1$ + 1.7% $^4F3$ + 1.2% $^4G3$ + 1.0% $^2F6$ + 1.0% ^6F + 1.0% $^2F2$ + ⋯.

eigenvector. Since most of the calculations have been done for Pu^{3+} systems, these will be summarized, but the same arguments hold for Am^{4+} in sites of the same symmetry.

Electron paramagnetic resonance has been reported for Am^{4+} diluted in ThO_2 and CeO_2 single crystals [15,16]. The site symmetry for the Am^{4+} ion is cubic. For a J = 5/2 state in cubic symmetry the crystal field will split this term into two states, a Γ_7 doublet and Γ_8 quartet state. For this symmetry the splitting of these two levels depends on two crystal field parameters B_0^4 and B_0^6, and angular factors which depend on the intermediate coupled wavefunction. These angular factors have been calculated for Sm^{3+} and Pu^{3+}, and it has been shown that the sign of B_0^4 (which is the dominant term) for Pu^{3+} (or Am^{4+}) is opposite to that for Sm^{3+} [14]. This sign change is due to the large admixture of higher L-S states by the spin-orbit coupling interaction. The net result is that for Sm^{3+} the Γ_8 state would be lowest, but for Pu^{3+} or Am^{4+}, the Γ_7 state is the ground state. This is illustrated in Fig. 1.

For an isolated Γ_7 state the calculated g value should be equal to

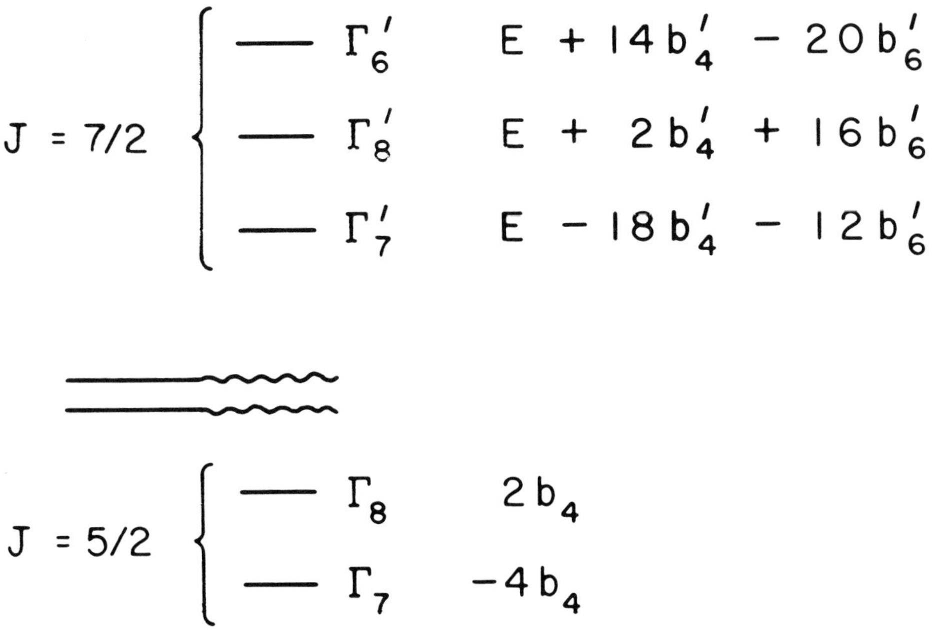

$J = 7/2$:
Γ_6' $E + 14b_4' - 20b_6'$
Γ_8' $E + 2b_4' + 16b_6'$
Γ_7' $E - 18b_4' - 12b_6'$

$J = 5/2$:
Γ_8 $2b_4$
Γ_7 $-4b_4$

Figure 1. Schematic energy level diagram for Pu^{3+}/CaF_2 and Am^{4+}/ThO_2, CeO_2. For Sm^{3+} (the 4f analogue), the Γ_7 and Γ_8 energy level ordering for the ground J = 5/2 state is reversed.

-0.700. The measured values for Am^{4+} in CeO_2 and ThO_2 are as shown in Table 3. Note that these two values are different with Am^{4+}/ThO_2 being larger. For Pu^{3+} the first excited state is a J=7/2 which is calculated to be at ~3200 cm^{-1}. In the actinide series, crystal field

Table 3. Measured g values for $5f^5$ Ions in Various Hosts (Ref. 16)

		$\|A\|$ (10^{-4} cm^{-1})		B_4'[a]
Matrix	$\|g\|$	$^{239}Pu^{3+}$	$^{241}Pu^{3+}$	(cm^{-1})
CeO_2	1.333 ± 0.001	22.4 ± 0.2		-5400
ThO_2	1.3124 ± 0.0005	65.4 ± 0.2		-5130
			46.1 ± 0.6	
CaF_2	1.297 ± 0.002	66.95 ± 0.03		-4945
			48.07 ± 0.10	
SrF_2	1.250 ± 0.002	84.6 ± 1.0		-4440
BaF_2	1.187 ± 0.004	102 ± 3		-3820
$SrCl_2$	1.1208 ± 0.0005	127.9 ± 0.4		-3190

		$\|A\|$ (10^{-4} cm^{-1})	
Matrix	$\|g\|$	$^{241}Am^{4+}$	$^{243}Am^{4+}$
CeO_2	1.3120 ± 0.0005		22.1 ± 0.2
ThO_2	1.2862 ± 0.0005	45.7 ± 0.1	45.3 ± 0.1

[a]Values of B_4' were calculated assuming that B_6'/B_4' = - 0.2.

effects are large and this interaction can mix excited states into the ground state. Using a model which considered only the mixing of these two states, Edelstein, et al. fit a series of g values for Pu^{3+} in CaF_2, SrF_2, and BaF_2 showing that the crystal field interaction decreased as the lattice parameter increased [14]. This same model has been applied to fit the g values for Am^{4+} in ThO_2 and CeO_2 [16]. The results are consistent with CeO_2 (the smaller lattice) having a larger crystalline field interaction at the Am^{4+} site than ThO_2. A plot of the g values vs. the crystal field parameter is shown in Fig. 2. A more complete

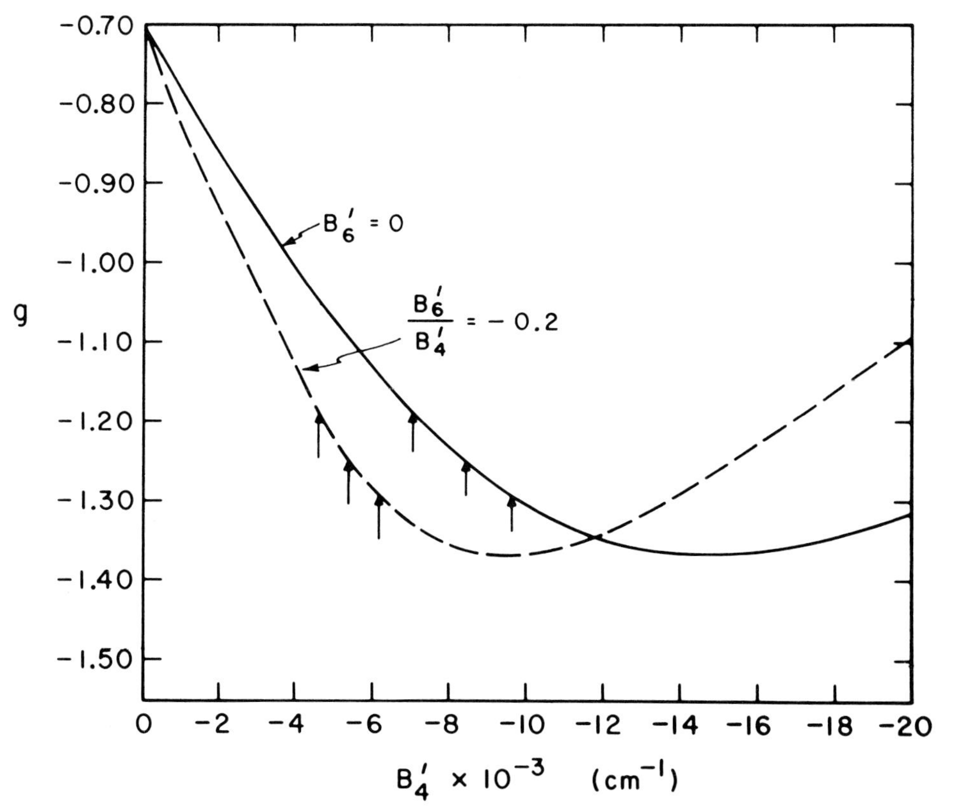

Figure 2. Plot of calculated g value vs. B_4'. The solid line describes the calculated values for $B_6' = 0$. The dotted line describes the values for $B_6'/B_4' = -0.2$. The arrows show where the experimental g values are located for Pu^{3+}/CaF_2, SrF_2, BaF_2. A similar plot may be drawn for Am^{4+}/CeO_2,ThO_2.

calculation for Pu^{3+} in cubic lattices has been given by Lam and Chan [17].

The magnetic susceptibility of $^{243}AmO_2$ as a function of temperature has been reported [18]. An antiferromagnetic transition was found at 8.5 ± 0.5 K. This transition was not found in a neutron diffraction study on $^{243}AmO_2$ at 20 K and 6.5 K nor was a hyperfine field detected for AmO_2 by a Mössbauer study [19]. The ground state g value of AmO_2 was found to be 1.51 which is 17% larger than for Am^{4+} in ThO_2, and the excited Γ_8 state was determined to be at ~35 K [18].

5. $5f^6$ - Am^{3+}, Cm^{4+}

The $5f^6$ configuration has a nominally $^7F_0$ ground term. Spin-orbit coupling will mix in other J = 0 states, but this state must be a singlet in all crystal fields. Consequently all magnetic effects will come from second order interactions and will result in temperature independent paramagnetism (TIP). The measured susceptibilities will be small in magnitude and impurities in the sample could drastically affect the measured values. With small radioactive samples (100 μg - 10 mg) containing the isotopes $^{241 \text{ or } 243}Am$ and $^{244 \text{ or } 248}Cm$, contamination of samples either with chemical impurities, or by radiation damage is a major problem. It is not surprising that the few results available vary greatly. If we assume the B_0^2 crystal field parameter does not change much from one compound to another (this is the only parameter which will split the J = 1 first excited state in compounds with C_3 or higher symmetry) then all Am^{3+} compounds should have about the same value for their temperature independent paramagnetism. Table 4 lists the values for a number of Am^{3+} compounds [20-25]. Am metal is the first actinide metal to exhibit a localized trivalent moment. Table 4 lists the magnetic susceptibility for Am, and whose value is in the range of trivalent Am compounds.

Cm^{4+} compounds show an anomalous temperature dependence [22,26-28]. Theoretically they should be similar to the Am^{3+} compounds and show

Table 4. Magnetic Susceptibility of Am^{3+} Compounds and Am Metal

Compound	TIP (10^{-6} emu/mole)	Reference
Am_2O_3	640 ± 40	20
$Cs_2NaAmCl_6$	5400 ± 400	21
AmF_3	714 ± 1%[a]	22
AmF_3	566 ± 25[b]	23
AmF_3	1040 ± 300[b]	24
Am^{3+} on bead	716 ± 100[b]	23
Am^{3+} in solution	720	25
Am metal	881 ± 46	23
Am metal	675	25a
Calculated	~500[c]	20

[a]Temperature dependent μ_{eff} = 0.63 BM.
[b]Slightly temperature dependent.
[c]From optical data on $Am^{3+}/LaCl_3$. The J=1 level is 2720 cm^{-1} above the ground J=0 level.

temperature independent paramagnetism. However they are temperature dependent and have effective magnetic moments between 2-4 Bohr magnetons. Kanellakopolus has followed the magnetic susceptibility of a $^{244}CmO_2$ sample as function of time and found the magnetism increases linearly [26]. He proposed that this increase was due to the formation of Cm^{3+} (a $5f^7$ ion with μ_{eff} ~ 7.9 BM) caused by the radiation damage due to the alpha particles from the decay of ^{244}Cm nuclei. He was able to fit his data by assuming a rate of production of Cm^{3+} of 2.09% per month. Goffart found a similar time dependence in his magnetic measurements on $^{244}CmO_2$ [27]. Presumably compounds synthesized with

^{248}Cm ($\tau_{1/2} = 4.7 \times 10^5$y compared with $\tau_{1/2} = 18.1$y for ^{244}Cm) should be much less sensitive to these effects. Hurray et al. have reported temperature dependent behavior for various $^{248}Cm^{4+}$ compounds. Goffart et al. and Morss et al. also found temperature dependent behavior for $^{248}CmO_2$ [27,28]. The lattice constants of these samples obtained by x-ray powder diffraction techniques showed no indication of an expanded structure [27,28]. Further studies are necessary for the Cm^{4+} compounds.

6. $5f^7$, Am^{2+}, Cm^{3+}

The half-filled shell configuration, $4f^7$, in the lanthanide series gives a ground state wavefunction for Gd^{3+} which is approximately 98% $^8S_{7/2}$ (see Table 5). This state will not be split by the crystalline field.

Table 5. Leading Terms in the Wavefunctions for Gd^{3+} and Cm^{3+}, J = 7/2 (Ref. 35)

Ion	Leading terms
Gd^{3+}	97.4% ^8S + 2.6% ^6P + .01% ^6D + •••
Cm^{3+}	79.4% ^8S + 17.1% ^6P + .81% ^6D + •••.

Nevertheless from epr measurements, splittings on the order of .1 to 1 cm^{-1} are observed for Gd^{3+} in various crystalline environments. These splittings arise from various higher order mechanisms which have been proposed [29]. The first report of the epr of Cm^{3+} was made by a group at Berkeley who found a group of seven lines at 4.2 K in $^{244}Cm^{3+}/LaCl_3$ characteristic of a J = 7/2 level with g = 1.991 [30]. This observation was confirmed by a group at Argonne National Laboratory [31]. The ground state of Cm^{3+} was supposed to be analogous to its $4f^7$ counterpart Gd^{3+}, and these data confirmed this similarity.

As more information became available about the electronic structure of the actinides, it became increasingly clear something was wrong with the above observations. The AmI ($5f^7 7s^2$) g value was measured by atomic beam methods, and it was found to be 1.937 [32]. Runciman [33] obtained wavefunctions for Cm^{3+} and from these wavefunctions the g value of the ground state was calculated to be 1.913 [34]. These deviations from g = 2.00 were due to the large extent of intermediate coupling for Cm^{3+}. Abraham, Judd, and Wickman [34] reexamined the $Cm^{3+}/LaCl_3$ epr spectrum and found a strong single line with $g_{||}$ = 1.925 ± 0.002 and g = 7.67 ± 0.02. A similar spectrum was found for Cm^{3+} diluted in lanthanum ethylsulfate. These spectra were readily explained on the basis of the zero field splitting for Cm^{3+} being large with respect to the microwave frequency. Thus the initial reports were spurious and the Cm^{3+} epr spectrum was consistent with other actinide data and theory.

The leading terms in the wave functions for Cm^{3+} and Gd^{3+} are shown in Table 5. Cm^{3+} is only about 79% pure $^8S_{7/2}$. The three leading terms will not split in a cubic crystalline field. An approximate calculation using the complete 50 term wavefunction for Cm^{3+} showed qualitatively that the effects of intermediate coupling can account for the much larger splitting in Cm^{3+} than in its $4f^7$ counterpart, Gd^{3+}.

The crystal field splittings for Cm^{3+} in cubic compounds are of the order of 5-50 cm^{-1}. The ground state is an isotropic Γ_6 state and the first excited state is an anisotropic Γ_8 state. If the splitting between these two states is of the order of magnitude of the magnetic splittings, these two states can be mixed by the magnetic field in an epr experiment [35]. This will result in the ground Γ_6 state showing some anisotropy. From the magnitude of the anisotropy, the Γ_6-Γ_8 splittings can be deduced. Some of the results are shown in Table 6.

Table 6. Zero-field Splittings of the $5f^7$ Ions in Various Crystals (Ref. 36)

Crystal	Ion	(cm^{-1})	(cm^{-1})	g_J
$SrCl_2$	Cm^{3+}	5.13 ± 0.05	15.3 ± 0.4	1.928 ± 0.002
SrF_2	Cm^{3+}	11.2 ± 0.4		1.9257 ± 0.001
CaF_2	Cm^{3+}	13.4 ± 0.5		1.926 ± 0.001
ThO_2	Cm^{3+}	15.5 ± 0.3		1.9235 ± 0.002
$SrCl_2$	Am^{2+}	5.77 ± 0.48		1.9283 ± 0.0008
SrF_2	Am^{2+}	15.2 ± 0.4		1.9254 ± 0.001
CaF_2	Am^{2+}	18.6 ± 0.5		1.926 ± 0.001

For Cm^{3+} in $SrCl_2$, both the Γ_6 and the Γ_7 resonances were observed. Both were anisotropic as shown in Fig. 3. Both the $\Gamma_7 - \Gamma_8$ and $\Gamma_6 - \Gamma_8$ splittings for Cm^{3+} in $SrCl_2$ are shown in Table 6 [36]. Finally, in Table 7 the zero field splitting for Gd^{3+} and Cm^{3+} in a number of cubic crystals are listed [38]. Note that the splitting for Gd^{3+}/CeO_2 is smaller than for Gd^{3+}/ThO_2. Based on an electrostatic model one would expect the crystal field splitting to be inversely proportional to the lattice constant (for anions of the same charge). The higher order mechanisms must predominate for Gd^{3+} in ThO_2 and CeO_2. However for Cm^{3+} with its larger splitting the electrostatic mechanism appears to be the predominant one.

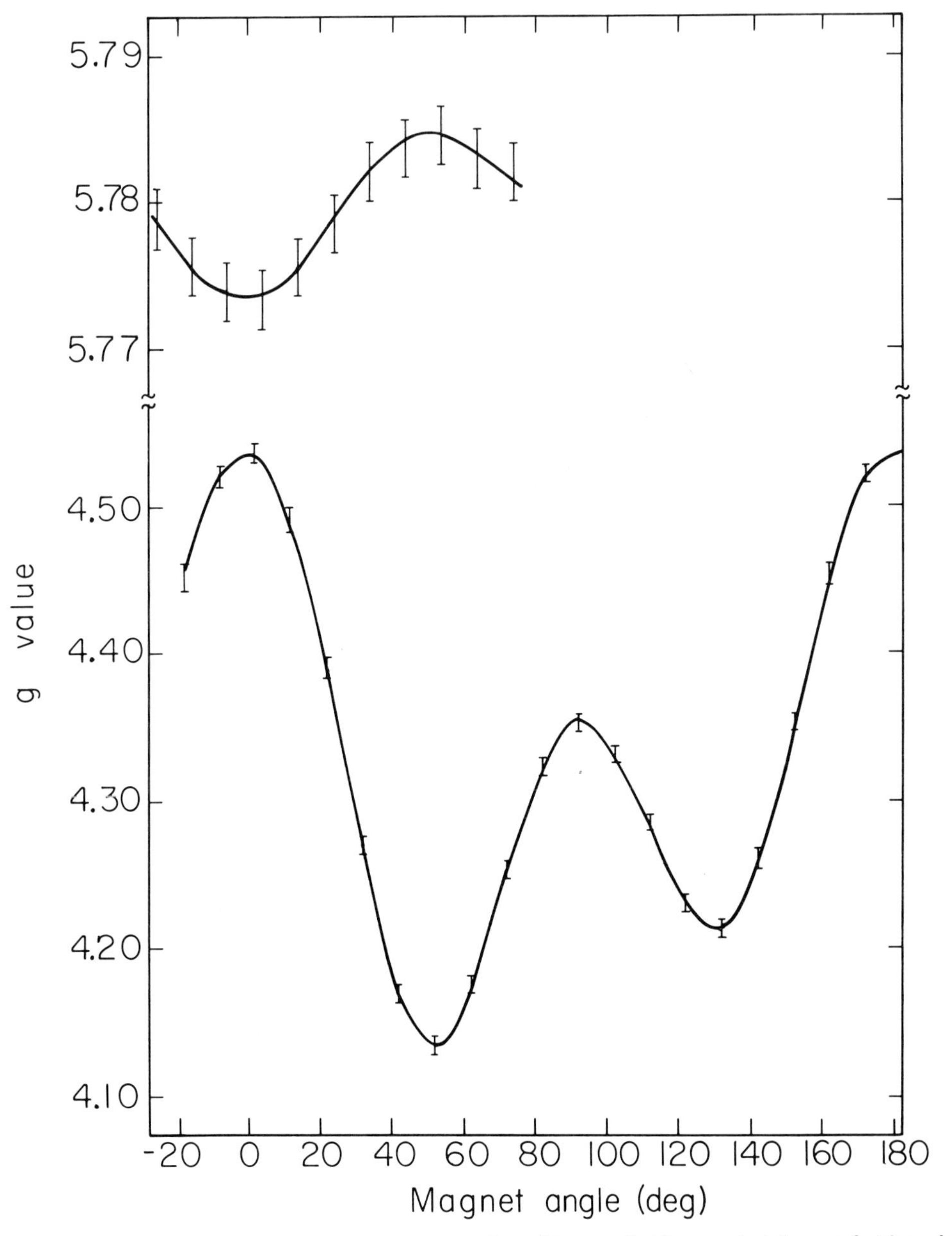

Figure 3. The measured g value as a function of the rotation of the dc magnetic field for $SrCl_2:Cm^{3+}$. The data were obtained at ~35 GHz and ~5500 G. The smooth curves are calculated values. Note the change in scale for the Γ_7 (upper curve) and Γ_6 (lower curve) states. The error bars for the Γ_6 state are ±0.005, for the Γ_7 state ±0.002.

Table 7. Zero-field Splittings for Gd^{3+} and Cm^{3+} in Various Host Crystals at 4.2 K (Ref. 37)

	Lattice Constant (Å)	Gd^{3+} (cm^{-1})	Cm^{3+} (cm^{-1})
CeO_2	5.41	0.0653 ± 0.0004	17.8 ± 0.3
ThO_2	5.60	0.06645 ± 0.00008	15.5 ± 0.3
CaF_2	5.46	0.0578 ± 0.0001	13.4 ± 0.5
SrF_2	5.80	0.0501 ± 0.0002	11.2 ± 4
BaF_2	6.20	0.0448 ± 0.0002	-
$SrCl_2$	7.00	0.01979 ± 0.0004	5.13 ± 0.5

The epr spectra of Cm^{3+} in a number of tetragonal crystals have been reported [38,39]. In D_{2d} symmetry the crystal field states will have eigenfunctions of the type

$$a|\pm 7/2\rangle + b|\mp 1/2\rangle$$

or

$$c|\pm 5/2\rangle + d|\mp 3/2\rangle$$

From the measured g values the admixture coefficients may be evaluated and, from measurements at different temperatures, it can be determined if the state observed is the ground state. For the system $Cm^{3+}/ThSiO_4$, no epr spectrum was observed [40]. In the hosts for which both Cm^{3+} and Gd^{3+} epr signals have been observed, the dominant crystal field term has had the same sign. For $Gd^{3+}/ThSiO_4$, B_0^2 is the dominant term with a sign < 0. Assuming that $B_0^2 < 0$ and dominant for the $Cm^{3+}/ThSiO_4$ system, the state with $a|\pm 7/2\rangle + b|\mp 1/2\rangle$ would be the ground state with the values $a \sim 1$ and $b \sim 0$. Some recent data of the various crystal field levels for the tetragonal system $Cm^{3+}/LuPO_4$ are shown in Table 8. The g values

Table 8. g Values for the Crystal Field Levels of $Cm^{3+}/LuPO_4$ (Ref. 39)

	Eigenvector	Calculated		Measured	
		$g_{\parallel}$	$g_{\perp}$	$g_{\parallel}$	$g_{\perp}$
77K	$0.9837\ |\pm 1/2\rangle - 0.1798\ |\mp 7/2\rangle$	1.424	7.436	1.424	7.436
	$0.9463\ |\pm 3/2\rangle - 0.3233\ |\mp 5/2\rangle$	4.155	4.07	4.15	4.07
	$0.9463\ |\pm 5/2\rangle + 0.3233\ |\mp 3/2\rangle$	8.0	4.07	8.0	4.07
	$0.9837\ |\pm 7/2\rangle + 0.1798\ |\mp 1/2\rangle$	12.95	0.25	-	-
4.2K	$0.98205\ |\pm 1/2\rangle - 0.1886\ |\mp 7/2\rangle$	1.373	7.402	1.373	7.402
	$0.9455\ |\pm 3/2\rangle - 0.3255\ |\mp 5/2\rangle$	4.13	4.096	4.12	4.10
	$0.9455\ |\pm 5/2\rangle + 0.3255\ |\mp 3/2\rangle$	7.976	4.086	7.98	4.096
	$0.98205\ |\pm 7/2\rangle + 0.1886\ |\mp 1/2\rangle$	12.9	0.27	-	-

measured at 77 K and 4.2 K are slightly different due to the change in the crystal field as a function of temperature. The net result of this change is a small difference in the energies and composition of the eigenvectors of the system as a function of temperature which is reflected in the g values. All the intensities of the epr lines for $Cm^{3+}/LuPO_4$ showed decreases as the temperature was lowered. This observation may be explained by again assuming that $B_0^2 < 0$ and is the dominant term. Thus the state with a ~ -1.0 must be the ground state.

Magnetic susceptibility measurements for various trivalent Cm compounds and AmI_2 are shown in Table 9 [21,22,41-44]. For a free ion

Table 9. Summary of magnetic data for Cm compounds

Compound	T (K)	μ_{eff} [a] BM	Θ (K)	References
$CmF_3 \cdot 1/2\ H_2O$	77-298	7.7	-5	41
CmOC1	77-298	7.6	-22	41
CmF_3 in LaF_3	77-298	7.7	-6	41
Cm^{3+} in	7.5-25	7.9	-4	21
$Cs_2NaLuCl_6$	25-45	7.5	-1	21
Cm_2O_3 [b,c]	20-80	8.20	-149	42
	100-300	7.89	-130	42
CmN	140-300	7.02	+109	43
CmAs	100-300	6.58	+88	43
CmF_3	~30-280	7.67	3.6	22
Cm_2O_3 [c]	50-300	7.74	-130	22
Cm_2O_3 [d]	4.2-300	7.51	-110	22
AmI_2	37-180	6.7 ± .7	e	44

[a] $\mu_{eff} = 2.828\ (T-\Theta)^{1/2}$ BM.

[b] Antiferromagnetic transition at T = 13 ± 2 K.

[c] Monoclinic phase.

[d] bcc.

[e] not given in Ref. 44.

μ_{eff} = 7.9BM. Thus most of the data at high temperatures agree with the expected theoretical value. Cm metal exhibits a localized moment characteristic of rare earth metals and the heavier actinide metals (starting at Am). Various magnetic measurements have been made on Cm metal with ^{244}Cm and ^{248}Cm [45]. The values given vary widely, and these measurements should be repeated. A transition to an antiferromagnetic phase has been reported below 52 K [46].

7. Conclusion

The magnetic properties of the elements Am and Cm, and their compounds can be explained on the basis of the wavefunctions obtained from optical spectra or from wavefunctions obtained from extrapolated parameters. For Cm^{4+} data the experimental data show an anomalous temperature dependence. Further work is needed on these compounds. There is a large scatter in the data for Cm metal, and this system should also be reexamined.

8. Acknowledgements

I would like to thank M.M. Abraham and L. Boatner for allowing me to quote their Cm^{3+} data before publication, and L. Soderholm and L.R. Morss for helpful comments about the manuscript. This work was supported by the Director, Office of Energy Research, Office of Basic Energy Sciences, Chemical Sciences Divison of the U.S. Department of Energy under Contract No. DE-AC03-76SF00098.

References

1. R.G. Pappalardo, W.T. Carnall, and P.R. Fields, J. Chem. Phys. 51, 1182 (1969).
2. J.G. Conway, J. Chem. Phys. 40, 2504 (1964).
3. J.B. Gruber, W.R. Cochran, J.G. Conway, and A.T. Nicol, J. Chem. Phys. 45, 1423 (1966).

4. J.P. Hessler and W.T. Carnall, in "Lanthanide and Actinide Chemistry and Spectroscopy," N. Edelstein, Ed., ACS Symposium Sseries 131, 349 (1980).
5. W.T. Carnall and K. Rajnak, J. Chem. Phys. 63, 3510 (1975).
6. T.K. Keenan, J. Am. Chem. Soc. 83, 3719 (1961).
7. L.B. Asprey and T.K. Keenan, J. Inorg. Nucl. Chem. 7, 27 (1958).
8. F.H. Krause and L.B. Asprey, Inorg. Chem. 1, 137 (1962).
9. L.B. Asprey and R.A. Penneman, Inorg. Chem. 1, 134 (1962).
10. J. Blaise, M.S. Fred, W.T. Carnall, H.M. Crosswhite, and H. Crosswhite in "Plutonium Chemistry," W.T. Carnall and G.R. Choppin, Eds., ACS Symposium Series 216, 173 (1983).
11. J.V. Beitz and J.P. Hessler, Nucl. Tech 51, 169 (1980).
12. For a comprehensive review of the epr of actinide ions see L.A. Boatner and M.M. Abraham Repts. Prog. Phys. 41, 87 (1978).
13. A review of the magnetic properties of the actinides is given by N. Edelstein and J. Goffart, "Magnetic Properties of the Actinides," in "Chemistry of the Actinide Elements," J.J. Katz, G.T. Seaborg, and L. Morss, Eds., Chapman and Hall, London, in press.
14. N. Edelstein, H.F. Mollet, W.C. Easley, and R.J. Mehlhorn, J. Chem. Phys. 51, 3281 (1969).
15. M.M. Abraham, L.A. Boatner, C.B. Finch, and R.W. Reynolds, Phys. Rev. B 3, 2864 (1971).
16. W. Kolbe, N. Edelstein, C.B. Finch, and M.M. Abraham, J. Chem. Phys. 60, 607 (1974).
17. D.J. Lam and S.K. Chan, Phys. Rev. B 6, 307 (1972).
18. D.G. Karraker, J. Chem. Phys. 63, 3174 (1975).
19. A. Boeuf, J.M. Fournier, J.F. Guegnon, L. Manes, J. Rebizant, and F. Rustichelli, J. Phys. Lett. 40, 335 (1979).
20. L.R. Morss, G. Shalimoff, and N. Edelstein, (to be published).
21. M.E. Hendricks, E.R. Jones, Jr., J.A. Stone, and D.G. Karraker, J. Chem. Phys. 60, 2095 (1974).
22. S.E. Nave, R.G. Haire, and P.G. Huray, Phys. Rev. B 28, 2317 (1983).
23. D.B. McWhan, Ph.D. thesis, UCRL-9695, 1961.

24. W.W.T. Crane, J.C. Wallmann, and B.B. Cunningham, UCRL-846, 1950.
25. J.J. Howland, Jr. and M. Calvin, J. Chem. Phys. 18, 239 (1950).
25a. W.J. Nellis and M.B. Brodsky, in "The Actinides: Electronic Structure and Related Properties, Vol. II," A.J. Freeman and J.B. Darby, Jr. Eds., Academic Press, New York, 1974, p. 266.
26. B. Kanellakopulos, "Magnetochemie," Lecture at the Institut für Radiochemie der Technischen Universität München, 1979.
27. J. Goffart, unpublished work, 1982.
28. L.R. Morss, E. Gamp, and N. Edelstein, unpublished work, 1983.
29. V.M. Malhotra and H.A. Buckmaster, Canad. J. Phys. 60, 1573 (1982) and references therein.
30. M. Abraham, B.B. Cunningham, C.D. Jeffries, R.W. Kedzie, and J.C. Wallmann, Bull. Am. Phys. Soc. 1, 396 (1956).
31. P. Fields, A. Friedman, B. Smaller, and W. Low, Phys. Rev. 105, 757 (1957).
32. R. Marrus, W.A. Nierenberg, and J. Winocur, Phys. Rev. 120, 1429 (1960).
33. W.A. Runciman, J. Chem. Phys. 36, 1481 (1962).
34. M. Abraham, B.R. Judd, and H.H. Wickman, Phys. Rev. 130, 611 (1963).
35. N. Edelstein and W. Easley, J. Chem. Phys. 48, 2110 (1968).
36. W. Kolbe, N. Edelstein, C.B. Finch, and M.M. Abraham, J. Chem. Phys. 56, 5432 (1972).
37. W. Kolbe, N. Edelstein, C.B. Finch, and M.M. Abraham, J. Chem. Phys. 58, 820 (1973).
38. M.M. Abraham and L.A. Boatner, Phys. Rev. B 26, 1434 (1982).
39. M.M. Abraham and L.A. Boatner, unpublished data, 1984.
40. M.M. Abraham, G.W. Clark, C.B. Finch, R.W. Reynolds, and H. Zeldes, J. Chem. Phys. 50, 2057 (1969).
41. S.A. Marei and B.B. Cunningham, J. Inorg. Nucl. Chem. 34, 1203 (1972).
42. L.R. Morss, J. Fuger, J. Goffart, and R.G. Haire, Inorg. Chem. 22, 1993 (1983).

43. B. Kanellakopulos, J.P. Charvillat, F. Maino, and W. Müller, in "Transplutonium Elements," W. Müller and R. Linder, Eds., North-Holland, Amsterdam, 1976, p. 181.

44. R.D. Baybarz, L.B. Asprey, C.E. Strouse, and E. Fukushima, J. Inorg. Nucl. Chem. 34, 3427 (1972).

45. For references to these measurements see P.G. Huray, S.E. Nave, J.R. Peterson, and R.G. Haire, Physica 102B, 217 (1980).

46. J.M. Fournier, A. Blaise, W. Müller, and J.C. Spirlet, Physica 86-88B, 30 (1977).

DELOCALISATION OF 5f ELECTRONS IN AMERICIUM METAL UNDER PRESSURE: RECENT RESULTS AND COMPARISON WITH OTHER ACTINIDES

U.Benedict, J.P.Itié, C.Dufour, S.Dabos, J.C.Spirlet
Commission of the European Communities, Joint Research Centre Karlsruhe Establishment, European Institute for Transuranium Elements, Postfach 2266, D-7500 Karlsruhe 1, Federal Republic of Germany

ABSTRACT. The high pressure behaviour of americium metal was reinvestigated by x-ray diffraction to 52 GPa. Phase Am III cannot be indexed in the monoclinic structure assigned to it previously. The resulting volumes are too large and would correspond to an expansion of the lattice by an increasing-pressure phase transition. This contradiction is lifted when Am III is indexed as a trigonal distortion of the cubic close packed Am II. A 6% volume decrease was observed on transition from Am III to Am IV. These findings complete the systematics of the high pressure transitions of the group of heavy actinide metals Am to Cf.

INTRODUCTION

Locating americium metal in the actinide series means first discussing the two main subgroups which are observed in the actinide metals. These subgroups can be defined on the basis of the dualism between the localised and the itinerant configuration of the 5f electrons which controls the solid state properties of these metals.

The first subgroup, protactinium to plutonium, has its 5f electrons in an itinerant (delocalised) state. This means they are of band type, hybridise with the conduction electrons and thus contribute to the metallic bonding. Magnetic order, which in the lanthanide metals is limited to the presence of localised 4f electrons, is consequently not observed in this subgroup. The strengthening of the metallic bond by the 5f participation leads to small atomic volumes (Fig.1), a high cohesive energy and low compressibility (Fig.2). Low symmetry (orthorhombic and monoclinic) crystal structures are found whose formation is probably related to the particular directional properties of the hybridised orbitals which include a contribution from 5f electrons.

The second subgroup, americium to californium, is characterised by localised 5 f electrons. In terms of electron energy, this means that these electrons have sharp energy levels and do not contribute to the metallic bond. In a spatial sense, it means that a particular 5f

N. M. Edelstein et al. (eds.), Americium and Curium Chemistry and Technology, 213–224.

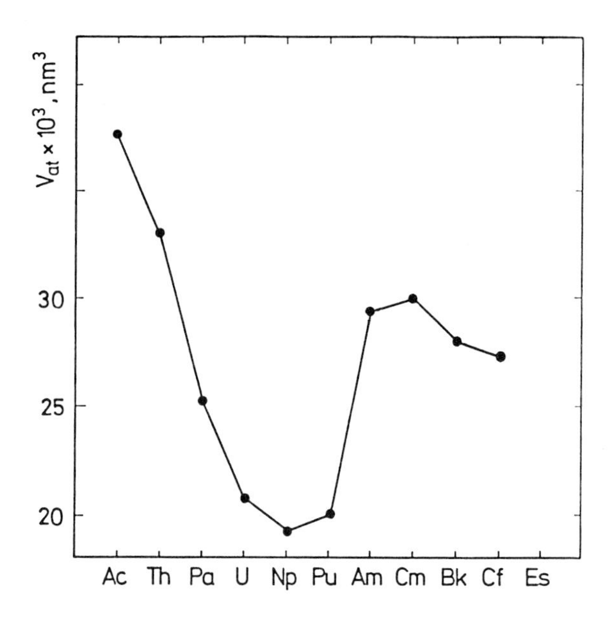

Fig.1 - Atomic volume of actinide metals.

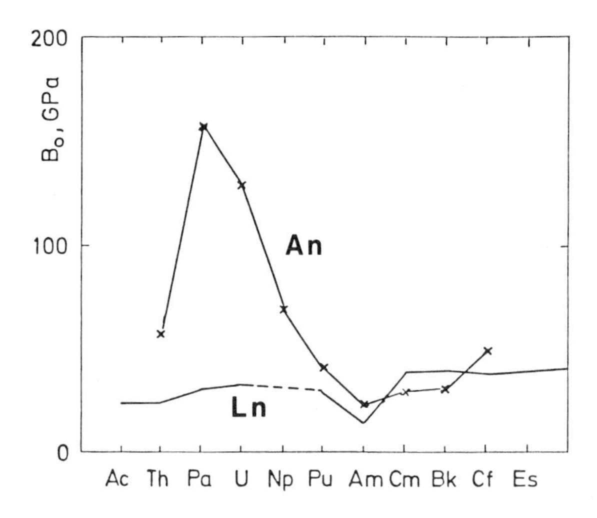

Fig.2 - Bulk moduli of the actinide metals, compared to those of the lanthanide metals.

electron is fixed ("localised") to a particular actinide atom. In contrast to the first subgroup, the localised 5f electrons contribute to the appearance of magnetic order in curium, berkelium and californium. The atomic volumes are larger (Fig.1), thus closer to those of the trivalent lanthanide metals. The cohesive energies are in general lower than those of the "itinerant" 5f metals.

The compressibilities are also of the same order as those of the trivalent lanthanides (Fig.2). The crystal structure of the four metals of the second subgroup is double-hexagonal close-packed (dhcp), thus of relatively high symmetry.

THE EFFECT OF PRESSURE

The most remarkable effect of pressure on the actinide metals is that due to closer contact between the lattice atoms, localised 5f electrons can become itinerant, hybridise with the conduction electrons, and participate in the metallic bond. The subgroup with localised 5f electrons can thus under pressure acquire properties which, at ambient pressure, are characteristic for the subgroup Pa to Pu. In the lighter actinides the 5f electrons participate in the bond at normal pressure; thus moderate pressure will not change their configuration and very high pressures are required to induce structural changes by other mechanisms.

In contrast with the lighter actinides, americium undergoes three phase transitions in the pressure range below 25 GPa. One of these is a consequence of the participation of the 5f electrons in bonding owing to the effects of pressure in this metal. The behaviour of americium under pressure is believed to be an important clue to the understanding of the properties of actinides.

The results of x-ray diffraction studies of Am metal up to 18 GPa were published in 1979-1982 by the Lawrence Livermore National Laboratory (LLNL) and by the Los Alamos National Laboratory (LANL in collaboration with the Oak Ridge National Laboratory (ORNL).

The two groups of researchers agree that with increasing pressure, the dhcp phase (Am I) transforms to a cubic close-packed (ccp) phase (Am II). But they give different structural descriptions for the low-symmetry phases Am III and Am IV. Both the LANL and the LLNL groups report a monoclinic structure for pressures between 10 and 15 GPa. The LANL group assigned to this structure space group no. 11 ($P2_1/m$) (1,2) but LLNL (3) proposed a c-centred structure with different lattice parameters. Roof (4) showed that these two structures are related. A similar situation exists for the allotrope described for the pressure range 15-18 GPa. The LANL group reported it (5) to have a structure of the α-U type. The data obtained by LLNL do not correspond to that structure type but are similar to it (6).

A preliminary x-ray diffraction study of Am under pressure was made in 1982 (17) using the energy-dispersive equipment of the European Institute for Transuranium Elements in Karlsruhe, FRG (7). Angular dispersive diffraction equipment with photographic film as the detecting device had been used in the preceding studies at LLNL and

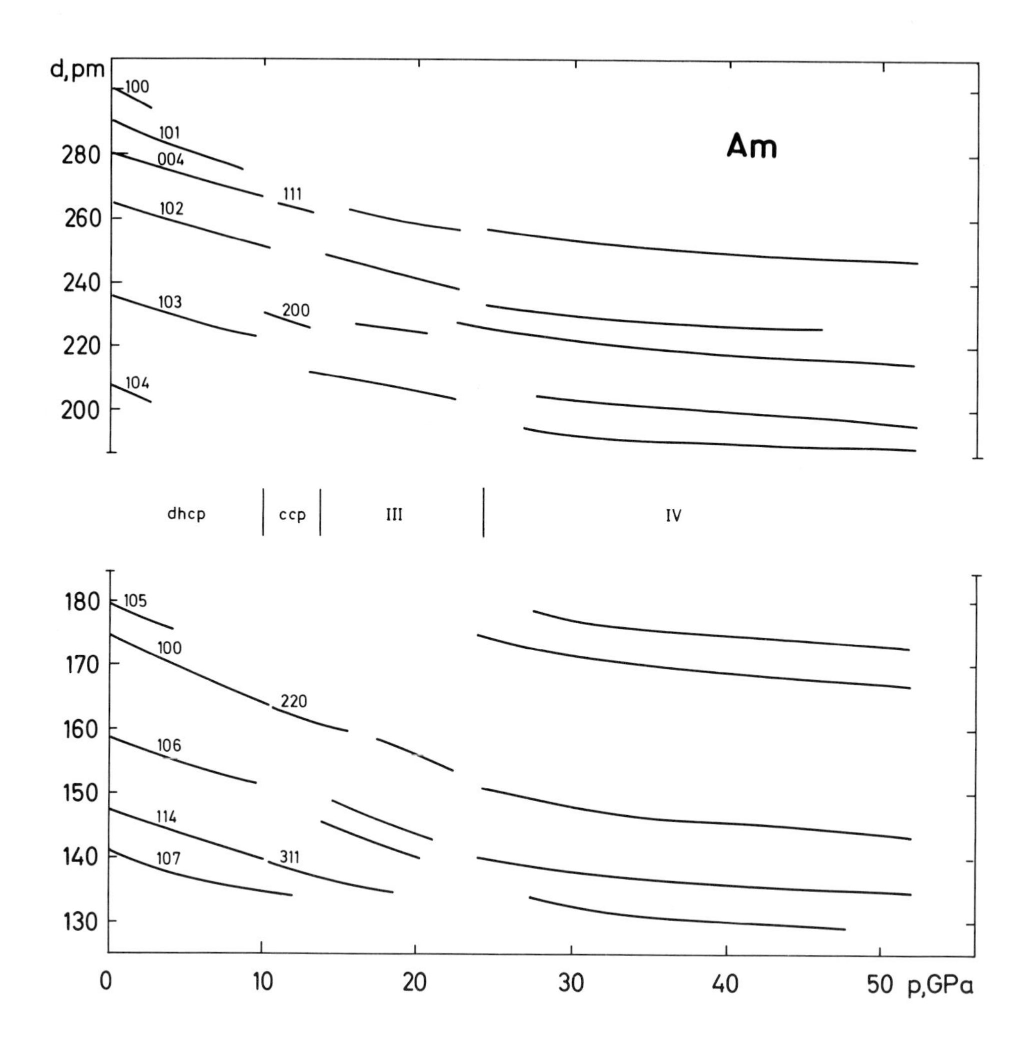

Fig.3 Interplanar distances in americium metal as a function of pressure. Miller indices hkl are indicated for the dhcp and ccp phases.

LANL. The Karlsruhe work yielded supplementary information, in part confirming the results of the preceding investigations,but did not yet allow definite conclusions on the structures of phases Am III and Am IV to be drawn.

The volume of Am as a function of pressure has been calculated using the atomic sphere approximation-LMTO (ASA-LMTO) method (8). If the relativistic, not spinpolarised volume calculated by Brooks (9) is employed for the lower limit of the volume collapse, which according to these calculations should occur when the 5f electrons go itinerant, a volume change of - 22% is obtained. None of the three above-mentioned experimental investigations confirmed such an important volume change upon formation of phases Am III or Am IV.

These differences between experimental results of different authors, on one hand, and between calculated and experimentally observed volume changes, on the other hand, led us to initiate further study of americium metal under pressure. Another important aim of this study was to obtain information on the structural behaviour of Am in the range 18-50 GPa, for which no data had been previously reported in the literature.

EXPERIMENTAL

The material studied was Am-243 containing 3.7% of Am-241. AmO_2 was reduced by lanthanum metal, the Am metal formed was distilled off and deposited on a tantalum surface. A second distillation was made to purify the metal. Small particles were scratched from the deposit in a pure argon atmosphere and loaded into a pressure cell of the Syassen-Holzapfel type. Silicone oil was used as a pressure medium. The pressure was determined by the ruby fluorescence method. Details of the experimental method were previously described (7,10). No chemical analysis could be carried out on the thin Am metal deposit. Material prepared by the same method was found to have about 5000 ppma nonmetallic (mainly O and N) and about 140 ppma metallic impurities (11).

RESULTS

The variation of interplanar distances of Am with pressure is shown in fig.3. This variation was studied up to 51.7 GPa. Diffraction spectra at 41 different pressure steps could be obtained in a reasonable time due to the use of the fast energy-dispersive, conical-slit method (7). 4 pressure ranges corresponding to 4 different crystallographic structures can be distinguished from this figure. The double-hexagonal close-packed (dhcp) phase (Am I) is stable up to 9 GPa and transforms at about 9.5 GPa to a cubic close packed structure (Am II). This in turn transforms to Am III at about 13.5 GPa. Am IV forms at about 23 GPa and remains stable up to the highest pressure applied in this experiment (51.7 GPa).

The relative volumes calculated from the d-spacings of Fig. 3 are plotted vs. pressure in Fig. 4. No major volume change occurs when the dhcp phase transforms to the ccp phase.

For Am III, the volumes were first calculated using the monoclinic indexing of Roof (2). The volumes thus obtained (dashed curve in Fig. 4) are too large with respect to the volumes of the ccp phase, indicating that this indexing cannot be applied to our data. In addition two of the strongest lines - (040) and (130) / (101) - of the monoclinic phase reported by Roof were not observed in our diffraction spectra.

Indexing Am III in the trigonally distorted ccp structure described by Vohra et al. (12) for Pr and other lanthanide metals gives a relatively good fit for line positions and intensities. The best intensity fit is obtained for x = 0.5, y = 0.33, z = o25. The volumes based on this indexing (full curve for Am III in Fig.4) are approximately in line with those of the ccp phase. This supports the view that Am III is indeed a trigonal distortion of the ccp structure, formed from Am II in a second order phase transition.

The volumes of Am IV were calculated on the basis of the orthorhombic α-uranium type structure which was reported for Am IV by Roof et al. (5). Indexing in this structure must be considered as provisional, because no line of our spectra could be fitted to hkl's (021) and (002), which theoretically correspond to intense diffraction lines. In addition, a shoulder on the high-d side of line (111), which is observed in all spectra, does not find a correspondence in the α-uranium structure. Further work will have to show whether another structure gives a better fit to the present data. The volume decrease on transition from the trigonally distorted ccp structure to Am IV is approximately 6%. This volume decrease probably corresponds to the onset of itinerancy of the 5f electrons.

By fitting the V (p) data for the dhcp phase with the Birch and Murnaghan equations (7), a bulk modulus B_o of 45 (1) GPa and a pressure derivative of B_o' = 6 (1) were determined (Fig.5). These values represent averages between the data for the Birch fit (B_o = 44.76, B_o' = 6.1, σ_B = 1.21) and for the Murnaghan fit (B_o = 45.14, B_o' = 5.6, σ_B = 1.24).

DISCUSSION

Significantly higher transition pressures were measured in this work, compared with work by LLNL and LANL. To exclude any doubt, the transition pressure of NaCl was measured under identical conditions and determined to be 31 ± 1 GPa, which is in good agreement with the values reported in literature. Good agreement was also observed between pressures determined by the ruby fluorescence method and from the lattice parameters of the NaCl sample using Decker's (18) equation of state.

The differences in transition pressures seem too large (8 GPa in the case of the transition Am III - Am IV) to be accounted for by the differences in impurity content which exist between the samples used in the three independent investigations. But such differences in impurity content can explain why different groups of workers report different crystallographic structures for the same pressure range. It is likely that several low symmetry structures have quite similar

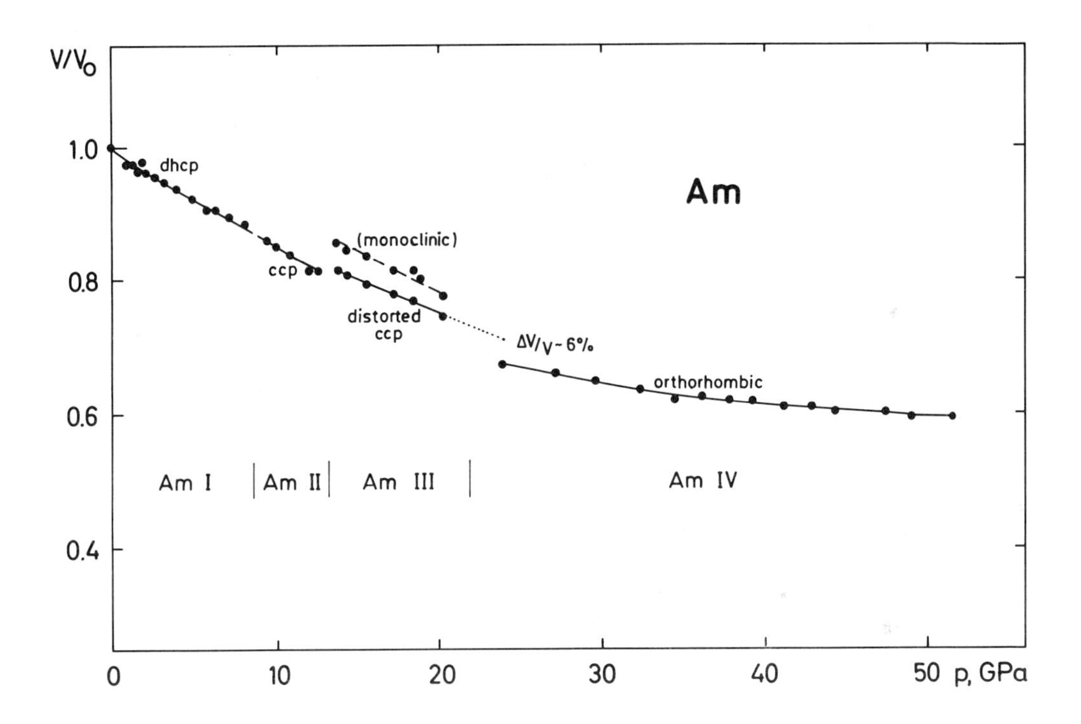

Fig.4 Relative volume of americium metal as a function of pressure.

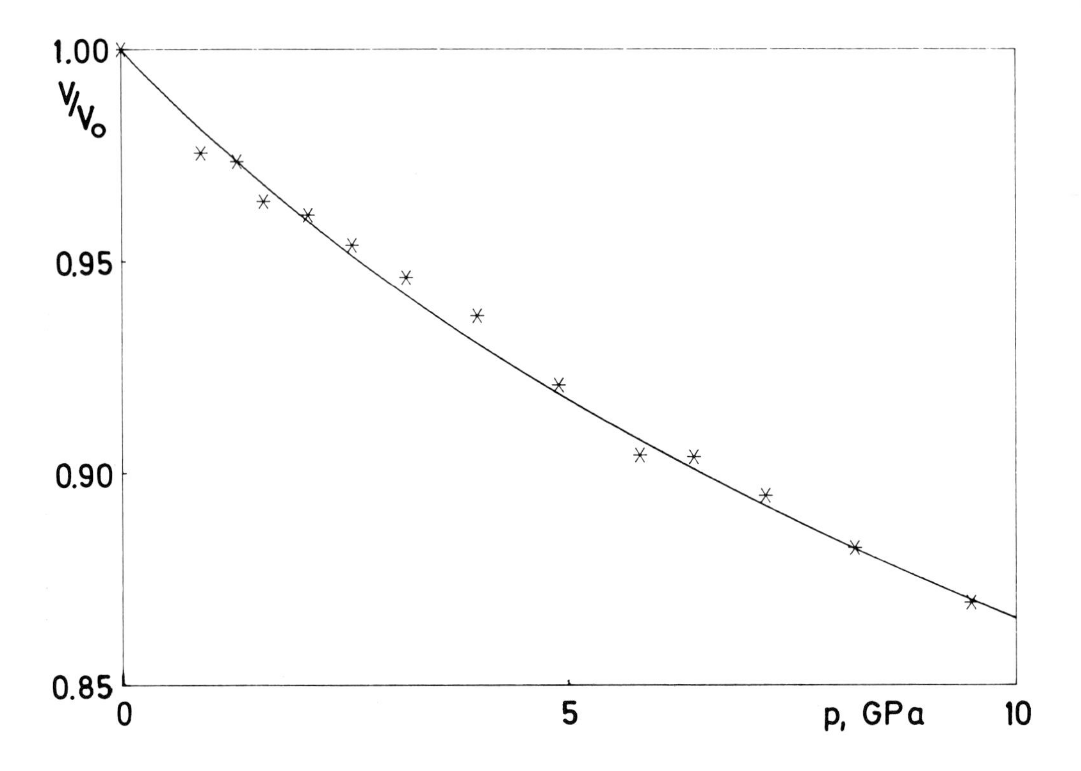

Fig.5 Fit of the Birch and Murnaghan equations to the experimental data for dhcp Am I. B=45 GPa, $B_o' = 6$. The curve represents the (coincident) Birch and Murnaghan equations.

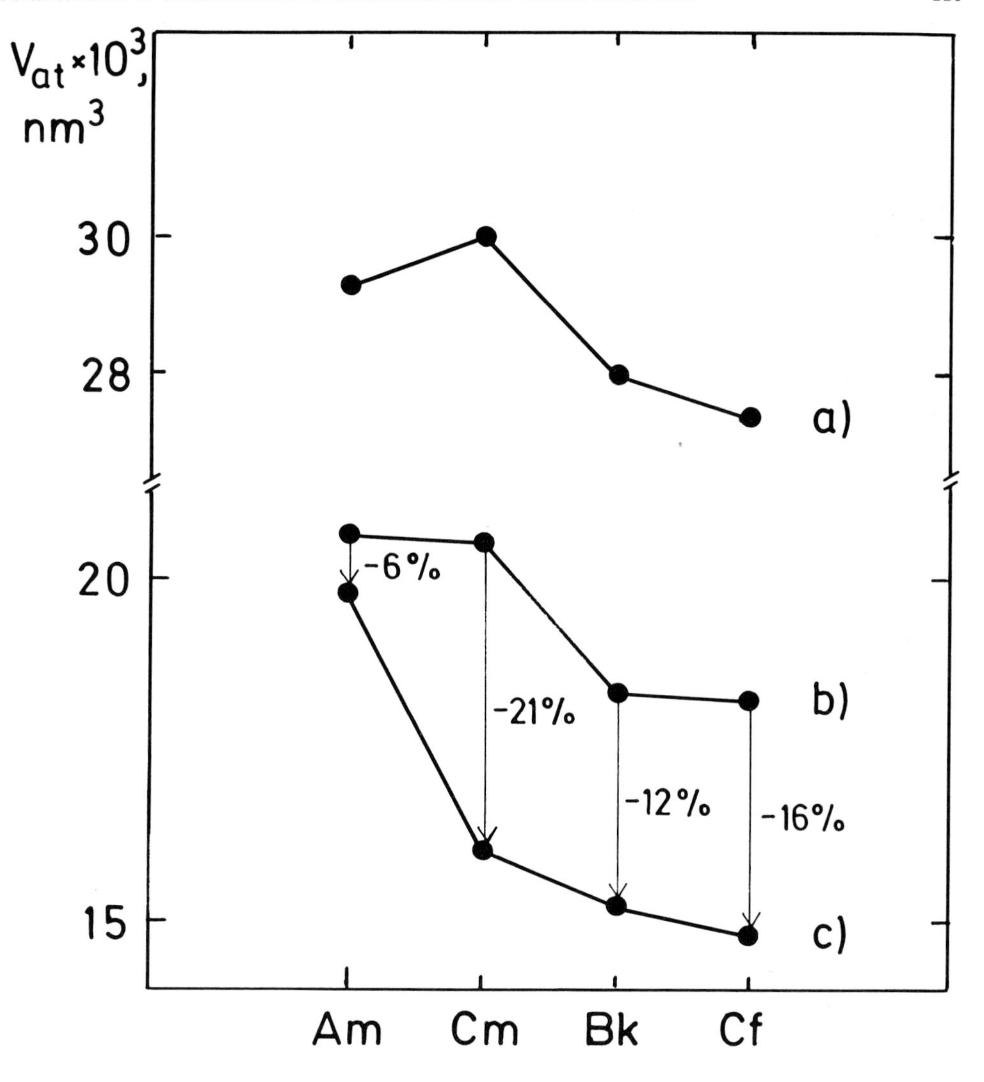

Fig. 6 Atomic volumes of Am, Cm, Bk and Cf.

a. at ambient pressure

b. ccp or distorted ccp phase at delocalisation pressure

c. α-U phase at delocalisation pressure

Volume decrease on delocalisation is indicated between curves b and c. (Volume in curve b taken as 100%).

thermodynamic stability, in a given pressure range, and that relatively small changes in conditions, such as differing impurity content, alternatively favor one or the other of these structures. Indications for this type of behaviour also exist for cerium metal. An α-uranium - type phase was reported as the main phase for Ce in the pressure range 5-12 GPa (13). Recent study, using cerium of different origin, did not furnish clear evidence for the α-uranium type structure (α'Ce), but revealed the existence of the monoclinic α'-Ce in the same pressure range (14).

The data of fig.4 exclude a major volume collapse as it was expected from theoretical considerations, and in this respect confirm the previous experimental work. It must be kept in mind that the theoretical calculations which predicted an important volume decrease on beginning itinerancy (8,9) were made for T = 0. The example of plutonium shows that the atomic volume varies considerably with temperature in such calculations. A correct comparison could thus only be made to calculations for room temperature (15). The present results also indicate that the itinerancy of 5f electrons in Am sets in at 23 GPa, a pressure much higher than that predicted from theoretical calculations (8).

The results obtained allow us to complete the systematics established in previous work on Cm (16,19), Bk (20,21) and Cf (20,22) for the pressure behaviour of this group of heavy actinides.

The transition pressures of Am, Cm, Bk and Cf are compared in Table 1. The transition pressures increase in

Table 1. Transition pressures of actinide metals which have localised 5f electrons at ambient pressure (16,19-22)

	Am	Cm	Bk	Cf
dhcp ⟶ ccp	9.5	23	9-10	17
ccp ⟶ distorted ccp	13.5	-	-	∿ 30
onset of itinerancy	23	43	32	41

principle with increasing 5f occupation. Curium is an exception, having higher transition pressures than the other three metals. The high spin polarisation energy of curium ($5f^7$) possibly contributes to the fact that particularly high pressures are needed to delocalise the 5f electrons in this metal. But the reasons for curium to transform from dhcp to ccp at considerably higher pressures than its neighbours of lower and higher Z are not clear (16).

Fig. 6 compares the atomic volumes of the four metals at different pressures. Curves b and c give the volume immediately before and immediately after the onset of itinerancy. Curve a,volume at ambient pressure,has a maximum at curium.This maximum corresponds to a minimum in 5f participation in the metallic bonding, curium having the most stable localised configuration among the actinide metals. It is seen that in the itinerant configuration (curve c) the maximum is at americium, re-establishing the trend of volumes decreasing with increasing Z ("actinide contraction").

The absolute volume of Am in the assumed α-uranium type structure at 46 GPa is $17.6 \cdot 10^{-3}$ nm^3. A volume of $\sim 18.10^{-3}$ nm^3 was recently estimated by Johansson (15) for Am at 46 GPa.

LITERATURE CITED

1. Roof R.B., Z. Kristallographie 1982 (158) 307-312.

2. Roof R.B., J. Appl. Crystallography 1981 (14) 447-450.

3. Smith G.S., Akella J., Reichlin R., Johnson Q., Schock R.N., Schwab M., Actinides 81, Asilomar, CA, September 1981.

4. Roof R.B., 1982, personal communication.

5. Roof R.B., Haire R.G., Schiferl D., Schwalbe L.A. Kmetko E.A., Smith J.L., Science 1980 (207) 1353-1355.

6. Akella J., 1982, personal communication.

7. Benedict U., Dufour C., High Temp., High Press, 1984, in press.

8. Johansson B., Skriver H.L., Andersen O.K., in Schilling J.S. and Shelton R.N., eds., Physics of Solids under High Pressure, North-Holland, Amsterdam, 1981, pp.245-262.

9. Brooks M.S.S., J. Phys. F., 1983 (13) 103-108.

10. Bellussi G., Benedict U., Holzapfel W.B., J. Less-Common Metals 1981 (78) 147-153.

11. Spirlet J.C., in "Actinides in Perspective", Edelstein N., ed., Pergamon 1982, pp. 361-380.

12. Vohra Y.K., Vijayakumar V., Godwal B.K., Sikka S.K., submitted to Phys. Rev. B., 1984.

13. Zachariasen W.H., Ellinger F.H., Acta Cryst. 1977, (A33) 155-160.

14. Benedict U., Gerward L., Staun Olsen J., Acta Cryst. 1984, (A 40) C-413.

15. Johansson B., J. Magn. Magn. Mat., 1984, in press.

16. Benedict U., Haire R.G., Peterson J.R., Itié J.P., J. Phys. F.: Metal Phys. 1985 (15) 29-35

17. Peterson J.R., Haire R.G., Benedict U., Spirlet J.C., Dufour C., Birkel I., Asprey L.B., p. 9 of: Rep. Workshop "Actinides under Pressure", Karlsruhe, April 21-22, 1983, Eur. Inst. Transuranium Elements Karlsruhe.

18. Decker D.L., J. Appl. Phys. 1971 (42) 3239-3244

19. Haire R.G., Benedict U., Peterson J.R., Dufour C., Itié J.P., J.Less-Common Metals, submitted.

20. Benedict U., Peterson J.R., Haire R.G., Dufour C., J. Phys. F: Metal Phys. 1984 (14) L43-L47

21. Haire R.G., Peterson J.R., Benedict U., Dufour C., J. Less-Common Metals 1984 (102) 119-126

22. Peterson J.R., Benedict U., Dufour C., Birkel I., Haire R.G., J. Less-Common Metals 1983 (93) 353-356.

PREPARATION, CHARACTERIZATION AND SOLUBILITY PRODUCT CONSTANT OF $AmOHCO_3$

Robert J. Silva*
Lawrence Livermore National Laboratory
P.O. Box 808
Livermore, CA 94550
USA

ABSTRACT. An investigation into the nature and solubility of a stable solid phase formed by a trivalent actinide, $^{243}Am^{3+}$, in dilute aqueous carbonate solutions was conducted. The compound exhibited an x-ray powder diffraction pattern which was nearly identical to that reported for $NdOHCO_3$ - type A. The pattern could be indexed in the orthorhombic system with unit cell parameters a = 4.958, b = 8.487, and c = 7.215 Å. The steady state solubility of the compound was determined from the results of both dissolution and precipitation experiments. The average solubility product quotient for 0.1M ionic strength, 25 ± 1°C and 1 atmosphere pressure was found to be 583 ± 206. The solubility product constant for zero ionic strength was estimated to be 335 ± 120.

1. INTRODUCTION

Computer calculations indicate that the actinides U-Am will be the major contributors to the radioactivity of high-level nuclear waste after storage periods of several hundred years or longer [1, 2], the design criteria for the lifetime of the metal canisters [3]. In the event that the waste packages fail to contain the radioactive materials, they will enter the local groundwater system and, if moving, the groundwater is expected to provide the most likely mechanism by which the radionuclides could be transported from an underground storage facility to the accessible environment [4]. The actinides can react with various components in the altered groundwater to form insoluble phases that can control their solution concentrations and, thus, their migration rates. The exact nature of the solid phases will depend on a number of factors, e.g., the composition and redox properties of the groundwater system and the oxidation states of the actinides. Therefore, a knowledge of the nature and solubilities of actinide compounds that could form in groundwater is one of the pieces of information that is needed to predict migration rates.

N. M. Edelstein et al. (eds.), Americium and Curium Chemistry and Technology, 225–238.

Thermodynamic data on trivalent lanthanide elements [5], good analogs for actinides in the 3+ state [6], suggest that carbonate compounds of the latter may be quite insoluble in carbonate bearing groundwaters; however, no thermodynamic data have been reported. The object of this work was to investigate the nature and solubilities of compounds of a trivalent actinide, $^{243}Am^{3+}$, in an aqueous carbonate system.

2. EXPERIMENTAL PROCEDURES

2.1. Preparation and Characterization of Solid Phase

The preparation of rare-earth carbonates via the formation and subsequent hydrolysis of the trichloroacetate complex in aqueous solution is a well established method for producing a pure, easily filterable and crystalline material [7, 8]. The exact nature of the rare-earth carbonate depends on the atomic number of the rare earth and is quite sensitive to the preparative conditions, e.g., temperature, CO_2 pressure and the washing procedure. The octahydrate, $R_2(CO_3)_3 \cdot 8H_2O$, is usually formed by the first members of the rare-earth series, e.g., La, while the dihydrate, $R_2(CO_3) \cdot 2H_2O$, is usually formed by the middle and end members of the series, e.g., Nd-Yb [9]. Charles has also reported that, even using the same procedures and conditions, some rare earths form the dihydrate, e.g., Nd, while others form the basic carbonate, $ROHCO_3$, e.g., Pr [10]. The compounds are, however, usually pure. Because of these complications, it seemed prudent to first test the preparative method and the stability of the resulting solid phase with neodymium, a good analog element for americium.

Approximately 10 mgs of neodymium carbonate were prepared by the trichloroacetate method. An x-ray powder diffraction pattern of the material was obtained with a Debye-Scherrer camera using CuK_α radiation (λ = 1.5418 A) and a Ni filter. The d-spacings and relative intensities derived from the powder pattern are given in Table I under Nd Solid (I). Head and Holley have investigated the preparation and thermal decomposition of the rare-earth dihydrates [8, 9]. These elements were stated to form dihydrates that were isostructural and that had similar x-ray powder diffraction patterns. Unfortunately, these patterns were not published. However, an original x-ray film containing the powder diffraction pattern of the $Nd_2(CO_3)_3 \cdot 2H_2O$ (assigned from the results of chemical analysis) was recently obtained from C. E. Holley of LANL and analyzed. The d-spacings and relative intensities are given in Table I. Our material appeared to be pure, microcrystalline $Nd_2(CO_3)_3 \cdot 2H_2O$.

The $Nd_2(CO_3)_3 \cdot 2H_2O$ solid was placed in a polypropylene cell containing an aqueous solution composed of 0.1$\underline{M}$ $NaClO_4$ and $2x10^{-4}$$\underline{M}$ HCO_3^- at a pH of 6.12. After a contact time of three weeks, an x-ray powder diffraction pattern was again obtained on the solid. The resulting powder pattern is given in Table I under Nd Solid (II). The pattern was distinctly different from that of the Nd Solid (I) but nearly identical to the powder pattern reported for $NdOHCO_3$ - Type A [11].

TABLE I. X-RAY POWDER DIFFRACTION PATTERNS OF Nd SOLID PHASES

$Nd_2(CO_3)_3 \cdot 2H_2O$		Nd Solid (I)		$NdOHCO_3$ [a]		Nd Solid (II)	
d(A)	I[b]	d(A)	I[b]	d(A)	I[c]	d(A)	I[b]
7.563	70	7.694	70	5.500	30	5.500	25
5.680	85	5.800	85	4.280	100	4.283	100
4.665	100	4.726	100	4.235	15	-	-
3.934	85	3.991	85	3.675	35	3.673	50
3.834	25	3.875	25	3.650	25	-	-
3.615	50	3.659	70	3.315	35	3.308	25
3.092	25	3.129	25	2.940	25	-	-
3.015	85	3.046	85	2.910	50	2.921	50
2.976	10	-	-	2.748	3	-	-
2.873	10	-	-	2.630	30	2.634	25
2.751	25	2.767	25	2.475	25	2.482	25
2.582	70	2.607	50	2.400	10	2.404	10
2.249	10	-	-	2.323	40	2.324	50
2.170	10	-	-	2.310	18	-	-
2.082	10	-	-	2.138	7	-	-
2.016	50	2.031	35	2.122	9	-	-
1.982	25	1.994	25	2.097	1	-	-
1.872	35	1.882	35	2.050	40	2.050	35
1.825	35	1.835	35	2.030	13	-	-
1.779	35	1.789	35	1.984	40	1.988	35
1.737	25	1.745	35	1.925	25	1.929	10
1.562	35	1.566	25	1.880	10	1.885	10
-	-	-	-	1.830	19	1.828	25

(a) Reference 11; (b) relative intensities visually estimated; (c) relative intensities by diffractometer.

There was no longer any evidence in the pattern for $Nd_2(CO_3)_3 \cdot 2H_2O$. The starting material had converted to the basic carbonate during the three week contact time.

An attempt was made to prepare $Am_2(CO_3)_3 \cdot 2H_2O$ by the same procedure as was used to prepare the $Nd_2(CO_3)_3 \cdot 2H_2O$. However, the x-ray powder diffraction pattern of the resulting Am compound was very similar to that of the $NdOHCO_3$ and not $Nd_2(CO_3)_3 \cdot 2H_2O$. The preparation of the solid was repeated with the Am, and the x-ray diffraction analysis of the solid gave a pattern again similar to the Nd basic carbonate. The d-spacings and relative intensities obtained from the pattern are given in Table II under Am Solid (I). Since the results with Nd suggested that the basic carbonate is more stable than the dihydrate in our aqueous carbonate system, no further attempts were made to produce the normal carbonate of Am.

2.2. Solubility Measurements

The solubility studies with trivalent americium were conducted in two parts. The first involved following to steady state the concentration of Am in a solution initially free of the element, i.e., undersaturation with respect to precipitation, but which had been placed in contact with the solid phase $AmOHCO_3$. The second involved following to steady state the concentration of Am in a solution initially supersaturated in Am with respect to the precipitation of a solid phase but with no solid initially present. Much of the equipment and techniques used in these measurements have been described elsewhere [12].

In order to obtain a reliable value for the solubility product quotient for the reaction,

$$AmOHCO_3(s) + 2H^+ = Am^{3+} + H_2O + HCO_3^-$$

the pH range available for measuring the solubility was somewhat limited. It was necessary to hold the pH to less than about 6.5 to avoid significant hydrolysis of the Am^{3+} ion but greater than about 5 to avoid possible dissolution of all of the available starting solid (8 mgs). In addition, the CO_3^{2-} concentration was set sufficiently low so as to avoid significant carbonate complexation of the Am^{3+} ion, i.e., $\sim 10^{-8}\underline{M}$. A pH of 6.12 was selected as it is within a reasonably well buffered region where the concentrations of H_2CO_3 and HCO_3^- are equal. A computer controlled pH-stat was used to maintain a pH near this value. The bicarbonate, and thus carbonate, ion concentration was fixed by maintaining a gas mixture of $0.792 \pm .010\%$ CO_2 and 99.3% Ar over the aqueous phase. The equilibrium concentrations of bicarbonate and carbonate ions in the aqueous phase under these conditions were calculated to be $2.05 \times 10^{-4}\underline{M}$ and $2.25 \times 10^{-8}\underline{M}$, respectively, using reported carbonate equilibrium quotients for $0.1\underline{M}$ ionic strength [13, 14]. To fix the ionic strength, an aqueous solution of $0.1\underline{M}$ $NaClO_4$ was used as supporting electrolyte. Two titrations of 50 mls of the aqueous phase at a starting pH value of 6.14 with $0.100\underline{M}$ HCl to the end point of the HCO_3^- to H_2CO_3 conversion were made prior to the initiation of the solubility experiments. From the results, the

TABLE II. X-RAY POWDER DIFFRACTION PATTERNS OF AMERICIUM SOLID PHASES

Am Solid (I)		Am Solid (II)		Am Solid (III)		Average (I, II, III)	Calculated	
d(A)	I[a]	d(A)	I[a]	d(A)	I[a]	d(A)	d(A)	hkl
5.500	25	5.503	25	5.505	25	5.503±0.003	5.497	011
4.283	100	4.279	100	4.281	100	4.281±0.002	4.281	110
3.673	50	3.671	50	3.666	50	3.670±0.004	3.682	111
-	-	-	-	-	-	-	3.658	021
3.308	25	3.314	35	3.326	35	3.316±0.009	3.320	012
-	-	2.942	25	-	-	-	2.943	121
2.921	50	2.919	50	2.912	50	2.917±0.005	2.917	102
2.634	25	2.636	25	2.635	25	2.635±0.001	2.634	031
2.482	25	2.480	25	2.475	25	2.479±0.004	2.479	200
2.398	10	2.402	10	2.406	25	2.402±0.004	2.404	122
2.324	50	2.335	35	2.331	50	2.330±0.006	2.326	131
-	-	2.322	10	-	-	-	2.314	013
-	-	2.134	10	2.132	25	2.133±0.002	2.141	220
2.050	35	2.054	35	2.051	50	2.052±0.002	2.052	221
1.988	35	1.986	25	1.985	25	1.986±0.002	1.986	212
1.929	10	1.929	25	1.924	25	1.927±0.003	1.928	123
1.885	10	1.889	10	-	-	1.887±0.003	1.883	141
1.828	25	1.835	10	-	-	1.832±0.005	1.832	033

(a) Relative intensities visually estimated.

bicarbonate ion concentration was calculated to be $2.67 \pm .16 \times 10^{-4}$M. This value is in reasonable agreement with the concentration of $2.15 \pm .17 \times 10^{-4}$M calculated for the CO_2 partial pressure in the gas mixture at this pH.

2.3. Solubility Measurements from Unsaturation

Approximately 8 mgs of the crystalline $AmOHCO_3$, Am Solid (I), were placed in a polypropylene cell with 50 mls of the aqueous phase at $25 \pm 1°C$ and 1 atmosphere pressure. Two 1 ml aliquots of the aqueous phase were taken after 1, 3, 7, 14, 21, and 30 days equilibration time. After centrifugation at 15,000 RPM for 15 minutes, the Am concentrations were determined from the results of liquid scintillation counting of the supernates. The beta-emitting daughter, ^{239}Np, was found not to be in nuclear decay equilibrium with the ^{243}Am in solution at the times of sampling. Because the ^{239}Np added to the measured total count rates of the aliquots, the samples required counting periodically for 3-4 weeks to allow secular equilibrium to be established so the appropriate counting corrections for the Np could be made.

At the end of the 30-day period, two additional 5 ml aliquots of the aqueous phase were taken. After centrifugation, the aliquots were split into two fractions. One fraction, 2 mls, was passed through a a 0.2μm acrodisc polysulfone filter while the other fraction, 3 mls, was passed through a 0.015μm Nuclepore polycarbonate filter. The first one-half of the fraction through the filters was discarded and a measured portion of the remaining filtrate analyzed for Am by liquid scintillation counting.

Also, at the end of the 30 day period, the solid phase remaining in the cell was subjected to x-ray diffraction analysis. The results obtained from the powder pattern are given in Table II under Am Solid (II). The powder pattern was identical to that of the starting material, Am Solid (I). There was no change in the solid phase during these dissolution measurements.

2.4. Solubility Measurements from Supersaturation

At the conclusion of the first part of the studies described above, the aqueous phase in the cell was made 0.1M in $HClO_4$ to dissolve the solid $AmOHCO_3$. The volume of the aqueous phase was adjusted to 50 mls by the addition of water and solutions of $NaHCO_3$ and NaOH in such a manner as to again produce an aqueous phase 0.1M in $NaClO_4$, 2×10^{-4}M in HCO_3^- and at a pH of 6.12. The remainder of the experiment proceeded as in the first part of these studies.

Aliquots of the aqueous phase were taken after 1, 3, 7, 14, 22, 31, 37, 44 and 51 days equilibration time in order to follow the Am solution concentration to steady state. At the 51 day period, in addition to separation of phases by centrifugation, aliquots were filtered as before. At this time, the solid phase that had precipitated was subjected to x-ray diffraction analysis. The information obtained from the powder pattern is given in Table II under Am Solid (III). The powder pattern was identical to that of Am Solid (I) and Am Solid (II).

3. RESULTS

3.1. Crystal Data

The corresponding d-values obtained from the x-ray powder diffraction patterns of the three solids, Am Solid I, II and III, were least square averaged; the results are given in Table II. Guided by the work on $NdOHCO_3$ type A [11], the averaged values could be indexed in the orthorhombic system with the unit cell parameters a = 4.958, b = 8.487 and c = 7.215Å. The calculated d-values and Miller indices are also given in Table II. The values of the parameters agree to within two decimal places with those given for $NdOHCO_3$ - type A [11]. Thus, the Am solid was identified as $AmOHCO_3$ and it is isostructural with $NdOHCO_3$ - type A.

3.2. Solubility from Undersaturation

The results of the measurements of the solubility of the crystalline $AmOHCO_3$ are given in Table III. The pH values given are the average values observed during each equilibration period. The errors in the pH values were estimated from the inherent reproducibility of measurements with the pH meter and electrodes, both on the dissolver solution and buffers, and from the root mean square deviations of pH values (measured every two hours) from the averages. No corrections were made for liquid junction potential differences.

The values and errors assigned to the Am solution concentrations are the average and deviations from the average of the two aliquots taken at each sampling period. The values of the bicarbonate concentrations were calculated from the partial pressure of CO_2 over the aqueous phase, i.e., 0.00792 $\pm$ .0004 atmospheres and the average pH value during the period.

Since the Am solution concentrations did not vary appreciably after the 7 day equilibration time, it was assumed that steady state had been achieved in the dissolution reaction. Some variation in the Am concentrations between equilibration times is due to differences in the pH. The Am concentrations of the solutions passed through the two different pore size filters at the 30 day period agreed to within experimental error. However, these values are only about 65% of the Am concentration of the centrifuged sample. Whether this difference was due to insufficient separation of solid and aqueous phases by our centrifugation system or due to loss of Am via absorption of soluble species by the filters was not determined. Since the source of this discrepancy was not known, the three values were given equal weight in subsequent data analysis.

3.3. Solubility from Supersaturation

The results of the measurements of the solution concentrations of Am as a function of equilibrium time during the precipitation of $AmOHCO_3$ are given in Table IV. The values for the average pH, the Am concentrations and bicarbonate concentrations and their associated errors were obtained as described in the previous section.

TABLE III. RESULTS OF SOLUBILITY MEASUREMENTS BY DISSOLUTION OF $AmOHCO_3$.

Equil. Time (days)	pH	Conc. Am (moles/l)	Conc. HCO_3 (moles/l)	Sol. Prod. Quot., Q_{sp}	log Q_{sp}
1	$6.13 \pm .03$	$2.86 \pm .25 \times 10^{-5}$	$2.09 \pm .18 \times 10^{-4}$	$6.32 \pm .10 \times 10^{3}$	$3.80 \pm .06$
3	$6.13 \pm .03$	$1.19 \pm .04 \times 10^{-5}$	$2.09 \pm .18 \times 10^{-4}$	$2.63 \pm .36 \times 10^{3}$	$3.42 \pm .06$
7	$6.15 \pm .03$	$2.17 \pm .54 \times 10^{-6}$	$2.19 \pm .19 \times 10^{-4}$	$5.51 \pm 1.56 \times 0^{2}$	$2.74 \pm .11$
14	$6.13 \pm .03$	$2.39 \pm .21 \times 10^{-6}$	$2.09 \pm .18 \times 10^{-4}$	$5.28 \pm .85 \times 10^{2}$	$2.72 \pm .06$
21	$6.14 \pm .03$	$2.02 \pm .04 \times 10^{-6}$	$2.14 \pm .19 \times 10^{-4}$	$4.79 \pm .65 \times 10^{2}$	$2.68 \pm .06$
30	$6.14 \pm .03$	$2.29 \pm .03 \times 10^{-6}$(a)	$2.14 \pm .19 \times 10^{-4}$	$5.43 \pm .73 \times 10^{2}$	$2.73 \pm .06$
		$1.54 \pm .13 \times 10^{-6}$(b)		$3.65 \pm .58 \times 10^{2}$	$2.56 \pm .06$
		$1.40 \pm .13 \times 10^{-6}$(c)		$3.32 \pm .55 \times 10^{2}$	$2.52 \pm .07$

(a) Centrifugation; (b) centrifugation plus 0.2 μm filtration; (c) centrifugation plus 0.015μm filtration.

TABLE IV. RESULT OF SOLUBILITY MEASUREMENTS BY PRECIPITATION OF $AmOHCO_3$.

Equil. Time (days)	pH	Conc. Am (moles/1)	Conc. HCO_3 (moles/1)	Sol. Prod. Quot., Q_{sp}	log Q_{sp}
0	.009	$6.61\pm.21\times10^{-4}$	-	-	
1	$6.09\pm.03$	$3.66\pm.13\times10^{-4}$	$1.91\pm.17\times10^{-4}$	$6.14\pm.82\times10^{4}$	$4.79\pm.06$
3	$6.11\pm.03$	$3.15\pm.05\times10^{-4}$	$2.00\pm.18\times10^{-4}$	$6.07\pm.82\times10^{4}$	$4.78\pm.06$
7	$6.12\pm.03$	$3.56\pm.13\times10^{-4}$	$2.05\pm.18\times10^{-4}$	$7.35\pm1.02\times10^{4}$	$4.87\pm.06$
14	$6.10\pm.03$	$7.86\pm1.67\times10^{-5}$	$1.95\pm.17\times10^{-4}$	$1.41\pm.35\times10^{4}$	$4.15\pm.10$
22	$6.13\pm.03$	$7.67\pm.72\times10^{-6}$	$2.09\pm.18\times10^{-4}$	$1.70\pm.28\times10^{3}$	$3.23\pm.07$
31	$6.14\pm.03$	$3.21\pm.14\times10^{-6}$	$2.14\pm.19\times10^{-4}$	$7.60\pm1.07\times10^{2}$	$2.88\pm.06$
37	$6.13\pm.03$	$4.42\pm.17\times10^{-6}$	$2.09\pm.18\times10^{-4}$	$9.77\pm1.36\times10^{2}$	$2.99\pm.06$
44	$6.13\pm.03$	$4.71\pm.30\times10^{-6}$	$2.09\pm.18\times10^{-4}$	$1.04\pm.15\times10^{3}$	$3.02\pm.06$
51	$6.11\pm.03$	$4.88\pm.11\times10^{-6}$(a)	$2.00\pm.18\times10^{-4}$	$9.40\pm1.33\times10^{2}$	$2.97\pm.06$
		$3.65\pm.06\times10^{-6}$(b)		$7.03\pm.95\times10^{2}$	$2.85\pm.06$
		$3.20\pm.32\times10^{-6}$(c)		$6.16\pm1.03\times10^{2}$	$2.79\pm.07$

(a) Centrifugation; (b) centrifugation plus 0.2 μm filtration; (c) centrifugation plus 0.015 μm filtration.

As can be seen in Table IV, the Am concentrations decreased during the first month of the measurements but then remained relatively constant for the following 20 days. Therefore, it was assumed that steady state had been achieved.

As in the dissolution experiments, the Am concentrations measured for the filtered samples at the 51 day equilibration time agreed reasonably well but were only about 60% of the value measured for the centrifuged sample. The three values were given equal weight in subsequent data analysis.

3.4. Evaluation of the Solubility Product Quotient and Constant

Solubility product quotients, Q_{sp}, were calculated from the measured concentrations given in Table III and IV for the following reaction

$$AmOHCO_3(s) + 2H^+ = Am^{3+} + HCO_3^- + H_2O \qquad (1)$$

The concentrations of hydrogen ion were derived from the measured pH values using an activity coefficient of 0.78. This value was estimated from trends in mean activity coefficients for 0.1M HCl [15], 0.1M $HClO_4$ [15] and HCl in 0.1M NaCl [16].

Since the Am^{3+} ion can undergo a small amount of hydrolysis and carbonate complexation under the solution conditions of the experiment, estimates of the degree of these reactions were made. The first hydrolysis constant for Am^{3+} has not been measured in a noncomplexing medium at 0.1M ionic strength; however, a log value of -7.7 ± 0.3 has been reported for Cm^{3+} [17]. Since Am^{3+} and Cm^{3+} are adjacent actinides of similar ionic radii, that of Am being slightly larger, a value of -7.8 ± 0.3 was assumed for Am^{3+}. From this quotient, the $AmOH^{2+}$ to Am^{3+} concentration ratio for a pH of 6.12 was calculated to be 0.016 ± 0.012.

A log value of 5.81 ± 0.04 has been reported for the formation constant, β_1, of the first carbonate complex of Am^{3+} in 1M $NaClO_4$ [18]. Unfortunately there are no data on mean activity coefficients of appropriate Am salts in mixed electrolyte systems which could be used to correct the quotients obtained in 1M $NaClO_4$ to 0.1M $NaClO_4$. As the next best approximation, it was decided to use mean activity coefficients of $LaCl_3$-HCl mixtures that have been measured experimentally [19]. The mean activity coefficient for appropriate carbonate species in a mixed electrolyte were estimated using Pitzer's equations [20]. Ion interaction parameters used in the calculations for carbonate species [21] and $NaClO_4$ [22] are reported in the literature. The single ion activity coefficient for $AmCO_3^+$ was taken to be equal to that of Na^+. The resulting estimate for log β_1(0.1M) was 6.11 ± 0.04. This quotient leads to a value of 0.027 ± 0.003 for the $AmCO_3^+$ to Am^{3+} concentration ratio.

From the above ratios, the concentration of free Am^{3+} was calculated as $95.9 \pm 1.3\%$ of the measured Am in solution. The total Am solution concentrations given in Tables III and IV were reduced by this percentage before calculation of the Q_{sp}'s given in the tables.

The Q_{sp} values calculated from the data obtained by the three different solid-solution separation schemes at the 30-day equilibration time for the dissolution experiment and the 51-day equilibration time for the precipitation experiment were averaged by the least squares method. The resulting mean and standard deviation of the solubility product quotient for 0.1M ionic strength was 583 $\pm$ 206. Estimates for the activity coefficients of Am^{3+}, HCO_3^- and H^+ at 0.1M ionic strength were used to obtain an approximate value for the solubility product constant, K_{sp}, at zero ionic strength from the solubility product quotient given in the preceeding paragraph. The sources of the estimated activity coefficients were given previously in this paper. Values of 0.46, 0.76, and 0.78 were used for Am^{3+}, HCO_3^- and H^+, respectively. The resultant value for K_{sp} was 335 $\pm$ 120.

4. CONCLUSIONS

The basic carbonate of americium, $AmOHCO_3$, was found to be a stable solid phase in dilute aqueous carbonate solutions at near neutral pH. The compound is isostructural with $NdOHCO_3$ - type A and its x-ray diffraction pattern can be indexed in the orthorhombic system. The solubility product quotient for 0.1M ionic strength, 25 $\pm$ 1°C and 1 atmosphere pressure was found to be 583 $\pm$ 206. The solubility product constant for zero ionic strength was estimated to be 335 $\pm$ 120.

5. ACKNOWLEDGEMENTS

The author wishes to thank C. E. Holley, Jr., of the Los Alamos National Laboratory for making available the x-ray film of the powder diffraction pattern of $Nd_2(CO_3)_3 \cdot 2H_2O$. The help of A. Zalkin of the Lawrence Berkeley Laboratory in the interpretation of powder patterns is gratefully acknowledged. Thanks are due J. J. Bucher (LBL) for helpful advise on estimating activity coefficients and H. Nitsche (LBL) for calculations involving Pitzer's equations. This work was supported by the office of Nuclear Regulatory Research, U.S. Nuclear Regulatory Commission, while the author was employed by the Lawrence Berkeley Laboratory.

*Work supported by the Division of Health, Siting and Waste Management, Office of Nuclear Regulatory Research, U.S. Nuclear Regulatory Commission under contract FIN No. B 3040-0 at Lawrence Berkeley Laboratory , and by the U.S. Department of Energy under Contract No. W-7405-Eng-48 at the Lawrence Livermore National Laboratory.

6. REFERENCES

1. Little, A. D., 1977. Technical Support of Standards for High-Level Waste Management, Vol. A, Source Term Characterization. EPA Report 520/4-79-007A. United States Environmental Protection Agency, Office of Nuclear Waste Isolation, Washington D.C., 20460

2. NRC, 1983. Staff Analysis of Public Comments on Proposed Rule 10CFR Part 60, Disposal of High-Level Radioactive Wastes in Geologic Repositories. U.S. Nuclear Regulatory Commission Report, NUREG-0804, Washington, D.C., p. 467.

3. U.S. Nuclear Regulatory Commission, 1983. 10CFR Part 60, Disposal of High-Level Radioactive Wastes in Geologic Repositories, Technical Criteria. Section 60.113, Performance of Particular Barriers after Permanent Closure, Federal Register, Vol. 48 (No. 120), p. 28224.

4. Klingsberg, C., and J. Duguid, 1980. Status of Technology for Isolating High-Level Radioactive Wastes in Geologic Repositories, DOE/TIC-11207, U.S. Department of Energy, Washington D.C., p. 15.

5. Smith, R. M., and A. E. Martell, 1976. Critical Stability Constants, Vol. 4, Inorganic Complexes, Plenum Press, New York, p. 37.

6. Kutz, J. J. and G. T. Seaborg, 1957. The Chemistry of the Actinide Elements, Wiley and Sons, New York, p. 406-475.

7. Salutsky, M. L., and L. L. Quill, 1950. 'The Rare-Earth Metals and their Compounds. XII. Carbonates of Lanthanum, Neodymium and Samarium', J. Amer. Chem. Soc., 72, p. 3306-3307.

8. Head, E. L., and C. E. Holley, Jr., 1963. 'The Preparation and Thermal Decomposition of Some Rare-Earth Carbonates'. Rare-Earth Research III, Gordon and Breach, N.Y. pp. 51-63.

9. Head, E. L., and C. E. Holley, Jr., 1964. 'The Preparation and Thermal Decomposition of the Carbonate of Tb, Dy, Ho, Er, Tm, Yb, Lu, Y and Sc.', Rare-Earth Research IV, Gordon and Breach, N.Y., pp. 707-718.

10. Charles, R. G., 1965. 'Rare-Earth Carbonates Prepared by Homogeneous Precipitation', J. Inorg. Nuclear Chemistry, 27, pp. 1489-1493.

11. Dexpert, H. and P. Caro, 1974. 'Determination de la Structure Crystalline de la Variete a des Hydroxycarbonates de Terres Rares $LnOHCO_3$', Mat. Res. Bull., 9, pp. 1577-1586.

12. Silva, R. J., and A. W. Yee, 1982. Geochemical Assessment of Nuclear Waste Isolation, 'Testing of Methods for the Separation of Solid and Aqueous Phases - a Topical Report', NUREG/LBL-14696, Lawrence Berkeley Laboratory, Berkeley, CA.

13. Harned, H. S. and R. Davis, Jr., 1943, 'The Ionization Constant of Carbonic Acid in Water and the Solubility of Carbon Dioxide in Water and Aqueous Salt Solutions from 0 to 50°'. J. Am. Chem. Soc., 65, pp. 2030-2037.

14. Phillips, S. L., 1982. Hydrolysis and Formation Constants at 25°C. Lawrence Berkeley Laboratory report LBL-14313, Berkeley, CA, p. 50.

15. Robinson, R. A. and R. H. Stokes, 1959. Electrolyte Solutions, Appendix 8.10, Butterworth, London, pp. 491-492.

16. Harned, H. S. and B. B. Owen, 1958. The Physical Chemistry of Electrolyte Solutions. Table 14-2-1A, Reinhold, N.Y., p. 748.

17. Edelstein, N. J. Bucher, R. Silva, and H. Nitsche, 1982. Thermodynamic Properties of Chemical Species in Nuclear Waste: Topical Report. ONWI/LBL-14325, Office of Nuclear Waste Isolation, Battelle Memorial Institute, Columbus, Ohio, p. 50.

18. Lundquist, R., 1982. 'Hydrophilic Complexes of the Actinides. I. Carbonates of Trivalent Americium and Europium', Acta Chem. Scand., A36, p. 741-750.

19. Khoo, K. H., T. K. Lim and C. Y. Chan, 1981. 'Activity Coefficients in Aqueous Mixtures of Hydrochloric Acid and Lanthanum Chloride at 25°C'. J. Sol. Chem., 10, pp. 683-691.

20. Pitzer, K. S., 1973. 'Thermodynamics of Electrolytes. I. Theoretical Basis and General Equations'. J. Phys. Chem., 77 (No. 2), pp. 268-277.

21. Pitzer, K. S., and G. Mayorga. 1973. 'Thermodynamics of Electrolytes. II. Activity and Osmotic Coefficients for Strong Electrolytes With One or Both Ions Univalent', J. Phys. Chem., 77 (No. 19), pp. 2300-2308.

22. Peiper, J. L. and K. S. Pitzer, 1982. 'Thermodynamics of Aqueous Carbonate Solutions Including Mixtures of Sodium Carbonate, Bicarbonate and Chloride', J. Chem. Thermodynamics, 14, pp. 613-638.

Part IV

Nuclear Studies, Environmental Studies, Separations, and Other Technological Aspects

HEAVY ION REACTIONS ON CURIUM TARGETS

Darleane C. Hoffman
Lawrence Berkeley Laboratory
University of California
Berkeley, California

Abstract: The availability of relatively large quantities of curium isotopes has made them available as targets for both the preparation of new elements (californium from curium-242 and nobelium from curium-246) and for heavy ion reactions to make new neutron-rich isotopes for study. In addition, curium-248 has been used as a target in a wide variety of attempts to make superheavy elements. The availability of such target materials is a tribute to Professor Glenn Seaborg's vision in establishing the Transplutonium Production Program in the early 1960's when he was Chairman of the U. S. Atomic Energy Commission. Some recent results of heavy ion experiments with curium-248 will be discussed briefly.

The availability of relatively large quantities of curium isotopes has made it possible to use them as target materials, both for the synthesis of new elements, for example, californium from ^{242}Cm (α,n) ^{245}Cf(44 m) in 1950 and nobelium from ^{246}Cm (^{12}C, 4n) ^{254}No (1 m) in 1958, and for heavy-ion reactions to make new neutron-rich isotopes and to study reaction mechanisms. More recently, ^{248}Cm has been widely used as a target for a variety of attempts[1,2] to make superheavy elements, and in other heavy element studies because of its high neutron-to-proton ratio. It is especially convenient to use because of its long half-life (3.4 x 10^5 years), and its availability in milligram amounts, nearly isotopically pure, by milking from ^{252}Cf. The availability of such target materials in the U.S. is a tribute to Professor Glenn Seaborg's vision in establishing the Transplutonium Production Program at the High Flux Isotope Reactor (HFIR) at Oak Ridge National Laboratory in the early 1960's when he was Chairman of the U.S. Atomic Energy Commission. In fact, as early as 1957 he had written a letter to then Chairman Lewis Strauss pointing out that research on the new transuranium elements in this country would be dependent on "substantial weighable quantities (say milligrams) of berkelium, californium and einsteinium". He further pointed out that this would require irradiation of substantial quanitities of ^{239}Pu and re-irradiation of the products in a proposed "very high flux reactor". And in fact, the HFIR began

N. M. Edelstein et al. (eds.), Americium and Curium Chemistry and Technology, 241–250.

operation in 1964, the Transplutonium Production Program was established, and milligram quantities of ^{252}Cf began to be produced and became available to researchers.

Because its neutron-to-proton ratio (1.58) is nearly the highest of any heavy element isotope feasible for use as a target, ^{248}Cm has been used in many attempts to make superheavy element (SHE) isotopes. In a recent attempt[1], we bombarded ^{248}Cm with ^{48}Ca ions at energies near the Coulomb barrier in order to try to produce SHE's with low excitation energies so they would not be completely lost by prompt fission. The resulting compound nucleus $^{296}_{180}116$ is near the island of stability which has been predicted to exist in the region of 114 protons and 184 neutrons. However, no evidence for SHE's with half-lives in the range from 1 microsecond to 10 years with cross sections larger than 10^{-34} to 10^{-35} cm^2 was found. (See Figure 1.)

Yields of isotopes of elements with atomic numbers both larger and smaller than that of the curium target were also measured[3,4] for reactions of ^{248}Cm with ^{40}Ca and ^{48}Ca ions at energies from near the Coulomb barriers to 60 to 80 MeV above the barriers. It is of interest to compare reactions of these two projectiles to assess the effect of the additional 8 neutrons in ^{48}Ca. A comparison of the mass-yield curves for these two systems for Bk, Cf, Es, and Fm is shown in Figure 2 for energies about 5 to 10 MeV above the Coulomb barrier. The maxima of the mass-yield distributions are only about 2 mass units larger for the ^{48}Ca than for the ^{40}Ca reactions; thus the 8-neutron difference in the projectiles is only partially reflected in the heavy product yields. This is in contrast to our results for ^{16}O and ^{18}O and ^{20}Ne and ^{22}Ne where the full two mass unit increase was observed in the mass-yield distributions for the heavier projectile of each pair. The maxima for these yield curves for reactions of ^{248}Cm with ^{48}Ca ($N/Z = 1.4$) occur at about the same mass numbers as for reactions with ^{18}O ($N/Z = 1.25$) and ^{22}Ne ($N/Z = 1.2$), and the half-widths for all the distributions are from 2 to 2.5 μ. The maximum cross sections for production of the Bk and Cf isotopes are about the same for ^{40}Ca and ^{48}Ca reactions, but the maximum yields for Es and Fm are about 5 times larger for ^{40}Ca than for ^{48}Ca reactions although the yields for the most neutron-rich isotopes are much higher for ^{48}Ca reactions. A comparison of the cross sections for actinide production from other neutron-rich projectiles including ^{136}Xe and ^{238}U indicates that cross sections for transfer of the same combinations of neutrons and protons from projectile to heavy actinide targets are about the same provided the reaction energy is not negative.

Excitation functions for these systems were also measured and are shown for the production of Es and Fm isotopes from ^{40}Ca and ^{48}Ca reactions in Figures 3 and 4, respectively. It is particularly interesting to note that the yields are near their maxima at the Coulomb barrier for the ^{40}Ca reactions, consistent with the calculated[5] positive excitation energies for these reactions. However, the observation that the yields decrease only slowly with increasing projectile energy is somewhat surprising because the fission barriers are only of the order of 5 to 6 MeV. Apparently only a small fraction of the projectile kinetic energy appears as excitation energy of the heavy product. Maxima in the mass-yield distributions for Es and Fm from ^{48}Ca shown in Figure 4 are observed at about 20 MeV above the Coulomb barrier, consistent with calculations showing that these reactions are endothermic by 10 to 15 MeV. Again, however, the cross sections decrease only slowly with increasing projectile energies up to 80 MeV above the barrier. Additional investigations are needed to try to ascertain how this excess energy is dissipated. In any case, these observations are encouraging for the production of new isotopes of the heaviest elements using similar "transfer" reactions in which the heavy product is formed with a low enough excitation energy that it is not entirely lost by prompt fission or particle emission.

The yields of some isotopes of elements lighter than the target were also measured and are shown in Figure 5 for $^{48}Ca + ^{248}Cm$ reactions. The cross sections are remarkably high with maxima of the order of 200 microbarns or more for elements down to Rn (Z = 86). The half-widths of these isotopic distributions are 4.5 to 5 mass units, considerably larger than those observed for the above target distributions. Yields were measured for only a few isotopes of Th, U, and Pu for the ^{40}Ca reactions, but the yields were much lower than for ^{48}Ca. A comparison of the integrated elemental yields for ^{40}Ca and ^{48}Ca are shown in Figure 6. As can be seen, the yields of the below target elements are lower by factors of 10 to 100 for the ^{40}Ca reactions. This may be a result of the tendency toward N/Z equilibration in which protons would tend to flow from the proton-rich ^{40}Ca to the neutron-rich (proton-deficient) ^{248}Cm. The data for the yields of Th, Ac, and Ra from the ^{48}Ca reactions imply the formation of the very neutron-rich Fe, Co, and Ni isotopes, ^{65}Fe, ^{68}Co, and ^{72}Ni, as complementary fragments (assuming emission of only one neutron from the light fragment) with similar cross sections of the order of 200 micobarns.

In summary, a variety of different projectiles and projectile energies have been used with ^{248}Cm targets to measure cross sections and excitation functions for elements both heavier and lighter than the target. Neutron-rich projectiles enhance the formation of neutron-rich heavy products. Cross sections for above target elements decrease rapidly with the number of nucleons transferred. Excitation functions indicate that the excess kinetic energy of the projectile does not go in to excitation energy of the target-like product and that "cold" heavy products are produced. In general, the maxima of the excitation functions are consistent with product excitation energies calculated assuming binary transfer reactions. The yields of below target elements for the $^{48}Ca + ^{248}Cm$ system imply formation of very neutron-rich complementary fragments of Fe, Co, and Ni.

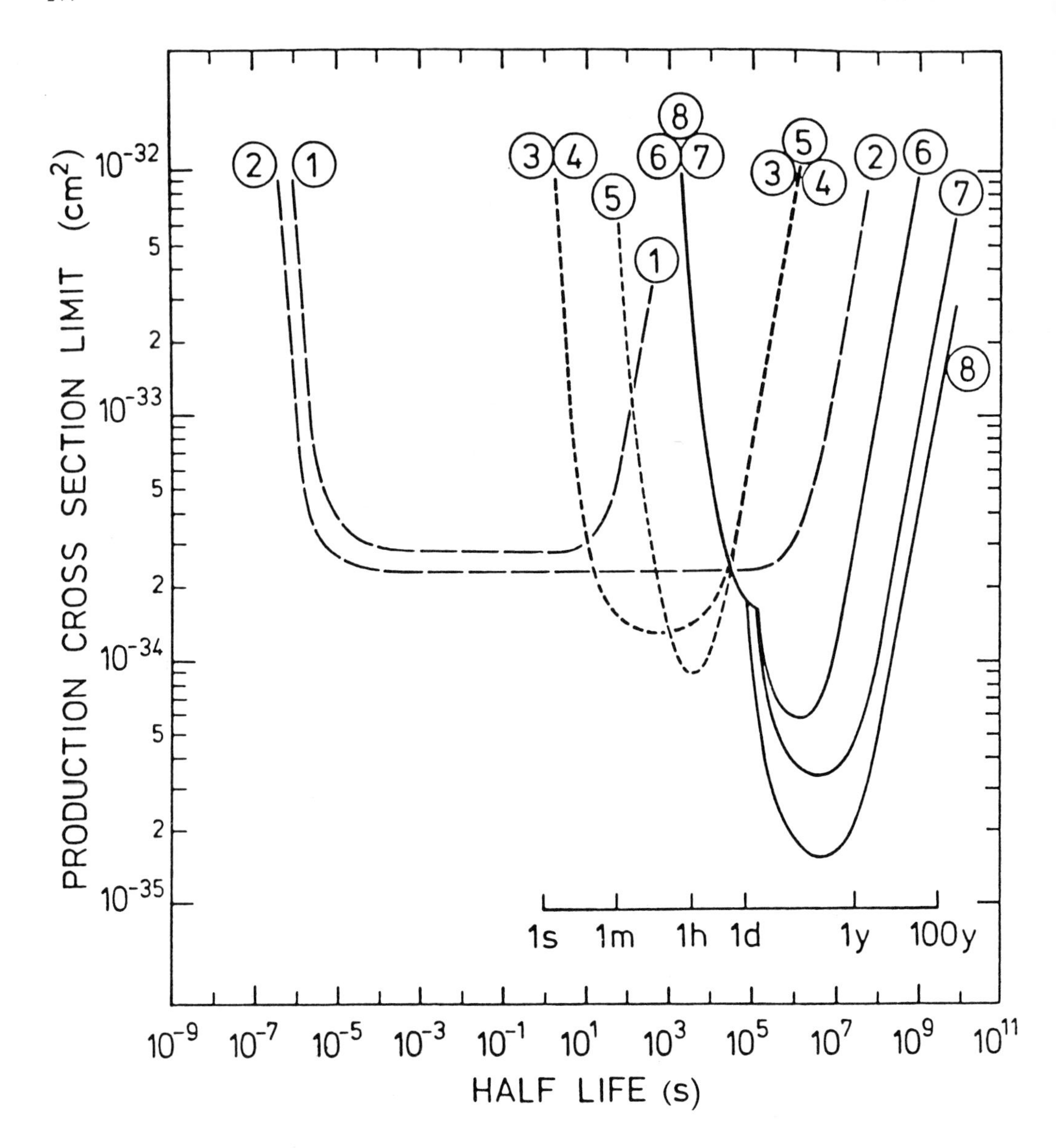

Figure 1. Upper limits (at 95% confidence level) for the production cross sections of superheavy nuclei in the reaction of ^{48}Ca with ^{248}Cm in the energy range 4.5 to 5.2 MeV/μ as a function of the half-life. The data result from experiments with recoil fragment separators (curves 1,2), with fast on-line chemical separations (3,4,5), and off-line chemistry (6,7,8). (Taken from Ref. 1.)

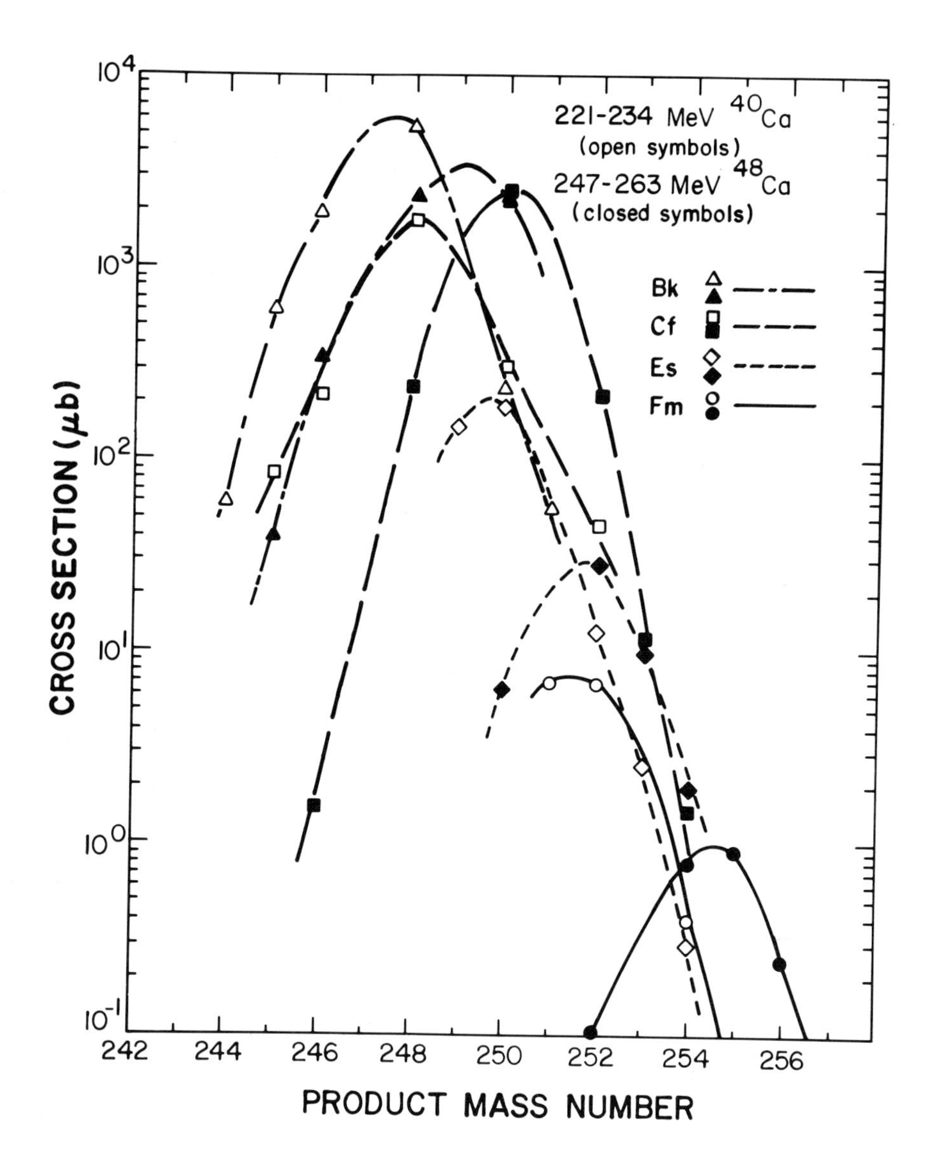

Figure 2. Comparison of heavy actinide yields from ^{40}Ca and ^{48}Ca reactions with ^{248}Cm. Projectile energies in the target are from 5 to 10% above the Coulomb barrier. (Taken from Ref. 3.)

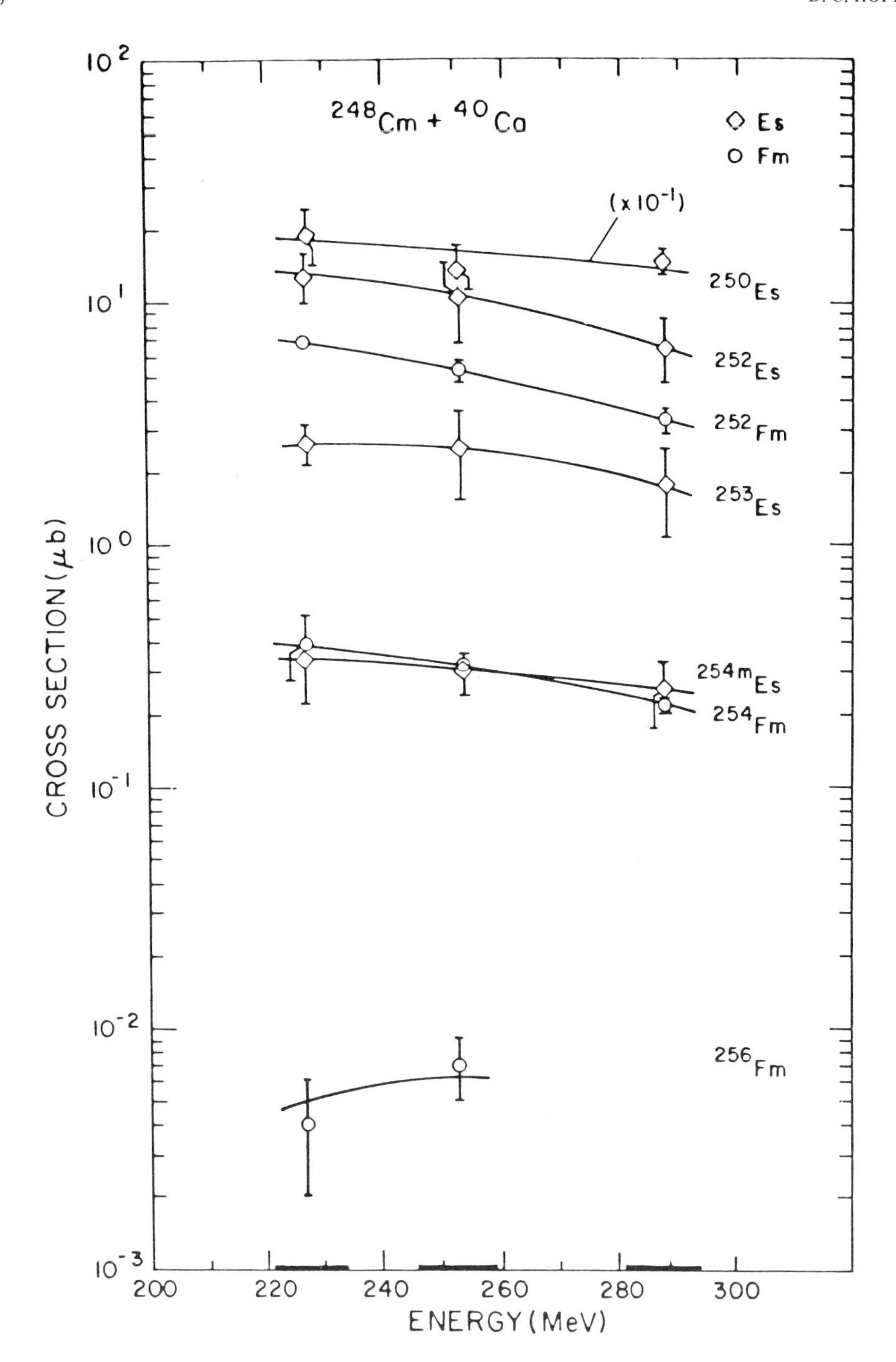

Figure 3. Excitation functions for some Es and Fm isotopes produced in the bombardment of ^{248}Cm with ^{40}Ca. (Taken from Ref. 3.)

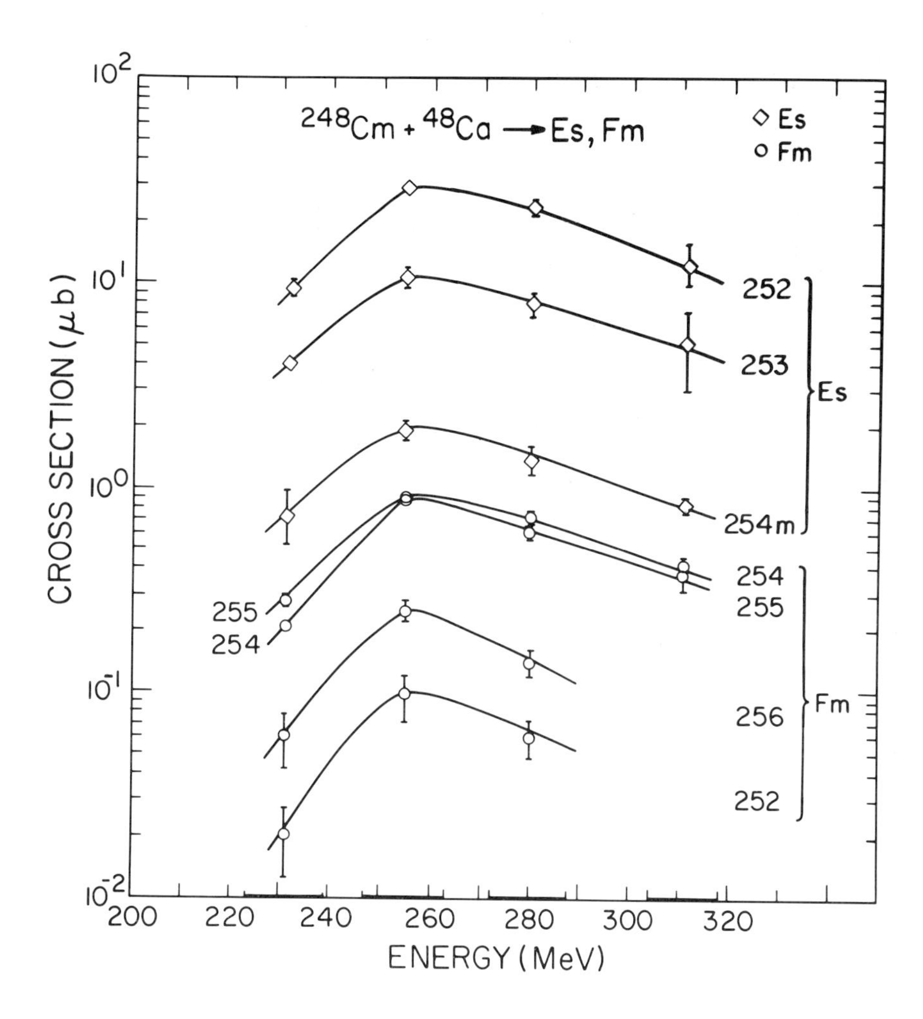

Figure 4. Excitation functions for some Es and Fm isotopes produced in the bombardment of ^{248}Cm with ^{48}Ca. (Taken from Ref. 3.)

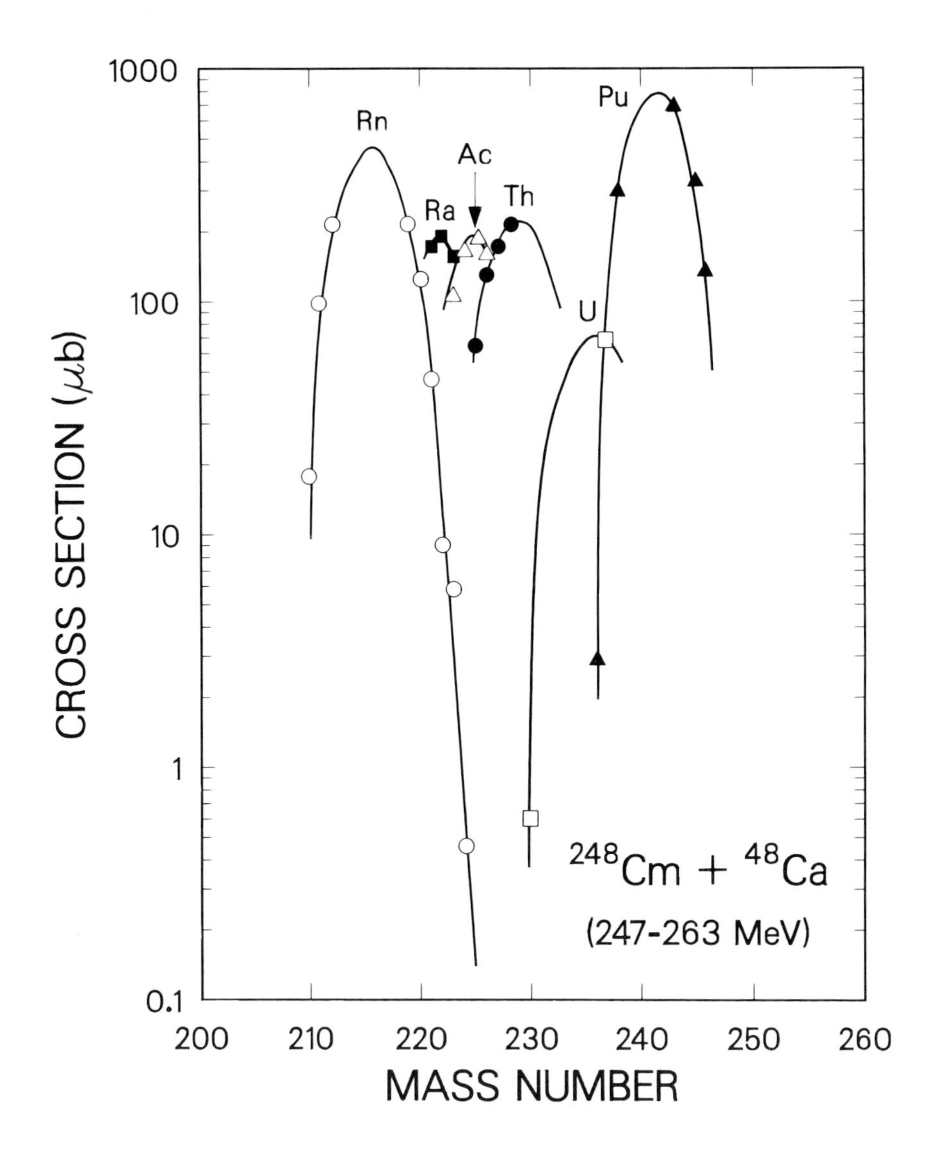

Figure 5. Mass-yield distributions for some lighter than target elements from the reaction of ^{48}Ca with ^{248}Cm. (Data from Ref. 4.)

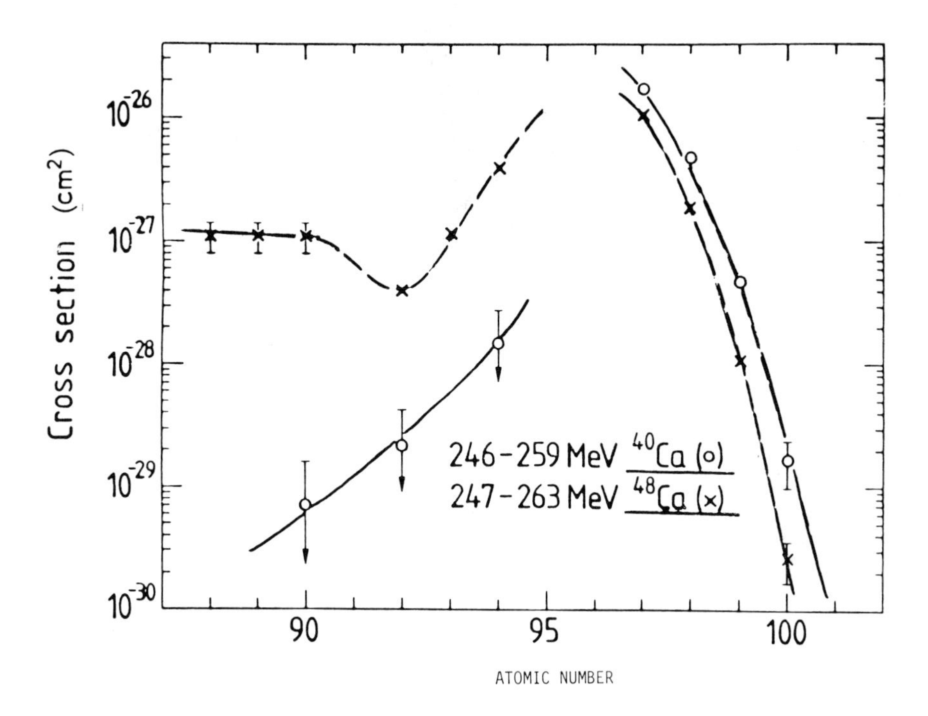

Figure 6. Comparison of elemental yields for ^{40}Ca and ^{48}Ca reactions with ^{248}Cm. (Data from Ref. 4.)

REFERENCES

1. P. Armbruster et al., Phys. Rev. Letters (submitted) 1984.
2. E. K. Hulet et al., Phys. Rev. Lett. 39, 385 (1977); J. V. Illige et al., Phys. Lett. 78B, 209 (1978); R. J. Otto et al., J. Inorg. Nucl. Chem. 40, 589 (1978); Yu. Ts. Oganessian et al., Nucl. Phys. A291, 213 (1978).
3. D. C. Hoffman et al., Phys. Rev. C (submitted) 1984.
4. H. Gäggeler et al., in preparation, 1984.
5. D. C. Hoffman and M. M. Hoffman, Los Alamos National Laboratory Report LA-UR-82-824, rev., March 1984.
6. D. Lee et al., Phys. Rev. C 25, 286 (1982).
7. D. Lee et al., Phys. Rev. C 27, 2656 (1983).

This work is supported by the Director, Office of Energy Research, Division of Nuclear Physics of the Office of High Energy and Nuclear Physics of the U.S. Department of Energy under Contract No. DE-AC03-76SF00098.

Synthesis of Transuranium Nuclides from Interaction of ^{16}O with ^{238}U

N.Shinohara, S.Ichikawa, S.Usuda, T.Suzuki, H.Okashita, T.Sekine, K.Hata
Japan Atomic Energy Research Institute,
Tokai-mura, Naka-gun, Ibaraki-ken, 319-11 Japan

T.Horiguchi, Y.Yoshizawa
Hiroshima University,
1-1-89, Higashi-Senda-cho, Naka-ku, Hiroshima-ken, 730 Japan

S.Shibata
Institute of Nuclear Study, University of Tokyo,
3-2-1, Midori-cho, Tana-shi, Tokyo, 188 Japan

I.Fujiwara
Institute of Atomic Energy, Kyoto University,
Gokasho, Uji-shi, Kyoto-fu, 611 Japan

Many studies of actinides, especially of transuranium elements, have been performed mainly owing to scientists' endeavors to discover new elements and nuclides.(1) In spite of these efforts, the mechanism of the formation of transuranium nuclides by heavy-ion bombardment and the nuclear and chemical properties of the synthesized actinide nuclides have not been clearly elucidated, and a great deal remains to be examined in more detail. In this context, we have begun the investigation of heavy-ion nuclear reactions(2): A system in which a uranium target is irradiated by oxygen ions (^{16}O +^{238}U), was chosen. For this reaction seventeen nuclides (^{250}Fm, $^{244,245,246}Cf$, ^{242}Cm, $^{238,239}Np$, $^{237,239}U$, $^{225,226,227,228}Th$, $^{224,225}Ac$, and $^{223,224}Ra$) were identified and the condition of formation for these nuclides was clarified. This paper deals with a novel method of mutual rapid separation of the synthesized transuranium elements by ion exchange and a discussion of the mechanism which governs the heavy-ion nuclear reactions.

Experimental

The uranium targets, 99.98 % ^{238}U, were prepared by electrodeposition from an isopropyl alcohol solution(3) onto an aluminum backing of 7-µm thickness. The targets varied in thickness

N. M. Edelstein et al. (eds.), Americium and Curium Chemistry and Technology, 251–260.

from 0.3 to 2.0 mg/cm^2 of uranium. A target assembly consists of aluminum foils for degrading the beam energy at the upstream side of the target and an aluminum foil (7 μm thickness) for catching the recoiling nuclei at the down stream side.

Irradiation by ^{16}O ions was performed by the JAERI Tandem accelerator. The beam energy was varied from 85 to 130 MeV and the beam intensity was up to 150 pnA. Typical irradiation durations were 0.5 to 3 hours.

After irradiation, both the target and the aluminum catcher foil or the catcher foil only were dissolved in aqua regia and the solution was heated to near dryness. The sample was then redissolved in a nitric acid and methyl alcohol mixture solution, and a rapid anion-exchange procedure for separation of transuranium elements was carried out using a high performance liquid chromatographic technique. Figure 1 shows the typical elution curves for the separation of transplutonium elements and rare-earths by anion exchange using an acid-alcohol mixture as eluant. Chemical separation of target-like nuclides was also carried out by ion exchange. After separation, the fractions were mounted on tantalum discs and dried to make the sources for radioactivity measurements. Time required for the separation is about 20 minutes including the source preparation.

Alpha- and gamma-activities of the recoiling nuclei, which were collected by the aluminum catcher foil, were measured with Si-surface-barrier and Ge(Li) detectors immediately after irradiation. The same measurements were performed for the chemically separated samples. Measurements were continued for about two months after irradiation.

Results and Discussion

Fermium

Fermium nuclides expected to be synthesized in ^{16}O + ^{238}U reactions are ^{248}Fm ($T_{1/2}$ = 36 sec.; E_{alpha} = 7.87 MeV), ^{249}Fm ($T_{1/2}$ = 2.6 min.; E_{alpha} = 7.53 MeV), and ^{250}Fm ($T_{1/2}$ = 30 min.; E_{alpha} = 7.43 MeV).(4,5) After bombardment for 30 minutes, the transplutonium fraction was separated from the irradiated sample by the ion-exchange method and subjected to alpha spectrometry (Fig. 2). The chemical behavior, the alpha energy, and the half-life observed indicate formation of the ^{250}Fm nuclides.

To study the conditions for the formation of ^{250}Fm, the dependence of the quantity of ^{250}Fm produced on projectile energy was investigated. The excitation function of ^{250}Fm thus obtained is given in Fig. 3. The ^{250}Fm nuclides are synthesized between about 84 MeV and 102 MeV of ^{16}O ion energy on the target (lab. system) and the formation cross section becomes maximum (about 2 x 10^{-30} cm^2) when the projectile energy is around 92 MeV.

Other fermium nuclides, $^{248,249}Fm$, could not be detected, but the upper limit of the formation cross section of these nuclides was determined to be 7 x 10^{-32} cm^2 for 100-MeV ^{16}O ions.

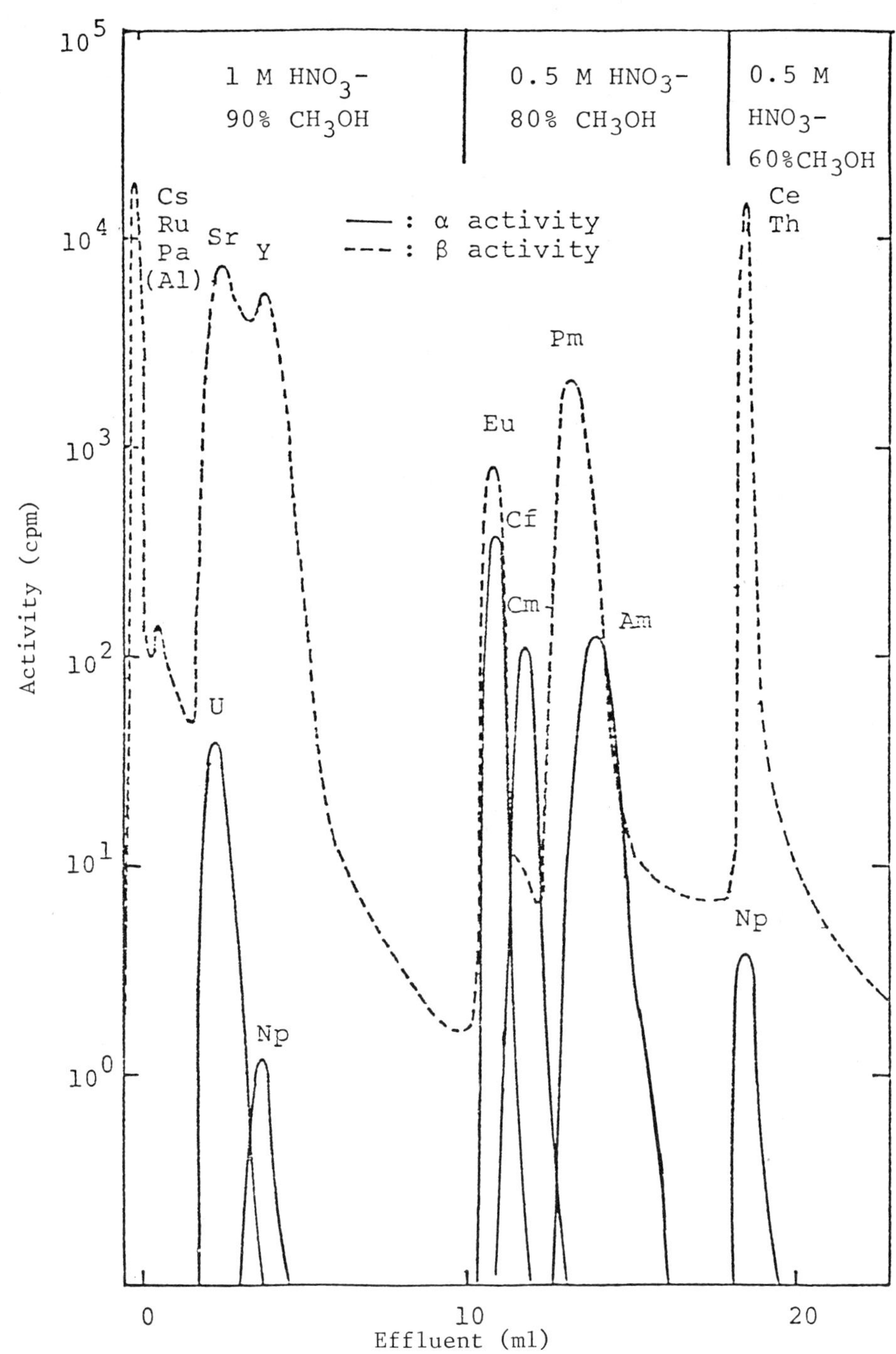

Fig. 1 Elution curves for the separation of transplutonium elements and rare-earths by anion-exchange using acid-alcohol mixtures as eluants.

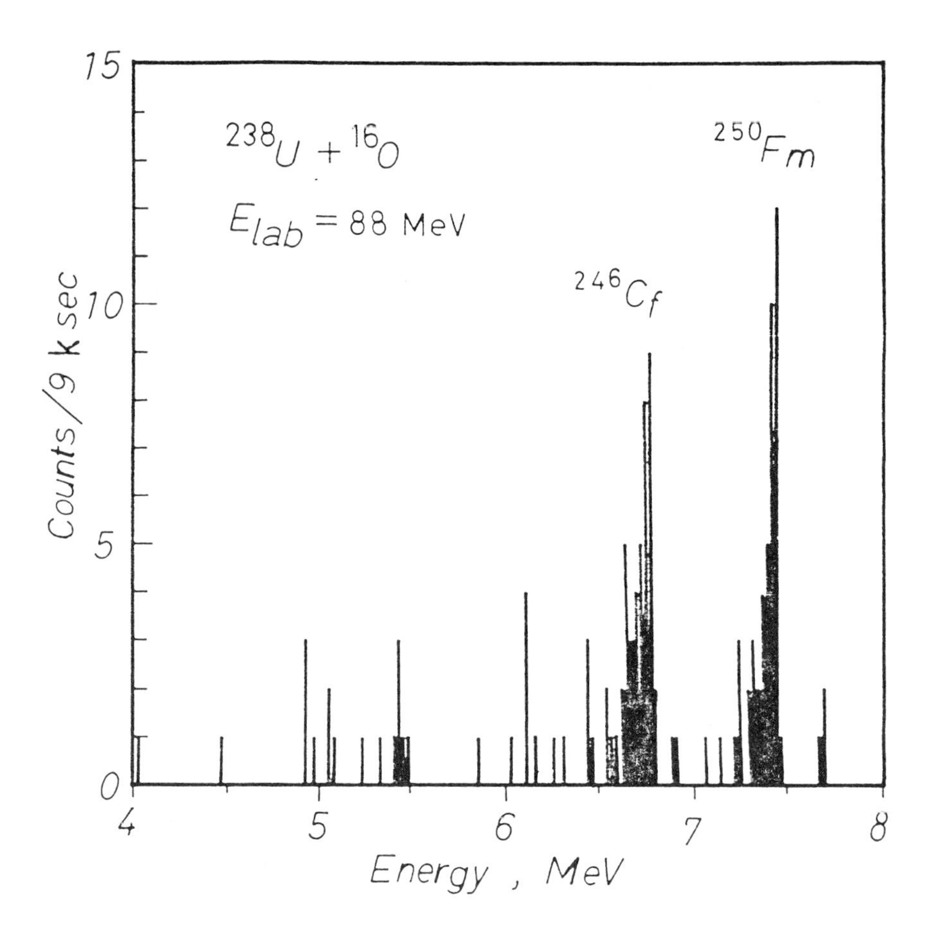

Fig. 2 Alpha-spectrum of the transplutonium fraction after the ion-exchange separation.

Californium and curium

By measuring alpha spectra of the aluminum catcher foils for long periods after irradiation and examining chemical behavior and half-lives, the formation of ^{246}Cf and ^{242}Cm was confirmed. The excitation functions are also shown in Fig. 3.

The formation of other californium nuclides, short-lived $^{244,245}Cf$, was also observed a short time after the bombardment. The alpha spectrum of the californium fraction just after the separation is given in Fig. 4.

Neptunium and uranium

The formation of the target-like nuclides emitting gamma-rays was investigated. After irradiation, the gamma-spectrum of the aluminum catcher foil was measured with a Ge(Li) detector and the chemical behavior of the gamma-emitter was studied. The formation of the nuclides $^{238,239}Np$ and $^{237,239}U$ was confirmed. The formation cross sections of these nuclides are also depicted in Fig. 3. Other target-like nuclides, e.g., $^{240,241}U$, could not be recognized despite prudent experiments.

Thorium, actinium and radium

When the projectile energy increased to 110-130 MeV, formation of the ^{250}Fm, $^{244-246}Cf$ and ^{242}Cm nuclides could not be observed and instead production of thorium, actinium and radium nuclides was detected. After measuring the alpha-activities and the half lives and examining the chemical characteristics, various nuclides such as $^{225,226,227,228}Th$, $^{224,225}Ac$ and $^{223,224}Ra$ were identified. Figure 5 gives the formation cross sections of these nuclides as a function of the number of neutrons and protons picked up from the target nucleus.

Formation mechanism of actinides

From the view of synthesizing actinides, the reactions are classified roughly into two processes: (i) complete fusion process and (ii) nucleon transfer process, dependent on the angular momentum carried into the system.(6) An example of the first mechanism is the reaction, $^{238}U(^{12}C,4n)^{246}Cf$ (7), and of the latter is $^{27}Al(^{16}O,^{15}N)^{28}Si$ (8).

(i) *complete fusion process (CF)*; this implies that two nuclei are fused in the bombardment to form one compound nucleus. Most of the compound nuclei decay by fission but a very small fraction of the nuclei undergoes nucleon evaporation reactions. The excitation functions of $^{238}U(^{16}O,4n)^{250}Fm$, $^{238}U(^{16}O,\alpha 4\text{-}6n)^{244-246}Cf$ and $^{238}U(^{16}O,2\alpha 4n)^{242}Cm$ reactions can be calculated by ALICE code assuming the particle evaporation from the compound nucleus.(9) The observed formation cross sections of ^{250}Fm have a nearly same feature as the calculated values by the ALICE code. The ^{250}Fm could be formed by evaporation of four neutrons, following the formation of the compound nucleus ^{254}Fm.

(ii) *nucleon transfer process*; nucleon transfer process can be classified into the following three mechanisms: quasi-elastic transfer (QET), deep-inelastic transfer (DIT), and massive transfer (MT). QET

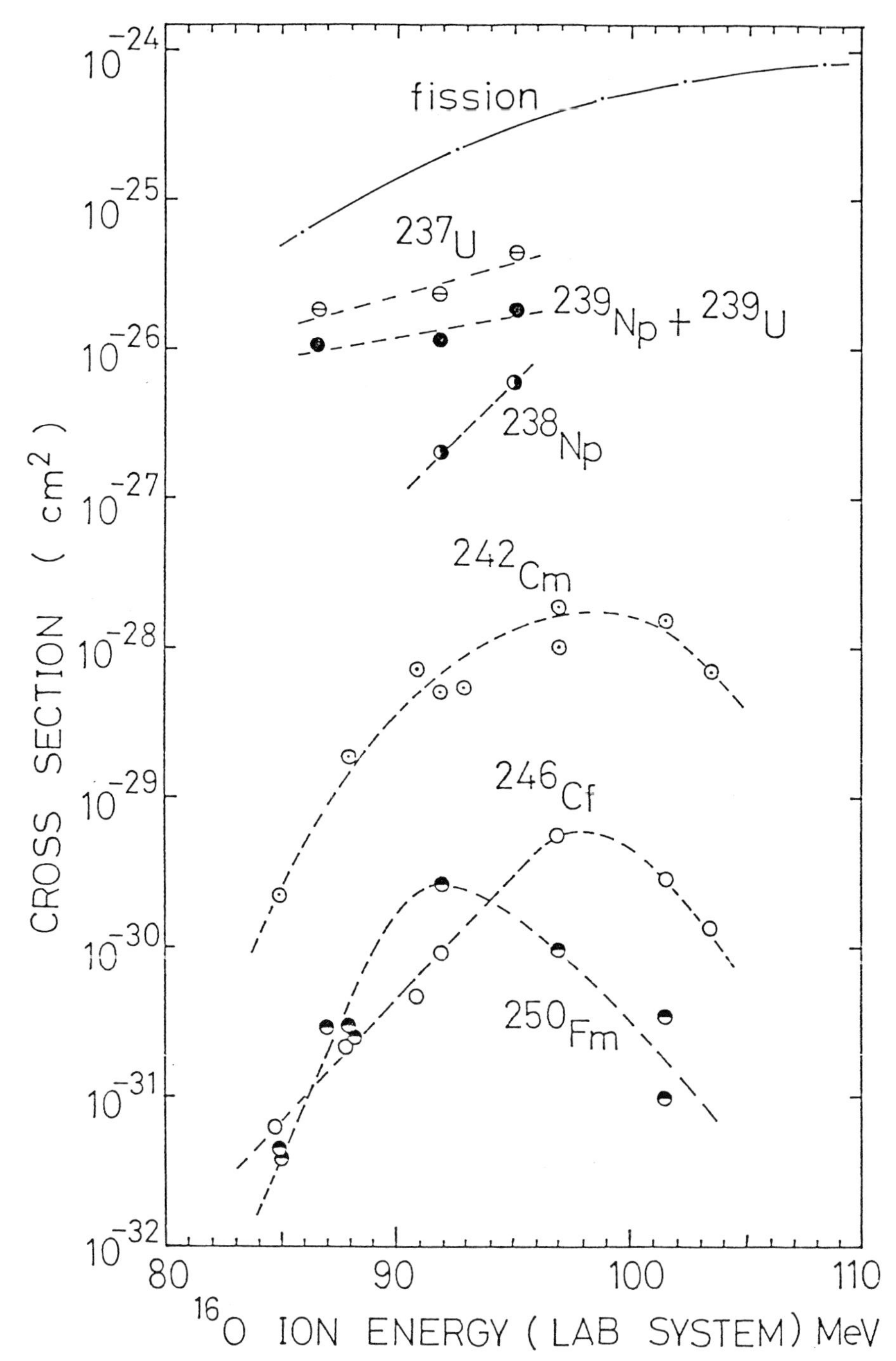

Fig. 3 Excitation functions of actinides synthesized in the ^{16}O + ^{238}U reaction. The fission data quoted are from V.E.Viola,Jr. et al. [Phys. Rev.,1962,128,767].

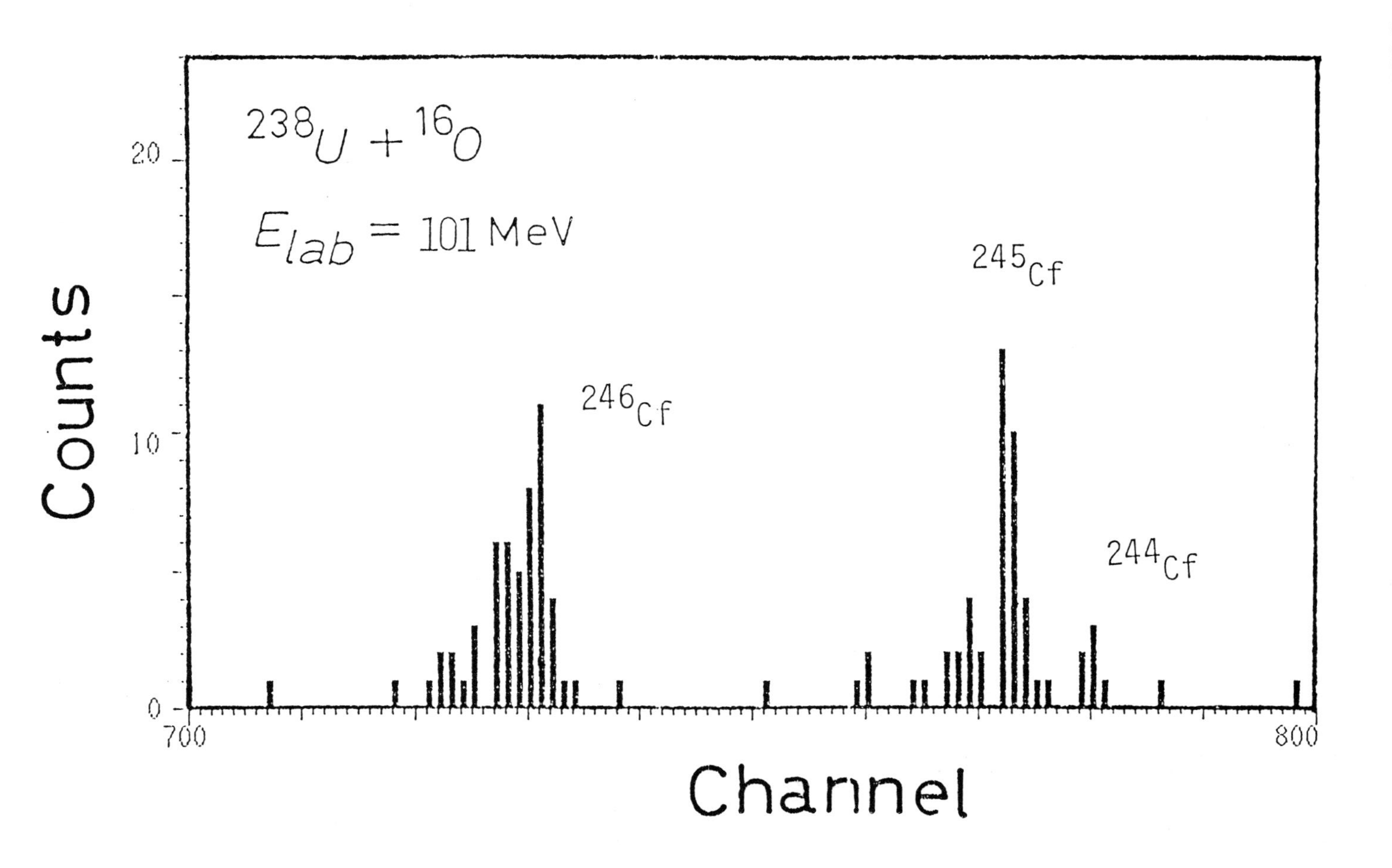

Fig. 4 Alpha-spectrum of the californium fraction after the ion-exchange separation.

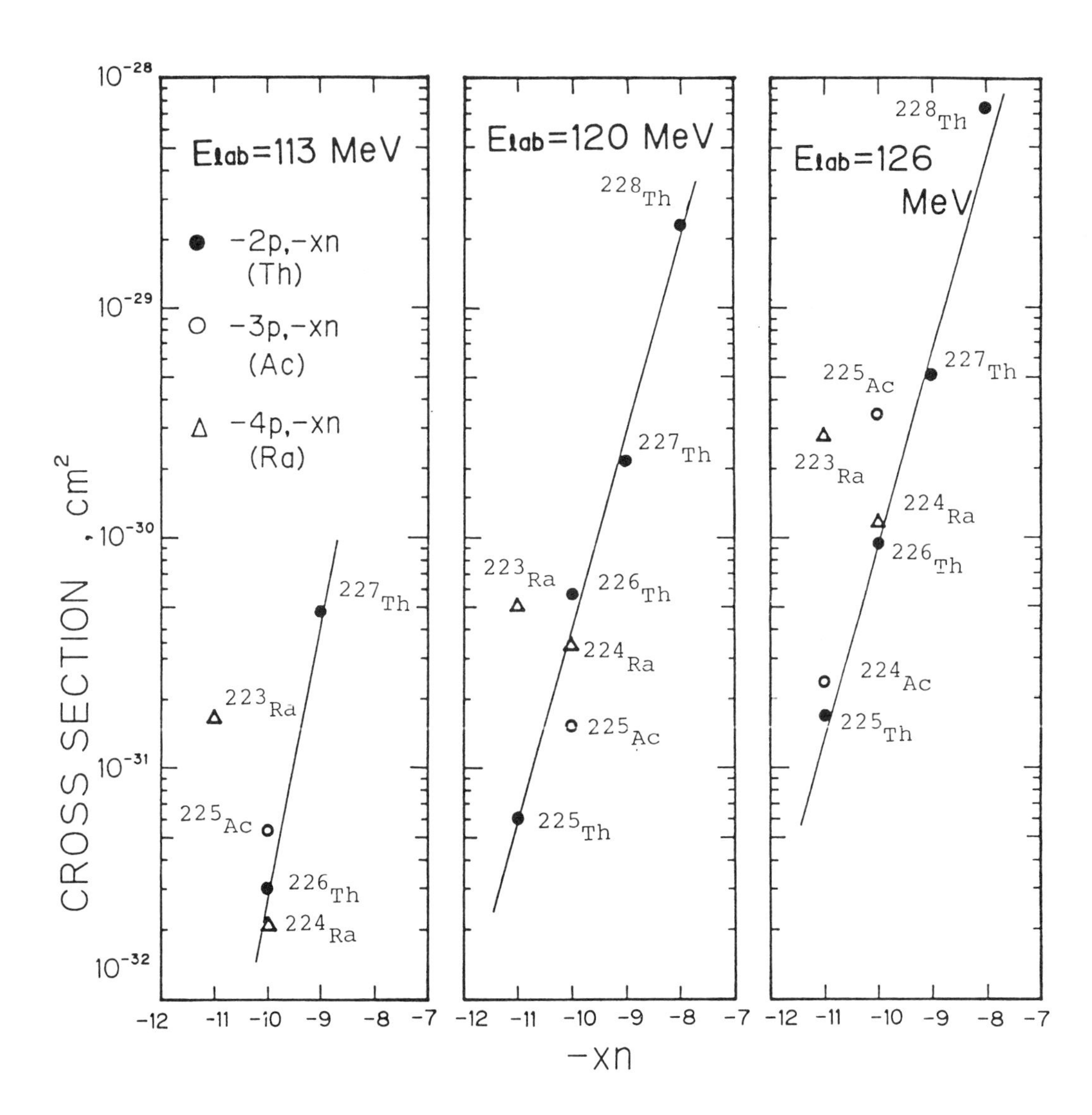

Fig. 5 Formation cross sections of the thorium, actinium and radium nuclides synthesized in the $^{16}O + ^{238}U$ reaction. The abscissa indicates the number of the picked up neutrons and the ordinate shows the measured cross sections.

takes place at the outermost shell of nuclei and DIT is the reaction in which the area of collision is larger than in the case of QET. MT means that the target nucleus and a part of the projectile are fused together. The obtained excitation functions of ^{246}Cf and ^{242}Cm differ apparently from the calculated ones by the ALICE code. The nuclides of $^{244,245,246}Cf$ and ^{242}Cm would be formed by the transfer reaction, probably QET or DIT, instead of the presumed particle evaporation reactions.

In the synthesized target-like nuclides, the ^{238}Np could be formed by neutron-proton exchange reaction between target and projectile nucleus, and the ^{239}Np or ^{239}U by a one proton or one neutron transfer reaction from the projectile to the target. The nuclide ^{237}U could be formed by a one neutron pick-up reaction, in which the projectile picks one neutron up from the target nucleus.

The nuclides of thorium, actinium and radium seem to be formed by picking up 2 to 4 protons and 8 to 11 neutrons from the ^{238}U target (Fig. 5); He, Li, or Be fragments of various masses could be transfered to the projectile. This reaction corresponds to either a multinucleon pick-up reaction or MT, and the Q_{gg} systematics (10) may be used to describe the reaction mechanism.

Conclusion

At least seventeen actinide nuclides were synthesized by ^{16}O + ^{238}U reactions and their formation conditions were investigated. We can classify those nuclides into four categories: the first is the reaction product ^{250}Fm formed by the complete fusion process; the second, the products $^{244-246}Cf$ and ^{242}Cm formed by quasi-elastic transfer or deep-inelastic transfer processes; the third, target-like nuclides $^{238,239}Np$ and $^{237,239}U$ formed by one-nucleon transfer processes; and the last, products such as $^{225,226,227,228}Th$, $^{224,225}Ac$ and $^{223,224}Ra$ formed by multinucleon pick-up or massive transfer processes. Because most of the radioactivity measurements were limited to alpha-spectrometry only, other nuclides may have been synthesized, which were not detected, and other mechanisms than those mentioned above could have occurred.

The authors are most greatful to the operating staff of the JAERI Tandem Accelerator for making this research possible by their considerable help in many ways.

References

1) Ghiorso,A.:"Actinides in Perspective", Proceedings of the Actinides 1981 Conference, edited by Edelstein, N.M., Pergamon Press, 1982, 23-56.
2) Shinohara,N. et al.: "JAERI TANDEM Annual Report 1982", JAERI-M 83-095 ,1983, 53-55.
3) Kono,N. et al.: private communication ,1982.

4) Lederer, C.M. and Shirley, V.S., Eds. : Table of Isotopes, 7th Ed., J.Wiley & Sons, New York ,1978.
5) Akap'ev,G.N. et al.: Atomnaya Energiya 1966,21,243.
6) Zucker,A. and Toth,K.S.: "Nuclear Chemistry", edited by Yaffe, L. Academic Press ,1968, 409-452.
7) Gerlet,Y.B. et al.: Zh. Eksperim. i Teor. Fiz. 1957, 33, 339.
8) Newman,E. et al.: Phys. Rev. 1963, 119, 1720.
9) Blann,M.: Report COO-3494-29, 1976.
10) Artukh.A.G. et al.: Nucl. Phys. 1973, A211, 299.

A STUDY OF THE PRODUCTION OF TRANSURANIUM ELEMENTS AND ITS APPLICATION TO THE SOLUTION CHEMISTRY IN TOHOKU UNIVERSITY

I. Satoh, T. Mitsugashira, M. Hara, M. Kishimoto, and S. Suzuki
The Research Institute for Iron, Steel and Other Metals,
Tohoku University
Katahira 2-1-1, Sendai 980
Japan

ABSTRACT. About 2.8 g of UO_2 were irradiated with neutrons(about 1.6x 10^{20} nvt) at the Japan Material Testing Reactor(JMTR). After 140 days cooling, the irradiated sample was processed chemically. The isotopic ratios of U and Pu were determined by α-ray spectrometry and mass spectrometry. A small amount of ^{241}Am(<100 μg) was also irradiated in JMTR. Am and Cm were purified mainly by ion-exchange chromatography. We have tried to estimate the irradiation conditions in JMTR. The thermal neutron flux(ϕ) and the epithermal ratio (the ratio of the epithermal to the pure thermal neutron density)(f) were chosen as adjustable variables for the calculation of the isotopic ratios of U, Pu, Am, and Cm. Then, the most reasonable values of ϕ and f were obtained which explained the values of the isotopic ratios obtained experimentally. Short-lived Bk tracers, $^{243-246}Bk$, were produced by the $^{241}Am + \alpha$ and $^{243}Am + \alpha$ reactions using the A.V.F. cyclotron at the Cyclotron and Radioisotope Center, Tohoku University. A compressed aluminium powder disk, in which Am was suspended, was used as the target. Bk tracers were isolated from Am and fission products by HDEHP or TTA extraction of Bk(IV). Stability constants(β_1 and β_2) of the chloro and bromo complexes of trivalent Sm, Eu, Gd, Tb, Ac, Am, Cm, Bk, and Cf were determined by a solvent extraction method using HDEHP with an ionic strength and acidity of 3.0 M and 0.15 M, respectively. Stability constants could be roughly estimated by assuming inner sphere complex formation.

INTRODUCTION

We have two methods to produce the transuranium elements. One is by the multiple neutron capture reaction in a nuclear reactor and the other is by nuclear reactions in an accelerator. For the application of the multiple neutron capture reaction, it is very effective to use a high flux reactor. In 1970, the Japan Atomic Energy Research Institute began irradiation service in the JMTR at the Oarai Reseach Establishment. We constructed the chemical processing equipment(1) in the 6th hot-cell of the Hot Laboratory of the Oarai Laboratory for Irradiation Experiment of the Research Institute for Iron, Steel and Other Metals, Tohoku Univer-

N. M. Edelstein et al. (eds.), Americium and Curium Chemistry and Technology, 261–273.

sity. First, the irradiated UO_2 was processed chemically, and then the irradiated AmO_2 was processed.

The cyclotron at the Cyclotron and Radioisotope Center, Tohoku University accelerates proton to 40 MeV and ^4He to 50 MeV. The Bk tracers were produced by bombardment of the AmO_2 target with ^4He.

The stability constants of the chloro and bromo complexes of several actinoid and lanthanoid elements, which were available or produced by the nuclear reactions, were determined.

MATERIALS AND METHODS

Neutron irradiated uranium

Uranium dioxide pellets, which contained about 2.8 g of natural uranium dioxide, were irradiated during three irradiation cycles of the JMTR at the N-10 irradiation hole in the Be reflector zone. The irradiated U was processed chemically about 140 days after the irradiation to reduce the radioactivity of fission products. The flow sheet of the separation scheme is shown in Figure 1. The pellets were dissolved in a 6 M HNO_3 solution and the Pu was adjusted to the tetravalent state by the addi-

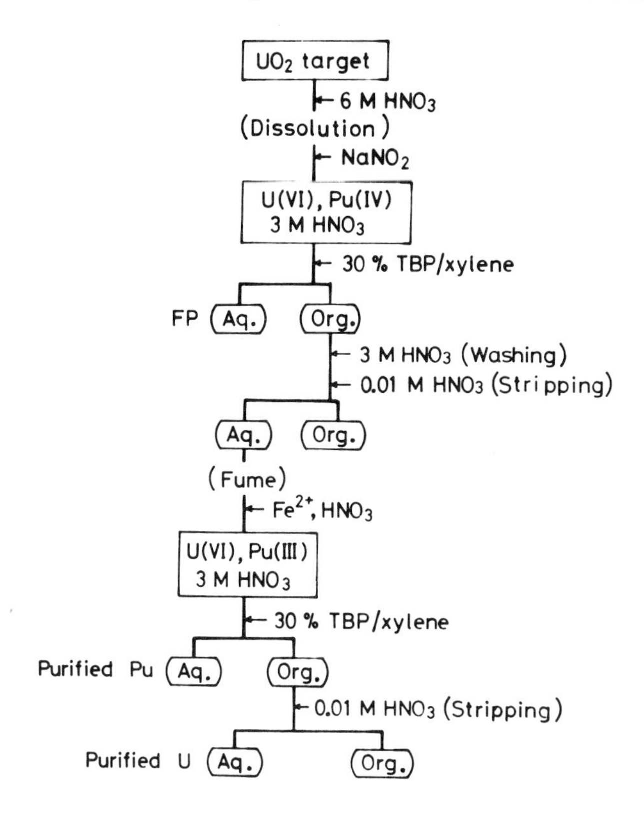

Figure 1. Separation scheme of neutron irradiated UO_2

tion of an equal volume of diluted $NaNO_2$ solution. Then, U(VI), and Pu(IV) were extracted into a 30 vol% TBP/xylene solution. The separation of U and Pu was carried out with the same system after the Pu was reduced to the trivalent state and remained in the aqueous phase.

The chemical yields of U, Pu, and some fission products were determined by α- and γ-ray spectrometry to examine the separation method. A Si surface-barrier detector and a Ge(Li) detector were applied to the the measurement of α-ray and γ-ray spectra, respectively. A 4096-channel PHA was used for the accumulation of the spectra.

A mass spectrometer(Atlas CH-4) was used for the determination of the isotopic composition of U and Pu. An aliquot of the purified soluions of U and Pu, which contained about 10 μg of U and 1 μg of Pu, was placed onto rhenium ribbon filaments and evaporated to dryness. Ionization cases of the double-filament type were used throughout the scanning of mass spectra. For the detection of weak ion beams, a secondary electron multiplier tube was used.

Neutron irradiated americium

$^{241}AmO_2$ was irradiated in the reflector area of the JMTR intermittently for 384 day.(2) The authors obtained a part of it from the Japan Atomic Energy Research Institute. The flow sheet of the separation procedure for the transuranium elements is shown in Figure 2. The chemical separations were carried out by remote control via ion-exchange methods in the 6th hot-cell. After the dissolution of the irradiated sample, the transuranium elements were separated from most of fission products and cladding materials on anion exchange resin using a 9.9 M LiCl-0.1 M HCl solution. Then the actinoids were eluted with 6 M HCl at 87 °C. The column was 80 mm in length and 8 mm in diameter; the flow rate was 0.25-0.35 $ml/cm^2/min$.

The transuranium elements, in the HCl solution containing a small amunt of LiCl, were coprecipitated with $Fe(OH)_3$. The iron and actinoids were separated by the anion exchange method. Subsequently, the transuranium elements were separated by the following two methods.

Method 1: cation exchange separation with 2-hydroxyisobutyric acid (2-HIB) was performed. An HCl solution of transuranium elements was dried and several drops of 0.4 M 2-HIB(pH 4) were added. This solution was placed and adsorbed on the top of the resin (maintained at 87 °C by the vapor of trichloroethylene) and eluted slowly. Suitable aliquots of the effluent were taken in polyethylene test tubes. Each fraction was analysed radiochemically. The exchange resin column was 117 mm in length and 4 mm in diameter; the flow rate was 0.3-0.6 $ml/cm^2/min$.

Method 2: anion exchange separation with a HNO_3-methanol solution was performed. The anion-exchange resin column was 70 mm in length and 4 mm in diameter; the flow rate was 0.25-0.35 $ml/cm^2/min$. One ml of a 1.0 M HNO_3-90 % methanol solution of the transuranium elements, in which ^{152}Eu and ^{137}Cs were added, was absorbed on the resin and eluted with a solution of 0.5 M HNO_3-methanol. The concentration of methanol was decreased from 90 to 0 % in steps according to the elution of the elements.

After the chemical separation, isotopic ratios of Cm and Am were determined by analysing their α- and γ-ray spectra obtained by the use

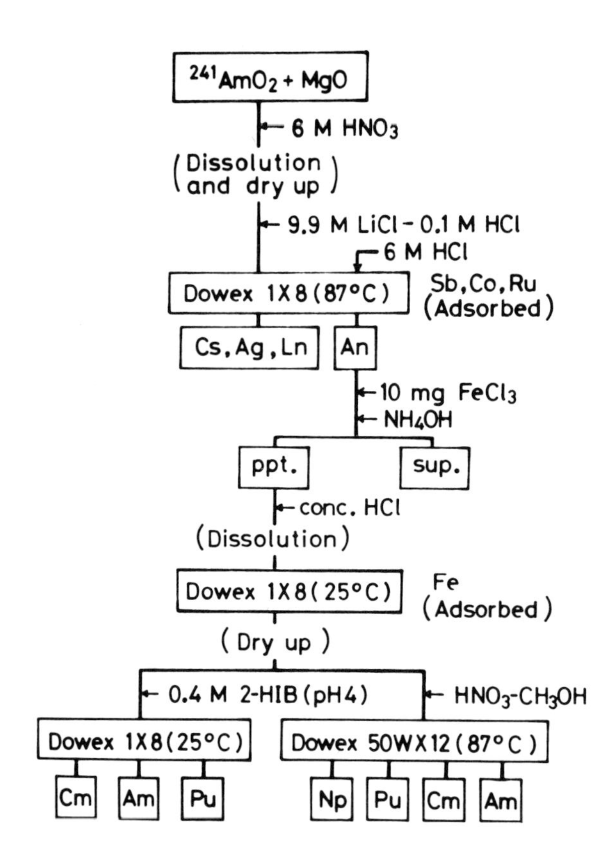

Figure 2. Separation scheme of neutron irradiated AmO_2
Ln: Lanthanoid elements, An: Actinoid elements

of Si- and Ge(Li)-detectors, respectively. The ^{242m}Am was quantified by analysing the growth curve of ^{238}Np, the α-decay product of ^{242m}Am. The measured isotopic ratios were compared with the values calculated from the irradiation history in JMTR to consider the irradiation of ^{241}Am.

Production of Bk tracers

The $^{243-246}Bk$, which were produced by the ^{241}Am + α or ^{243}Am + α reaction, were isolated rapidly from the Am target by the solvent extraction method shown in Figure 3. Bk was oxidized to the tetravalent state and extracted into a TTA/xylene solution(3) or HDEHP/heptane solution.(4) The γ-ray spectra of the Bk fraction were measured using an intrinsic Ge detector and a 4096-channel PHA.

Stability constants

The stability constants of the chloro and bromo complexes of trivalent Sm, Eu, Gd, Tb, Ac, Am, Cm, Bk, and Cf were investigated. The tracers of these elements were obtained as follows: ^{152}Eu, ^{153}Gd, ^{160}Tb, and ^{252}Cf

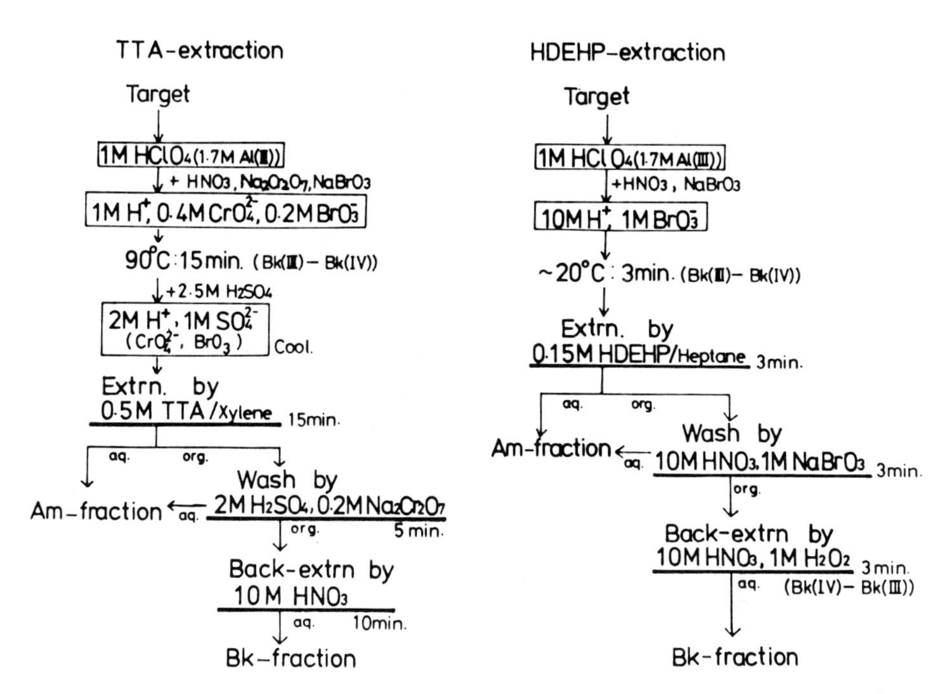

Figure 3. Separation scheme of Bk

were obtained from the Radiochemical Centre, Amersham, and ^{241}Am from Oak Ridge National Laboratory. ^{225}Ac was isolated from ^{229}Th which was obtained from ^{233}U. Cm and Bk tracers were prepared as described above. ^{145}Sm was produced by means of the Nd + α reaction and was purified by successive extractions with HDEHP. These tracers were prepared in 3.0 M $HClO_4$ solutions and stored in Pyrex vessels.

A 6 ml sample solution was prepared by mixing appropriate amounts of stock solution, 3.0 M LiCl or LiBr solution, and 3.0 M $LiClO_4$ solutions in a separatory funnel. The sample solution was equilibrated with an equal volume of an HDEHP/octane solution pre-equilibrated with a solution of the same composition as the sample solution. To determine the relation between the distribution ratio and ligand concentration, the acidity was fixed at 0.15 M and the concentration of HDEHP was controlled so that the distribution ratio in the absence of the ligand was to be 1.5-1.8. Extraction equilibrium was reached by shaking for 20 min at about 20 °C. From each phase, a 2 ml portion was placed in a polyethylene test tube and the radioactivity was counted. For bromide solutions, values were corrected for self-absorption of γ-rays using absorption data preliminarily determined.

RESULTS AND DISCUSSION

Neutron irradiated uranium

^{144}Ce, ^{106}Ru, $^{134,137}Cs$, and ^{95}Zr-^{95}Nb were observed in the γ-ray spectrum of irradiated U. All these nuclides were effectively removed from the Pu-U fraction. The chemical yield of Pu in the Pu-U fraction was greater than 90 %. The isotopic composition of U and Pu were determined by α-ray spectrometry and mass spectrometry, and are shown in TABLE I.

TABLE I. The isotopic ratios of U and Pu

Isotopic ratio	Measured (x100)	Calculated* (x100)
$^{235}U/^{238}U$	0.657 ± 0.010	0.657
$^{236}U/^{235}U$	1.86 ± 0.23	1.650
$^{240}Pu/^{239}Pu$	2.846 ± 0.045	2.873
$^{241}Pu/^{240}Pu$	3.29 ± 0.20	3.270

* $\phi=3.17 \times 10^{13}$ n/cm^2/sec and f=0.06

The yields and the isotopic ratios of actinoids were calculated with a computer program(5) adjusting two parameters, ϕ and f. We also took into account the fluctuation of the neutron flux during the irradiation assuming that the flux is proportional to the reactor power. As is shown in TABLE I, the calculated isotopic ratios agreed well with the measured values. Thus we could conclude that f=0.06 and $\phi=3.17 \times 10^{13}$ n/cm^2/sec at the N-10 irradiation hole when the reactor power is 30 MWt. As we did not consider the absorption of neutrons by the target, the actual f and ϕ will be 1.7-2.0 times greater than the fitted values. The maximum power of JMTR is 50 MWt, so we can say that the neutron flux of JMTR is high enough to apply JMTR to the production of transuranium elements through the irradiation of ^{241}Am and ^{243}Am.

Neutron irradiated americium

The α- and γ-ray spectra of the irradiated sample showed that it contained the transuranium elements ^{238}Pu, $^{241-243}$Am, and $^{242-244}$Cm, the fission products ^{134,137}Cs, ^{144}Ce, ^{106}Ru, and ^{125}Sb, and the cladding materials ^{110m}Ag and ^{60}Co. Via the anion exchange separation procedure with a LiCl solution, Cs, Ce, Sb, Co, and most of Ag and Ru were separated from the transuranium elements.

In the case of the cation exchange separation procedure with 2-HIB, the retention time of transuranium elements depends greatly on the pH value of the 2-HIB. The most appropriate pH value was found to be 4.0. Figure 4 shows the elution curve of the transuranium elements Np, Pu, Am, and Cm using 0.4 M 2-HIB of pH 3.98. Although Cm cannot be perfectly separated from Am, the separation factor of Cm/Am was found to be about 1.4. A similar value has been reported.(6)

The result of the anion exchange separation with the HNO_3-methanol solution is shown in Figure 5. The two peaks are attributed to Cm and Am. Cs was not adsorbed and eluted in the first peak; Eu eluted before Cm and Am. The ^{239}Np($t_{1/2}$=2.35 d), which is the daughter of ^{243}Am, was eluted while ^{243}Am remained on the resin but its elution behavior was not clearly known. In comparison with the cation exchange method with

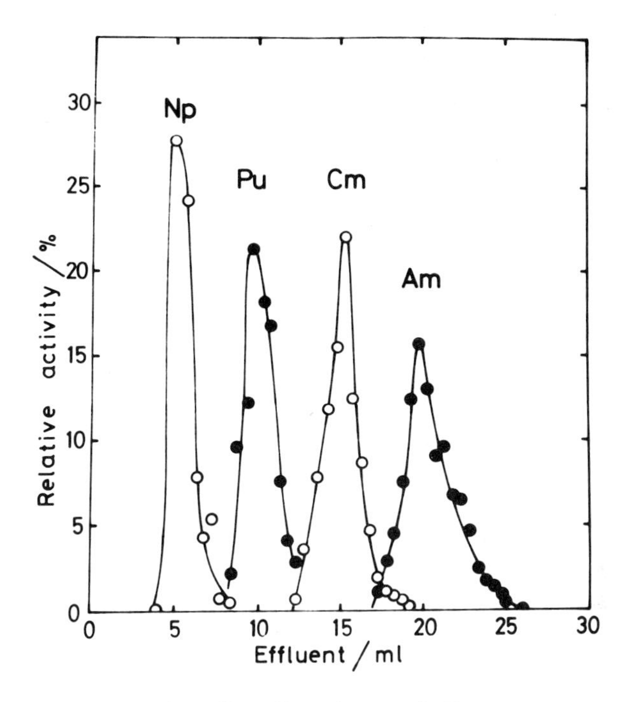

Figure 4. Elution curves for Np, Pu, Am, and Cm

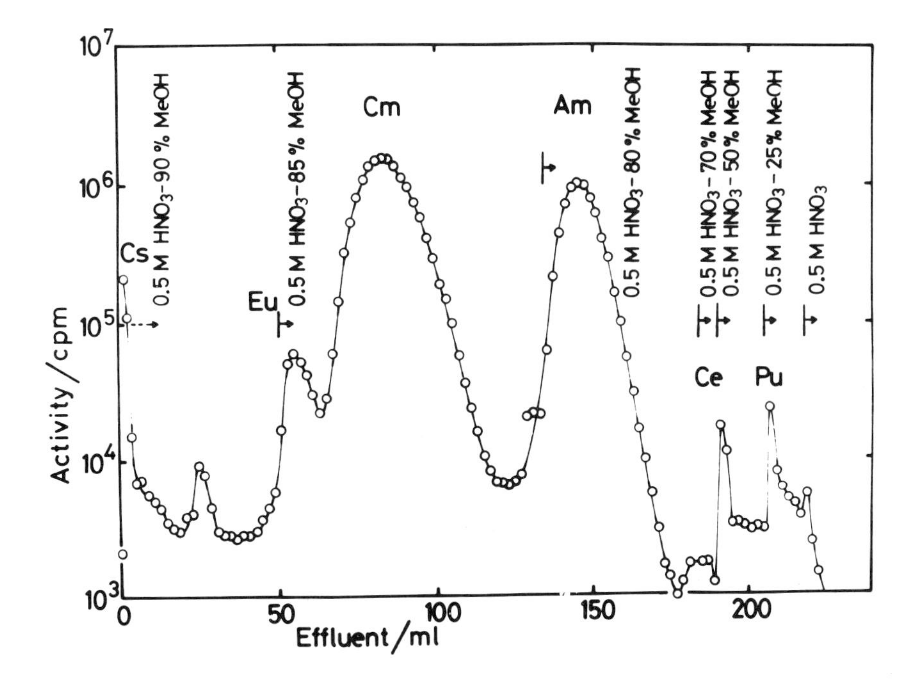

Figure 5. Elution curve of HNO_3-CH_3OH system

2-HIB, the anion exchange method with HNO_3-methanol is superior because of the large separation factor of Cm/Am reported to be about 3.(7) The anion exchange method has the additional merits that the separation can be carried out at room temperature and salt-free samples can be prepared easily for the α-ray measurements. But on the other hand, the latter is inferior to the former with respect to the low flow rate of eluent and the longer time needed for the separation. This disadvantage can be overcome by use of high pressure liquid chromatography.

The isotopic ratios of Cm and Am were determined by the α-ray and γ-ray spectrometry and the results are shown in TABLE II. The quantity of ^{242m}Am was determined by analysing the growth of ^{238}Np which was the daughter of ^{242m}Am.(8) The yields of the nuclides of Cm and Am at the time of separation were computed by a computer program.(5) For the calculation, ϕ and f were chosen as adjustable variables. Values of $\phi=(2.15 \pm 0.20)\times10^{14}$ n/cm^2/sec and f=0.084 $\pm$ 0.026 were found to reproduce the isotopic ratios measured within an error of 2σ.(8,9) The isotopic ratios, which were calculated using the values of $\phi=2.1\times10^{14}$ n/cm^2/sec and f=0.09, are shown in TABLE II.

TABLE II. Isotopic ratio of Cm and Am at the time of separation

Isotopic ratio	Measured	Calculated*
$^{242m}Am/^{241}Am$	0.0116 $\pm$ 0.0017	0.0112
$^{243}Am/^{241}Am$	0.245 $\pm$ 0.026	0.240
$^{243}Cm/^{242}Cm$	16.7 $\pm$ 2.5	16.5
$^{244}Cm/^{242}Cm$	17.6 $\pm$ 2.7	19.65

* $\phi=2.1\times10^{14}$ n/cm^2/sec and f=0.09

Production of Bk tracers

The two extraction methods give good results with respect to the removal of Am and Al for the recovery of Bk. However, some fission products as well as $^{141,143}Ce$ could not be removed. ^{105}Ru, ^{99}Mo, and ^{99m}Tc were the main impurities in the Bk fraction obtained by the HDEHP extraction. On the other hand, ^{97}Zr, ^{97}Nb, and ^{99m}Tc were the main impurities in the TTA extraction. In spite of the contamination of the fission products, the KX-rays of Cm were isolated well enough from the γ-rays of the fission products that measurement of $^{243,244}Bk$ was not disturbed when a Ge detector was applied to the measurement of the spectrum. Therefore, when $^{244}Bk(t_{1/2}=4.35$ h) and $^{243}Bk(t_{1/2}=4.5$ h) are used as radiochemical tracers, the HDEHP extraction method is more suitable than the TTA extraction because the former is more rapid.

A thick target yield was calculated assuming a homogenous mixture of 0.385 wt% of Am in Al. The result is shown in TABLE III. For ^{243}Bk and ^{244}Bk, the emission rate of γ-rays had never been determined. So, we temporarily assumed that the disintegration rates of ^{243}Bk and ^{244}Bk were equal to the emission rate of Cm KX-rays. For the preparation of Bk tracers, a ^{241}Am target seems to be more suitable than ^{243}Am

target because the former gives a higher yield of ^{243}Bk and ^{244}Bk. The radioactivity of the main product, ^{245}Bk, using the latter target is not high enough for application to radiochemical experiments.

TABLE III. Thick target yield of Bk

Nuclear reaction	Energy of α (MeV)	Yield (KBq/μAh)
$^{241}Am(\alpha,n)^{244}Bk$	28.2	0.53
$^{241}Am(\alpha,n)^{244}Bk$	44.7	2.1
$^{241}Am(\alpha,2n)^{243}Bk$	28.2	0.55
$^{241}Am(\alpha,2n)^{243}Bk$	44.7	2.7
$^{243}Am(\alpha,n)^{246}Bk$	48.2	0.07
$^{243}Am(\alpha,2n)^{245}Bk$	48.2	0.15
$^{243}Am(\alpha,3n)^{244}Bk$	48.2	2.0
$^{243}Am(\alpha,4n)^{243}Bk$	48.2	0.3

Stability constants

The distribution ratios were calculated from the solvent extraction data by an equation:

$$D = [M]_{org}/[M]_{aq}, \quad (1)$$

where $[M]_{org}$ and $[M]_{aq}$ mean the metal concentrations in the organic and aqueous phases, respectively. In the presence of a complexing anion, L^- (L=Cl or Br), in the aqueous phase, the complex formation reaction of a tripositive cation proceeds as follows,

$$M^{3+} + xL^- ===== ML_x^{3-x}, (x=1, 2, \cdots) \quad (2)$$

and the stability constant is given as $\beta_x=[ML_x^{3-x}]/[M^{3+}][L^-]^x$. Therefore, the distribution ratio can be expressed as follows:

$$D = [M]_{org}/\{[M^{3+}] + [ML^{2+}] + [ML_2^+] + \cdots\}$$

$$= D_o/\{1 + \beta_1[L^-] + \beta_2[L^-]^2 + \cdots\}. \quad (3)$$

The following equation can be derived:

$$(D_o/D) - 1 = \beta_1[L^-] + \beta_2[L^-]^2 + \cdots, \quad (4)$$

where D_o is the distribution ratio in the absence of the ligand.

The relation between D and $[L^-]$ are shown in Figure 6 and Figure 7. The values of β_1 and β_2, shown in TABLE IV, were obtained via equation 4 by least-mean-square fitting(10) to the values of $(D_o/D)-1$ vs $[L^-]$ calculated from the data in Figure 6 and Figure 7.

For the actinoids(III), β_1 of the chloro complexes are 0.4-0.6 and obviously larger than β_1 of the bromo complexes. For the lanthanoids-

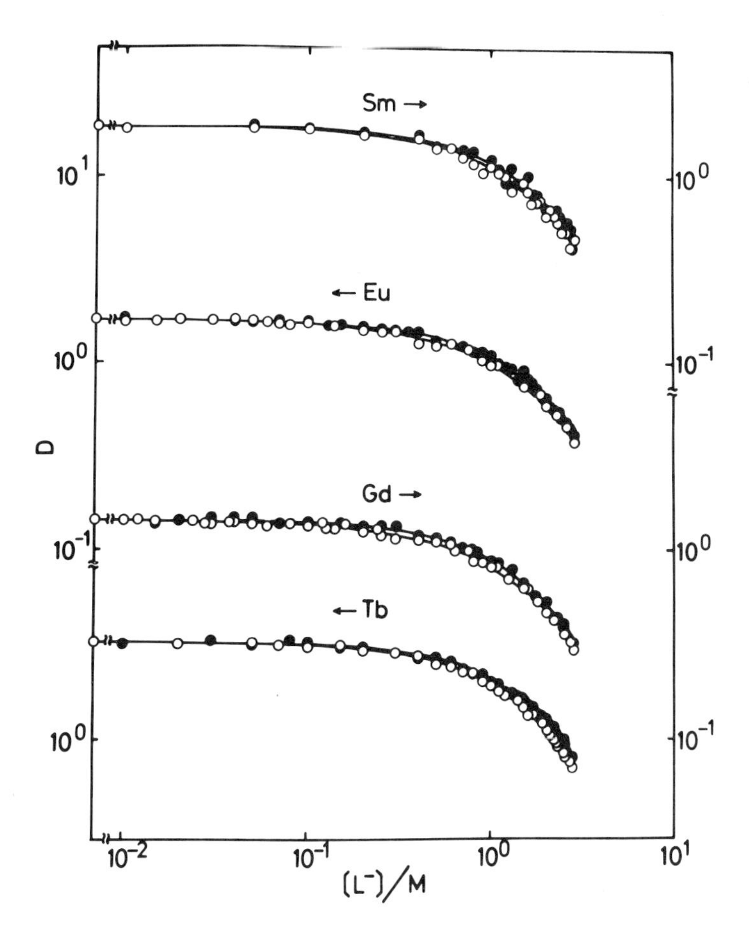

Figure 6. Dependence of D of some lanthanoids on chloride ion(O) and bromide ion(●)

(III), β_1 of the chloro complexes are also 0.4-0.6 and slightly larger than β_1 of the bromo complexes. But for β_2, no significant difference between chloro and bromo complexes were observed.

Yamada and Tanaka(11) proposed that the complex formation reaction proceeds by the Eigen mechanism(12) and the apparent formation constant K is expressed as follows:

$$K = K_{os}(1 + K_{is}), \quad (5)$$

where K_{os} is an outer sphere formation constant and K_{is} is a transformation constant of an inner sphere complex. Fuoss(13) proposed an equation which has been applied frequently to the calculation of K_{os}:

$$K_{os} = (NV/1000)\exp(-u/kT), \quad (6)$$

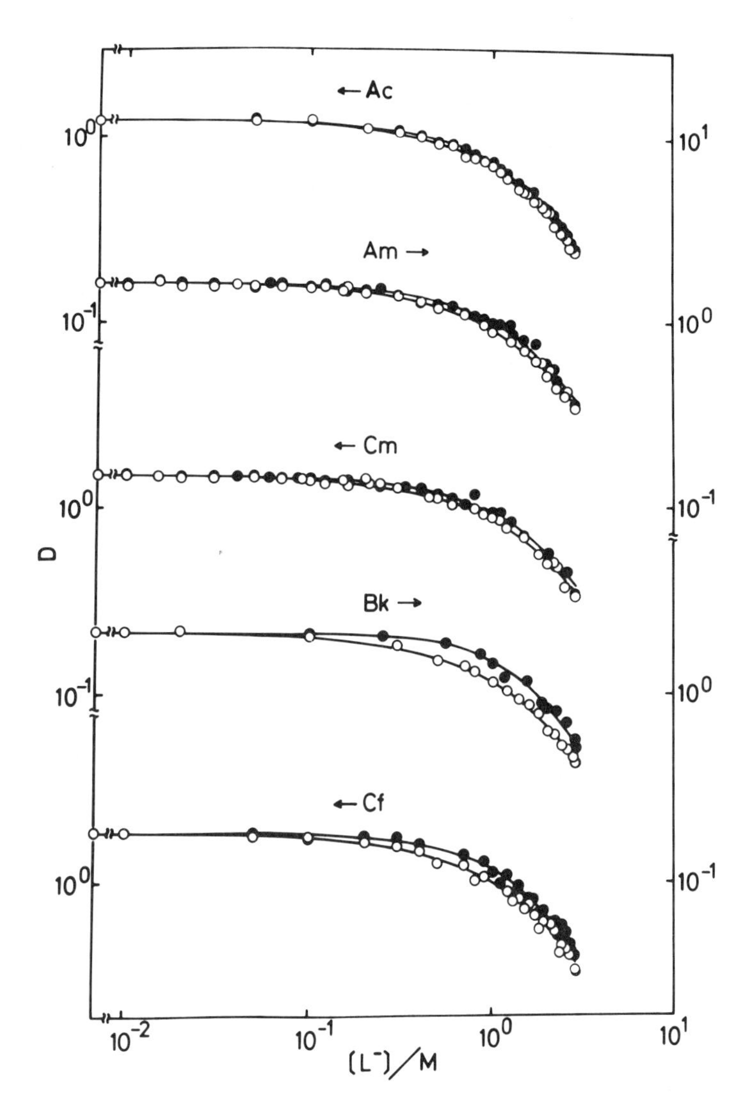

Figure 7. Dependence of D of some actinoids on chloride ion(O) and bromide ion(●)

where V is the exclusive volume of a cation and u is the electrostatic energy; N, k, and T mean Avogadoro constant, Boltzmann constant, and absolute temperature, respectively.

This equation gives calculated formation constants for the bromo complexes of actinoids(III) and lanthanoids(III) of 0.34-0.35 in good agreement with the experimental values of β_1. On the other hand for the chloro complexes of actinoids(III), the experimental values of β_1 are obviously higher than the values calculated, i.e. K_{os}=0.35-0.36. According to equation 5, this implies that an inner sphere mechanism can not be neglected for the formation of actinoid(III) chloro complexes.

TABLE IV. Stability constants of chloro and bromo complexes

Anion	Metal	β_1	β_2	K_{os}
	Sm	0.41 ± 0.04	0.25 ± 0.03	0.36
	Eu	0.52 ± 0.02	0.22 ± 0.02	0.35
	Gd	0.56 ± 0.02	0.21 ± 0.02	0.35
	Tb	0.45 ± 0.02	0.26 ± 0.02	0.35
Cl^-	Ac	0.44 ± 0.02	0.31 ± 0.02	0.38
	Am	0.55 ± 0.03	0.22 ± 0.02	0.36
	Cm	0.56 ± 0.03	0.20 ± 0.02	0.36
	Bk	0.59 ± 0.02	0.25 ± 0.02	0.36
	Cf	0.61 ± 0.04	0.25 ± 0.03	0.36
	Sm	0.33 ± 0.04	0.24 ± 0.03	0.34
	Eu	0.38 ± 0.02	0.23 ± 0.01	0.34
	Gd	0.37 ± 0.02	0.26 ± 0.02	0.34
	Tb	0.41 ± 0.03	0.22 ± 0.01	0.34
Br^-	Ac	0.42 ± 0.02	0.29 ± 0.01	0.37
	Am	0.30 ± 0.03	0.28 ± 0.02	0.35
	Cm	0.39 ± 0.02	0.22 ± 0.02	0.35
	Bk	0.15 ± 0.04	0.29 ± 0.03	0.35
	Cf	0.30 ± 0.04	0.30 ± 0.03	0.34

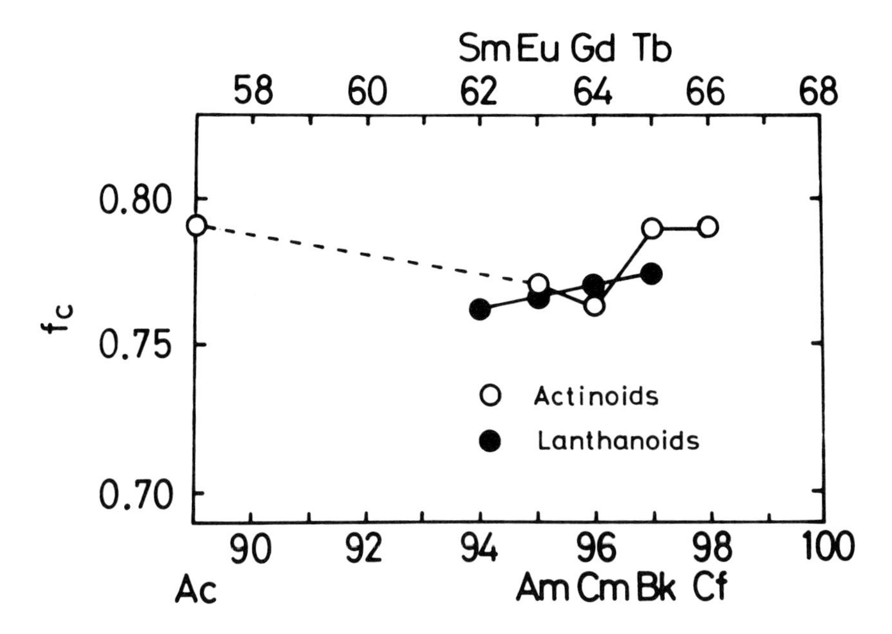

Figure 8. Variation of f_c with atomic number

In order to examine the difference in the complexing ability between actinoids(III) and lanthanoids(III), we calculated the fraction of complex, i.e. $f_c=f_1+f_2$, (f_x is a mole fraction of the species of ML_x^{3-x} in the aqueous phase) for a $[Cl^-]=2.85$ M. The results are shown in Figure 8; f_c shows a minimum at Cm, which may be due to the half-filled shell effect of the 5f electrons, while for lanthanoids(III), f_c increases continuously with atomic number. However, the large difference in the complex formation between the actinoids(III) and lanthanoids(III), which should be observed in the anionic complex formation and provides the experimental basis for the group separation, can not be observed in β_1 and β_2 values experimentally determined in the aqueous system of ionic stength 3.0 M and $[Cl^-] \leq 2.85$ M.

ACKNOWLEDGMENTS

We wish to thank Dr. K. Ueno of the Japan Atomic Energy Research Institute for providing the irradiated ^{241}Am. A part of the expenses for this investigation has been defrayed by a Grant-in-aid for Scientific Research No.505002 from the Ministry of Education, Science and Culture, Japan.

LITERATURES CITED

1. Suzuki, S.; Mitsugashira, T.; Hara, M.; Sato, A.; Kawasuji, I.; Shiokawa, Y.; Kishimoto, M.; Yamamura, Y.; Sci. Rep. RITU, 1979, **A-28**, 64.
2. Ueno, K.; Watanabe, K.; Sagawa, C.; Ishimori, T.; J. Nucl. Sci. Technol., 1975, **12**, 356.
3. Moore, F. L.; Anal. Chem., 1966, **38**, 1872.
4. Peppard, D. F.; Moline, S. W.; Mason, G. W.; J. Inorg. Nucl. Chem., 1957, **4**, 344.
5. Hara, M.; Mitsugashira, T.; Sato, A.; Suzuki, S.; Sci. Rep. RITU, 1979, **A-28**, 41.
6. Choppin, G. R.; Harvey, B. G.; Thompson, S. G.; J. Inorg. Nucl. Chem., 1956, **2**, 66.
7. Guseva, L. I.; Lebedev, I. A.; Myasoedov, B. F.; Tikhomirova, G. S.; Radiokhimiya, 1975, **17**, 321.
8. Kawasuji, I.; Fukasawa, T.; Suzuki, S.; Radiochem. Radioanal. Lett.,
9. Suzuki, S.; Sato, A.; Hara, M.; Mitsugashira, T.; Kawasuji, I.; Kikuchi, H.; Fukasawa, T.; Sci. Rep. RITU, 1979, **A-28**, 73.
10. Fukasawa, T.; Kawasuji, I.; Mitsugashira, T.; Sato, A.; Suzuki, S.; Bull. Chem. Soc. Jpn., 1981, **55**, 726.
11. Yamada, S.; Tanaka, M.; J. Inorg. Nucl. Chem., 1975, **49**, 55.
12. Eigen, M.; Wilkins, R.; Adv. Chem. Ser., 1965, **49**, 55.
13. Fuoss, R. M.; J. Am. Chem. Soc., 1958, **80**, 5059.

GEOCHEMICAL STUDIES ON AMERICIUM AND PLUTONIUM IN SOIL

Masayoshi YAMAMOTO, Kazuhisa KOMURA* and Masanobu SAKANOUE*

School of Pharmacy, Hokuriku University, Kanagawa-machi, Kanazawa-shi, Ishikawa 920-11, Japan
*Low-Level Radioactivity Laboratory, Kanazawa University, Tatsunokuchi-machi, Nomi-gun, Ishikawa 923-12, Japan

ABSTRACT. Fallout Am-241 and Pu isotopes in soils from various locations of Japan including Nagasaki and Hiroshima where the atomic bombs had exploded in 1945 have been measured for studying their contamination levels, distributions, and behaviors in soils.

From these results, it was estimated that the average deposited amount of Pu-239,240 in soil was 1.5-1.7 mCi/km^2, and the Am-241/Pu-239,240 activity ratio in soil was about 0.3 at present and its ratio would have a peak of 0.41 in the year of 2033, at which time the activity of Am-241 would become 0.62-0.70 mCi/km^2. The studies on the Nagasaki soils showed that the soils at around Nishiyama area in Nagasaki City were highly contaminated with Pu-239,240 brought from the Nagasaki atomic bomb. The contribution of atomic bomb Pu-239,240 to the total Pu-239,240 for the Nishiyama soils was 95-97% from the activity ratios of Pu-241/Pu-239,240 and Am-241/Pu-239,240 by comparing them with those of global fallout. The leaching studies of Am-241 and Pu-239,240 using the Nishiyama soil highly contaminated with Pu-239,240 suggested that plausible environmental leaching conditions might mobilize the Am much more easily than the Pu present in soil.

1. INTRODUCTION

Global contamination of the environment with americium (Am-241) and various plutonium isotopes (Pu-238,-239,-240 and -241) have been caused by atmospheric nuclear weapons testing [1]. Most of Am-241 (an alpha-decay isotope with a 433-yr half-life) detectable in the environmental samples is not a direct result of the atomic bomb, but is formed in the environment by the decay of the beta emitter Pu-241, which has a half-life of 14.4-yr.

The investigation of these transuranium elements in the environment are of importance not only from the geochemical point of view but also from the standpoint of health physics. In the terrestrial environment, soil is the principal reservoir for these artificial radionuclides and acts as the source of the transport system.

N. M. Edelstein et al. (eds.), Americium and Curium Chemistry and Technology, 275–292.

This paper presents the estimations of fallout levels and distributions of Am-241 and Pu-239,240 in surface soils of various areas of Japan and the variation of Am-241/Pu-239,240 activity ratio with time in soil. Furthermore, the study includes measurements of the Am-241 and Pu isotopes present in the soils of Nagasaki and Hiroshima where the atomic bombs had exploded in 1945. The depth profiles and leaching characteristics of Am-241 and Pu-239,240 in soil are also given together with other fallout radionuclide, cesium (Cs-137).

2. SAMPLES AND ANALYTICAL METHODS

Air-dried soil samples sieved through a 2-mm screen were used in this studies, unless otherwise noted.

After gamma-spectrometric measurement of Cs-137, a 20-50 g aliquot was subjected to analysis for plutonium isotopes (Pu-238 and Pu-239, 240) and americium (Am-241). The sequential separation of plutonium and americium was done according to the method described by Thein et al. [2] or Yamato [3], Pu-242 and Am-243 being used as yield monitors. The alpha activity of each electroplated sample was measured using an alpha-ray spectrometer with a Si(Au) surface barrier detector.

To determine Pu-241, the amount of Am-241 ingrowth from electroplated Pu-241 stored for 2-3 years was measured either by comparing the new alpha spectrum with the old one or by applying the radiochemical separation of Am-241 from the plate.

The determination of the Pu-240/Pu-239 isotopic ratio in some soil samples was performed using our new method based on the measurement of LX-/alpha-ray activity ratio using a Ge-LEPS (low-energy photon spectrometer) and a Si(Au) surface barrier detector [4].

3. RESULTS AND DISCUSSION

3.1. The levels and distributions of Am-241 and Pu-239,240 in surface soils of various areas of Japan

The soil samples were collected from the surface to a depth of 12-20 cm at rice-fields of 15 national and/or prefectural agricultural experimental stations in Japan during 1963-1966 (mostly in 1963) and during 1972-1976 (mostly in 1976) [5].

As shown in Fig.1 and Table I, considerable differences are found for the integrated amounts of deposited Am-241, Pu-239,240 and Cs-137. Higher values were obtained for the soil samples from Akita, Niigata, Ishikawa and Tottori Prefectures located along the Japan Sea coast, while the lowest values for the soil from Osaka. Generally, the Japan Sea coast region has 2.5 to 3 times higher values than the Pacific coast region. The average deposition of Pu-239,240 in Japan was estimated to be 1.5-1.7 mCi/km^2 (2.8-2.9 mCi/km^2 for the Japan Sea coast of Honshu and 1.0-1.2 mCi/km^2 for the Pacific coast).

The activity ratio, Am-241/Pu-239,240, in the "1963" soil samples ranged from 0.30 to 0.43 with a mean value of 0.34 ± 0.04, which

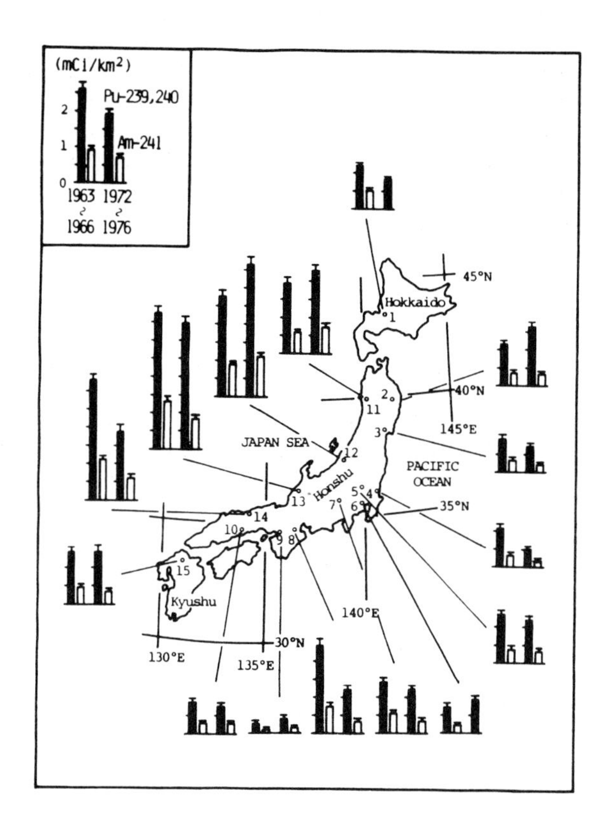

Figure 1. Levels and distributions of Am-241 and Pu-239,240 of integrated fallout in soils collected from Japanese rice-fields. The Am-241 data are as of the dates of analysis from the end of 1981 to the beginning of 1982. The propagated counting error of one sigma are given by the symbol "T" as the highest value on the top of each bar. (11:Akita,12:Niigata(Joetsu),13:Ishikawa,14:Tottori)

was slightly higher than the mean value of 0.28 ± 0.04 (each value ranged from 0.22 to 0.34) obtained for the "1976" soil samples. These values are consistent with the reported data for global fallout in the soil and the coastal sediment samples [6]. A slight decrease of Am-241/Pu-239,240 ratio may be explained by the contribution of recent fallout causing low Am-241/Pu-239,240 ratio.

3.2. The time variation of the Am-241/Pu-239,240 activity ratio

The activity of Am-241 in the environment is quite dependent on the

TABLE I. Regional average depositions of Pu-239,240, Am-241 and Cs-137 in soils collected from Japanese rice-fields, and their activity ratios

Year of sampling	Region	Concentration (pCi/kg)			Deposition (mCi/km^2)			Activity ratio	
		Pu*	Am-241	Cs-137#	Pu*	Am-241	Cs-137	Am-241 / Pu*	Pu* / Cs-137
"1963"	Japan Sea coast	22.9/3.8**	7.6/1.2	1.91/.52	2.9/.8	0.98/.30	242/58	0.33/.02	0.012/.001
	Pacific coast	9.2/3.8	3.0/1.2	0.68/.30	1.2/.6	0.38/.19	85/46	0.34/.03	0.014/.002
	All area	13.0/7.1	4.4/2.3	1.02/.65	1.7/.9	0.55/.33	128/84	0.34/.04	0.013/.002
	Japan Sea coast / Pacific coast	2.5	2.5	2.8	2.6	2.6	2.9		
"1976"	Japan Sea coast	21.2/6.7	5.9/1.7	1.23/3.9	2.8/.9	0.92/.33	163/52	0.28/.04	0.017/.001
	Pacific coast	7.6/4.3	2.0/.9	0.44/.25	1.0/.4	0.27/.10	55/30	0.29/.03	0.017/.002
	All area	11.3/7.7	3.2/2.1	0.65/.45	1.5/.9	0.43/.27	85/57	0.28/.04	0.017/.002
	Japan Sea coast / Pacific coast	2.8	2.9	2.8	2.9	3.3	3.0		

*) Pu: ^{239,240}Pu. #) pCi/g.

**) The number after slash denotes standard deviation of one sigma: 22.9/3.8 = 22.9 ± 3.8.

The ^{137}Cs data were corrected for decay to the dates of sampling, while the ^{241}Am data are as of the dates of of analysis from the end of 1981 to the beginning of 1982.

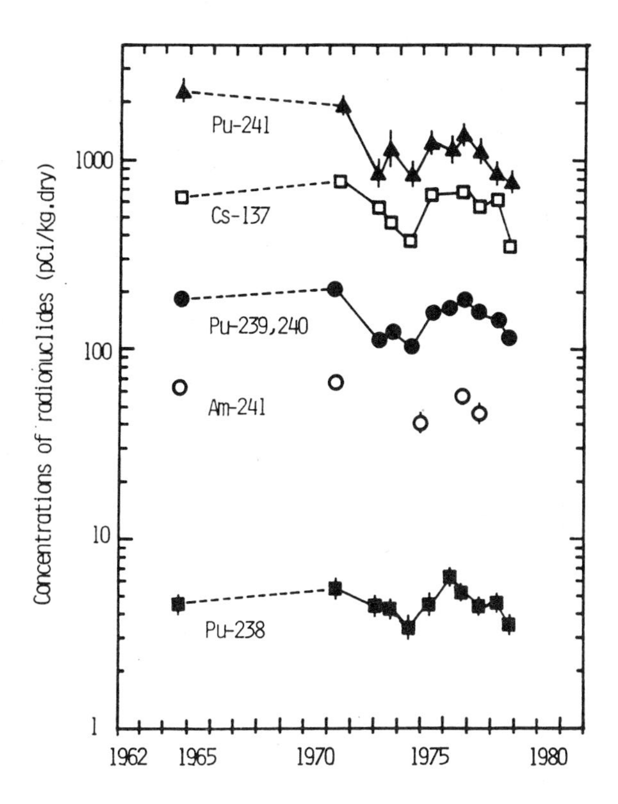

Figure 2. The time variation of Pu isotopes, Am-241 and Cs-137 in sediments periodically collected from the central area of Niu Bay, Fukui Pref.,Japan. The concentrations of Pu isotopes and Cs-137 were corrected for decay to the dates of sampling, while that of Am-241 is as of the dates of analysis,May 1981.

elapsed time since most of Am-241 is due to the decay of deposited Pu-241. To estimate the present or near future level of Am-241, it is fundamental to know the time variation of the activity ratios of both Am-241/Pu-239,240 and Pu-241/Pu-239,240 in the past. Therefore, this work was undertaken to study the activity ratios of Am-241/Pu-239, 240 and Pu-241/Pu-239,240 in the coastal sediment samples collected during 1964-1978 from Niu Bay, Fukui Pref., and the activity ratio of Am-241/Pu-239,240 in soil samples collected during 1957-1980 from Niigata and Akita Prefectures located along the Japan Sea coast [5].

Figure 2 shows the time variation of the concentrations of several plutonium isotopes, Am-241 and Cs-137 in the sediments from the central area (maximum depth: 13-15 m) of Niu Bay [7]. From these data, the Am-241/Pu-239,240 activity ratios at the dates of analysis (May 1981) were found to show the similar value (mean ratio of 0.30 ± 0.03) regardless to the different dates of their sampling. This fact indicates that the Am-241 in the different samples has grown in from its precursor Pu-241 from nearly the same date for all samples.

The Pu-241/Pu-239,240 activity ratios (R) at the date of the sampling are shown in Fig.3 for each sample along with their Am-241/Pu-239,240 activity ratios (R') corrected to the growth of Am-241 from Pu-241 during the period from the sampling to the analysis. The variations of these two activity ratios with time (t) can be expressed by the following equations considering decay and growth, respectively:

$$(R)_t = (R)_0 \cdot e^{-k_3(t-t_0)} \qquad (1)$$

$$(R')_t = (R)_0 \cdot \frac{k_2}{k_2-k_1} \cdot [e^{-k_1(t-t_0)} - e^{-k_2(t-t_0)}] \qquad (2)$$

where, k_1= the decay constant of Pu-241 ($\frac{0.693}{14.4}$ y^{-1})

k_2= the decay constant of Am-241 ($\frac{0.693}{433}$ y^{-1})

The subscripts t and 0 refer to the ratios at time t and t_0, respectively. The parameters $(R)_0$, k_3 and t_0 were calculated to be 14 ± 2, 0.693/(13.2 ± 0.5) y^{-1} and 1960 ± 1, respectively, by means of non-linear optimization techniques based on the activity ratios of Pu-241/Pu-239,240 and Am-241/Pu-239,240 for the Niu Bay sediment samples. The solid lines in Fig.3 show the regression curves calculated for this sediment. The estimated Am-241/Pu-239,240 ratios for the soil samples collected periodically during 1963-1980, as shown in Fig.3, lay on this regression curve. These regression curves are close to those reported by Bennett [8], as shown by dashed curves after the year of 1963.

It is estimated from these regression curves that the activity ratio of Am-241/Pu-239,240 in soil in 1984 is about 0.3, and will reach a maximum in the year of 2033 when its activity ratio becomes 0.41, assuming no further atmospheric nuclear testing. The maximum accumulation of Am-241 can be predicted to be 0.62-0.70 mCi/km^2 by using the average deposition of Pu-239,240 (1.5-1.7 mCi/km^2) in Japan.

3.3. Studies for the soils from Nagasaki and Hiroshima

Based on the above-mentioned experimental results, similar studies were extended to the soils from Nagasaki and Hiroshima. As is well known , plutonium was used for the atomic bomb dropped on Nagasaki, while U-235-enriched uranium was used for the atomic bomb dropped on Hiroshima.

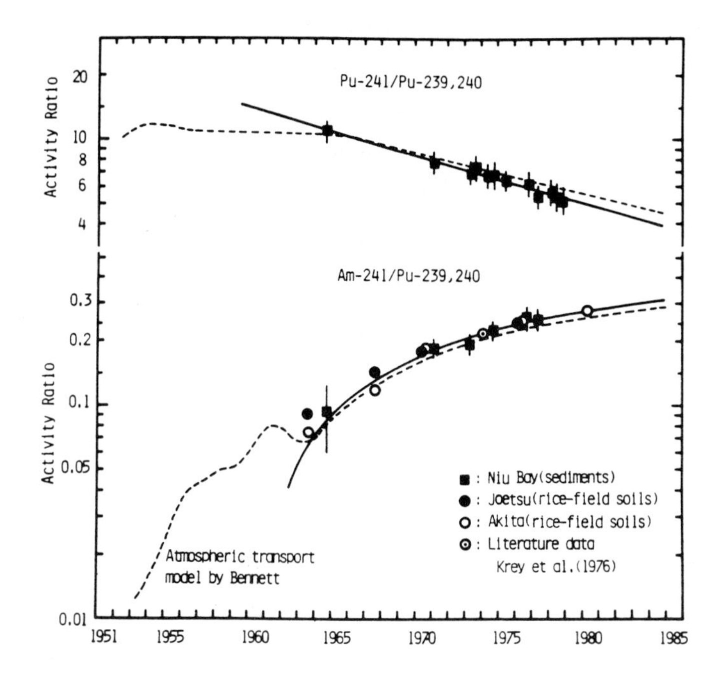

Figure 3. The variation with time of the Pu-241/Pu-239,240 and Am-241/Pu-239,240 activity ratios in soils or sediments. The regression lines for the Niu Bay sediment data are given in solid lines. Also are given in dashed lines the variation with time of these two ratios estimated by Bennett based on the atmospheric transport model.

The sampling points for the Nagasaki soils are shown in Fig.4 together with the accumulations of Pu-239,240 and Cs-137 per unit area. As seen in this figure, extremely high Pu-239,240 accumulation was found in the soils of the Nishiyama area, which is located 2-4 km from the epicenter. The accumulation level of Pu-239,240 ranging from 14.1 to 45.1 mCi/km^2 found in the Nishiyama area is higher on an average by a factor of about fifteen than the values of 1.0-3.7 mCi/km^2 in other areas. Whereas, the accumulation level of Cs-137 in the Nishiyama area (98.7-145 mCi/km^2) is not so largely different from the levels in the soils collected from other areas (59.8-135 mCi/km^2), indicating that most of Cs-137 is derived from global fallout and the contribution of the atomic bomb in 1945 is relatively small.

In order to clarify the contribution of the atomic bomb, it seems much better to use the activity ratios instead of comparing only the accumulation levels of the fallout radionuclides. Because the

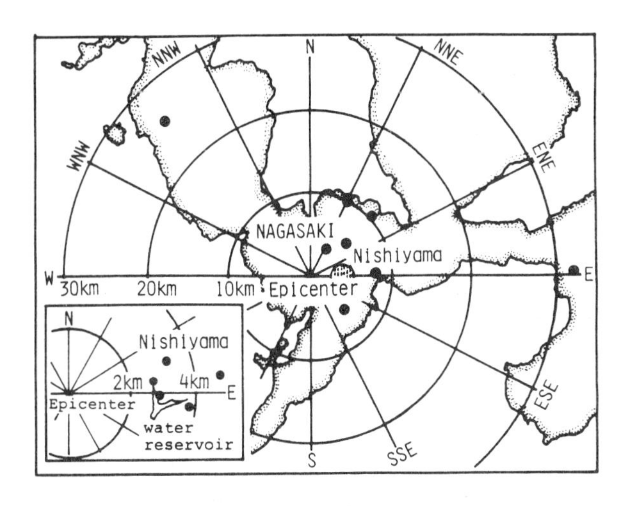

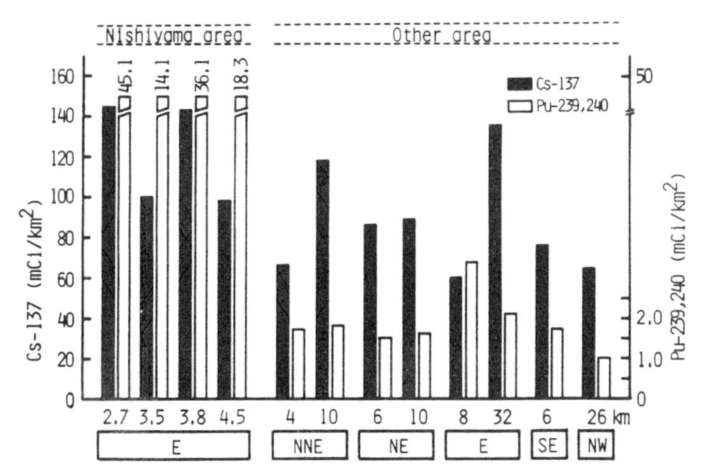

Figure 4. Sampling points for the Nagasaki soils and the integrated deposits of Pu-239,240 and Cs-137 in soils (0-10 cm depth). The Cs-137 values are corrected for decay to the dates of sampling (1976).

accumulation of fallout nuclides in a given area is influenced by many causes such as the difference of soil nature, configuration of the ground surface, local meterology such as the precipitation and so on. Fig.5 shows the activity ratios of Pu-238/Pu-239,240, Pu-239,240/Cs-137 and Am-241/Pu-239,240. As shown in this figure, the activity ratios of Pu-238/Pu-239,240 (0.050-0.062), Pu-239,240/Cs-137

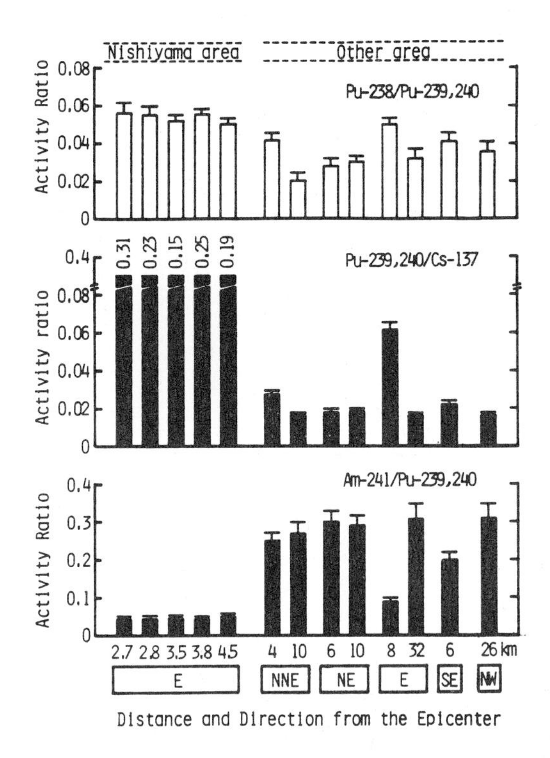

Figure 5. Activity ratios of Pu-238/Pu-239,240, Pu-239,240/Cs-137 and Am-241/Pu-239,240 in soils from Nishiyama and other area in Nagasaki. The propagated counting error of one sigma is given by the symbol"T" as the highest value on the top of each bar. The Cs-137 values were corrected for decay to the dates of sampling (1976), while the Am-241 values are as of the dates of analysis from the end of 1982 to the beginning of 1983.

(0.15-0.31) and Am-241/Pu-239,240 (0.04-0.05) of the Nishiyama soils are distinctly different from the corresponding global fallout values of 0.03-0.04 [1], about 0.02 [5] and about 0.3 [5],respectively. The Pu-240/Pu-239 atom ratios (0.02-0.04), shown in Table II, deviate significantly from the commonly accepted value of 0.18 for global fallout. These results indicate that fairly large amount of weapons-grade Pu having a low Pu-240/Pu-239 atom ratio due to the Nagasaki atomic bomb still remain in the soils at Nishiyama area. The contribution of atomic bomb Pu-239,240 to the total Pu-239,240 activity for the Nishiyama soil has recently been estimated to be 95-97% by the measurements of Pu-241/Pu-239,240 and Am-241/Pu-239,240 activity

TABLE II. Atom ratio Pu-240/Pu-239 and activity ratios of some nuclide pairs in soils collected from Nagasaki and Hiroshima

Location	Sample (Depth)	Sampling date	Pu-239,240 concentration (pCi/g)	Activity ratio			Atom ratio
				Pu-238 / Pu-239,240	Am-241 / Pu-239,240	Pu-239,240 / Cs-137	Pu-240 / Pu-239
"Nishiyama" district (Nagasaki)	E-2.8* (0-5 cm)	July 1979	1.380/.030	0.055/.005	0.041/.003*1	0.23/.01	0.031/.012
	E-3.5 (0-10 cm)	June 1976	0.337/.009	0.052/.003	0.048/.003*2	0.15/.01	0.034/.010
	E-4.5 (0-10 cm)	June 1976	0.462/.009	0.050/.003	0.051/.003*2	0.19/.01	0.020/.009
"Black rain" area (Hiroshima)	NWN-10 (0-5 cm)	July 1983	0.050/.002	0.033/.003	0.090/.010*3	0.015/.001	0.179/.023
	NWN-12 (0-5 cm)	July 1983	0.061/.004	0.036/.002	0.13/.01*3	0.020/.001	0.173/.023

*) Direction and distance from epicenter.

The number after slash denotes the propagated counting error of one sigma: 1.380/.030 = 1.380 ± 0.030. The $^{239,240}Pu/^{137}Cs$ ratios were decay corrected to the dates of sampling, while the $^{241}Am/^{239,240}Pu$ ratios are as of the dates of analysis: *1) January 1981, *2) March 1983, *3) August 1983.

ratios [10].

By contrast, the soils from Hiroshima did not show, in the activity ratios among the plutonium isotopes, Am-241 and Cs-137, any significant differences from global fallout. Therefore, the contribution of the transuranic elements derived from the Hiroshima atomic bomb was found to be negligibly small as compared with those from global fallout.

3.4. Depth profiles of Am-241 as compared with those of Pu-239,240 and Cs-137

The _in-situ_ gamma-spectrometric measurements of environmental radionuclides have been made extensively by using a potable Ge(Li) detector to evaluate the exposure rates along with the contents of natural and artificial radionuclides [11][12]. These results showed that the accumulation of fallout radionuclides was rather high in the natural forest or the area just under the eaves of roofs where snow and rainwater might potentially deposit fallout. In order to investigate the depth profiles of Am-241, Pu-239,240 and Cs-137, soil core samples were collected up to a depth of 20-30 cm from such three locations, Kohno (under the eaves of a temple), Oguroi and Kanagawa (from relatively flat area in the forests).

The depth profiles of Am-241, Pu-239,240 and Cs-137 in the soils from these locations are shown together with their activity ratios in Fig.6. As seen from this figure, nearly all of the Am-241,Pu-239,240 and Cs-137 were found in the soil layer up to a depth of 20-21 cm. The concentrations of these nuclides decreased smoothly with increasing depth. Over 80% of these nuclides, as seen in Table III, were retained in the top 10 cm layer of all soils.

A comparison of the activity ratios of Pu-239,240/Cs-137 or Am-241/Pu-239,240 among each depth fraction of core sample gives us information about the relative mobility of these nuclides in soils. As seen in the upper part of Fig.6, the Pu-239,240/Cs-137 activity ratios have rather constant values for Kohno and Kanagawa samples, indicating that the behavior of Pu-239,240 is similar to that of Cs-137 in the soil. On the other hand, this ratio has a tendency to increase with depth for the Oguroi sample. This fact may be explained by the different mobilities of these nuclides in the soil of this area, that is, more rapid leaching of Pu-239,240. Alternatively, if the soil of this area is low in K and clay minerals, enough plant (root) uptake of Cs-137 retards downward migration.

The Am-241/Pu-239,240 activity ratios in the soil profiles were found to be in the range of 0.25-0.40 for the Kohno sample and 0.27-0.38 for the Kanagawa sample. These values agree well with the Am-241/Pu-239,240 ratios obtained for the rice-field surface soil (see Fig.3). For the Oguroi sample, the Am-241/Pu-239,240 ratios increase from 0.25 to 0.62 with depth, though the measurement error is rather high in the deeper fractions. As pointed out by other investigator [13], the greater downward movement of Am-241 relative to Pu-239,240 may cause the increase of the Am-241/Pu-239,240 ratios with depth.

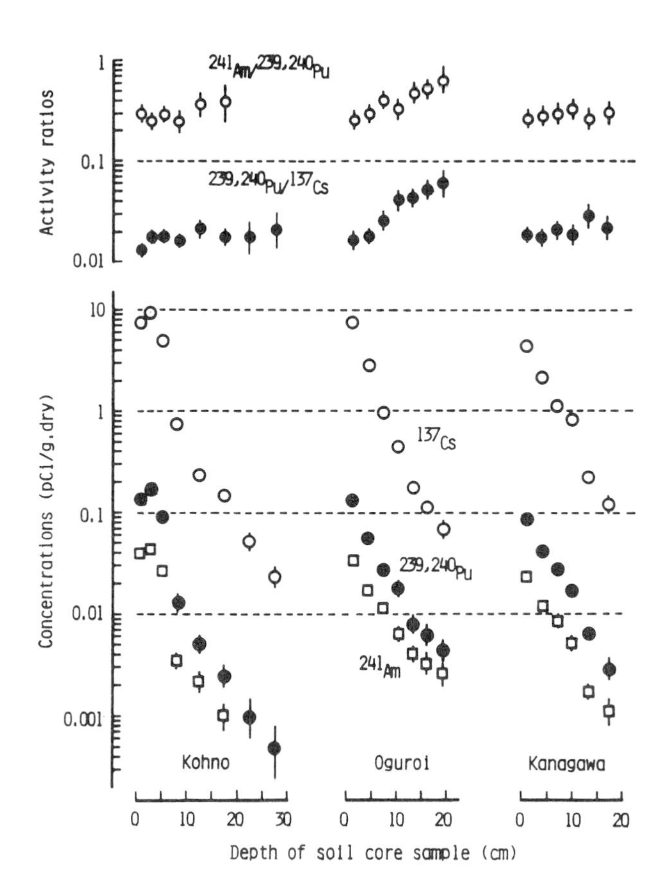

Figure 6. Depth profiles of Am-241, Pu-239,240 and Cs-137 in soils, and their activity ratios. The soils from Kohno,Oguroi and Kanagawa were collected on Sept. 1978, May 1980 and Oct. 1982, respectively. The Cs-137 data were corrected for decay to the dates of sampling, while the Am-241 data are as of the dates of analysis within three months after the sampling.

However, since most of the environmental Am-241 is grown up _in-situ_ by the decay of short-lived Pu-241, one must consider the larger amount of Am-241 derived from Pu-241 in the layer containing old plutonium. Here, the determination of Pu-241 is desired to clarify these circumstances.

TABLE III. Depth distributions of Pu-239,240, Am-241 and Cs-137 in soils (% of total)

Kohno (Sept.1978)				Oguroi (May 1980)				Kanagawa (Oct.1982)			
Depth (cm)	Pu*	Am-241	Cs-137	Depth (cm)	Pu*	Am-241	Cs-137	Depth (cm)	Pu*	Am-241	Cs-137
0-1.5	19.3	22.1	19.7	0-3	43.4	33.9	52.5	0-3	47.4	40.6	45.4
1.5-4.0	46.4	45.6	46.8	3-6	26.0	25.5	29.0	3-6	25.6	28.4	25.4
4.0-6.5	24.7	26.3	24.3	6-9	17.6	22.4	13.0	6-9	12.2	16.3	15.4
6.5-10	4.8	2.8	5.2	9-12	5.7	6.1	2.8	9-12	9.6	10.1	10.2
10-15	2.8	2.2	2.1	12-15	3.3	5.1	1.4	12-15	3.0	2.8	2.6
15-20	1.2	1.0	1.3	15-18	2.1	4.2	0.8	15-18	2.2	1.8	1.0
20-25	0.5		0.4	18-21	1.9	2.8	0.5				
25-30	0.3										
(mCi/km^2)	9.0	2.3	505		3.8	1.1	200		3.0	0.9	150

*) Pu: $^{239,240}Pu$.

The ^{137}Cs data were corrected for decay to the dates of sampling, while the ^{241}Am data are as of the dates of analysis within three months after the sampling.

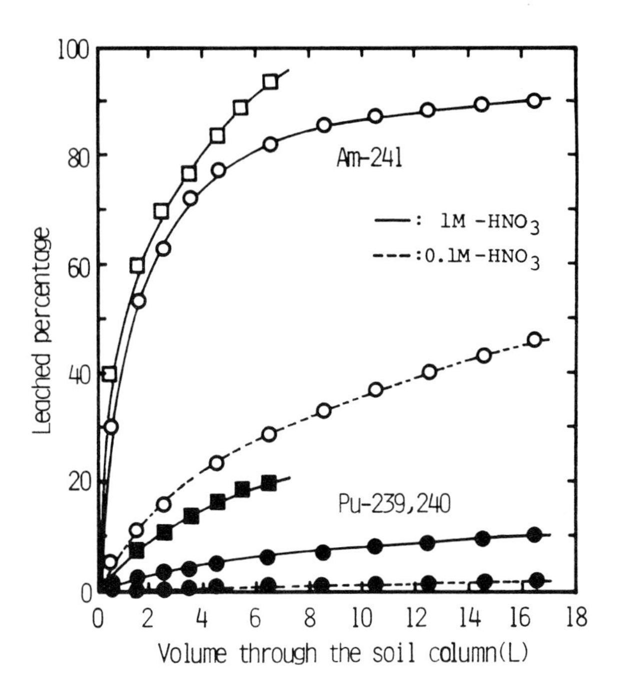

Figure 7. Dynamic leaching of Am-241 and Pu-239,240 from the soils collected at Nishiyama (circle symbol) in Nagasaki and Oguroi (square symbol) in Fukui using diluted nitric acid solution.

3.5. Leaching characteristics of Am-241 and Pu-239,240 in soil

To know whether there is any difference in the behavior of Pu and Am in soil, both column and batch leaching experiments were performed with diluted nitric acid solution using the Nishiyama soil highly contaminated with Pu-239,240.

For the column experiment, about 100 g of field-moist Nishiyama soil was packed into the cylindrical column, and the leaching solution (1M- and 0.1M-HNO_3) was fed upward through the column at flow-rate of 200 ml per hour. The effluent was collected in 1000 ml fractions, and analysed.

For the batch experiment, 10 g of the air-dried soil was suspended in nitric acid solutions (1000 ml) of various concentrations ranging from 1M to 0.0001M. After filtering (pore size of 0.45-μm), both the leaching solution and the residual soil on the filter were analysed for Am-241 and Pu-239,240.

The experimental results of the column leaching with 1M-HNO_3 are

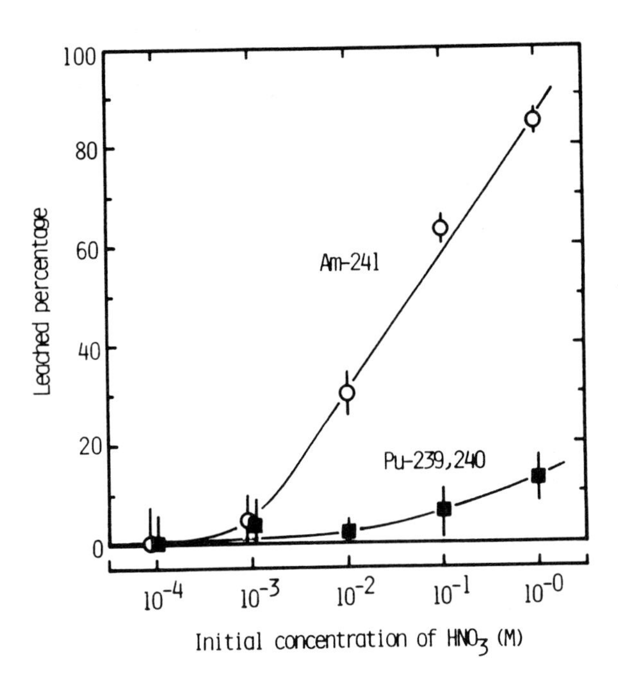

Figure 8. Batch leaching of Am-241 and Pu-239,240 from the Nishiyama soil using nitric acid solutions ranging in concentration from 1M to 0.0001M. (Soil: 10 g, Extractant: 1000 ml, Shaking time: Two months, Temp.: Room Temp.)

shown in Fig.7 together with those with 0.1M-HNO_3. The data are plotted in terms of the cumulative percentage of leaching for Am-241 and Pu-239,240. In this figure, the results for the Oguroi soil contaminated only with global fallout are also shown for comparison. The continued leaching with the 17 liters of 1M-HNO_3 solution showed the high solubility of Am-241 (about 90% of the total) and the low solubility of Pu-239,240 (about 10% of the total). These results are similar to the results for the Oguroi soil. Even in the case of the leaching with 0.1M-HNO_3, approximately 50% of the total Am-241 was removed, while the Pu-239,240 was only slightly leached (less than 2% of the total).

The batch leaching experiments with 1M-HNO_3 removed nearly 85% of the total Am-241 as shown in Fig.8, and the leaching percentage of Am-241 was reduced with decreasing concentration of nitric acid. Below 0.001M-HNO_3, Am-241 was not significantly leached. On the other hand, only an about 10% of the total Pu-239,240 was removed with

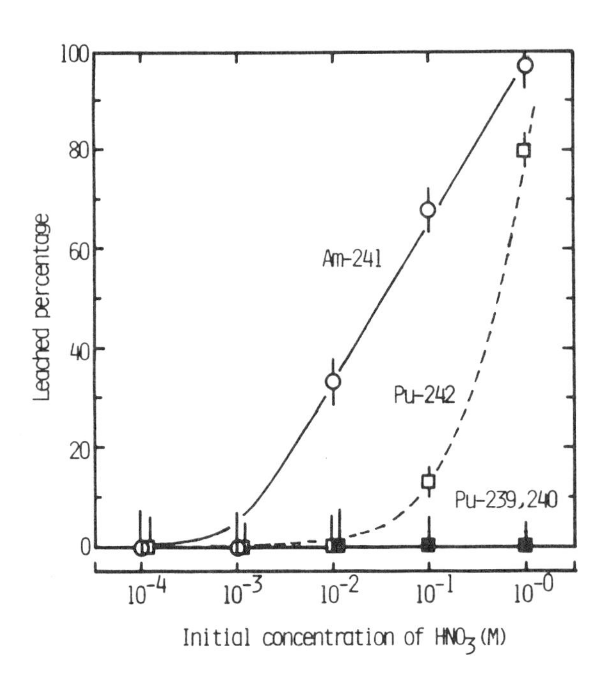

Figure 9. Batch leaching of Am and Pu from the soil prepared by adsorbing Am-241 and Pu-242 tracers from the solution of pH about 7 followed by air-drying, and the soil containing refractory Pu-239,240 oxide prepared by ignition at 1000°C, using nitric acid solutions ranging in concentration from 1M to 0.0001M. (Soil: 0.5 g, Extractant: 50 ml, Shaking time: Two months, Temp.: Room Temp.)

1M-HNO_3, and below 0.1M-HNO_3 Pu-239,240 was not leached.

A similar batch leaching experiment was performed using the following mixed soil samples which contained Am-241, Pu-242 and Pu-239,240 : a soil sample prepared by adsorbing Am-241 and Pu-242 tracers from the solution of pH about 7 followed by air-drying, and another one containing refractory Pu-239,240 oxide prepared by ignition at 1000°C.

The results obtained are shown in Fig.9. The results of Am-241 leaching are very close to the results for the Nishiyama soil shown in Fig.8, indicating that the Am-241 in the Nishiyama soil is likely to be present in the form similar to the Am-241 prepared here, probably adsorbed onto soil as the hydrolyzed or colloidal species. As for plutonium, high solubility of Pu-242 (about 80% of the Total) was observed in the leaching experiment with 1M-HNO_3 and its solubility was reduced rapidly with decreasing concentration of nitric acid. Whereas, no significant leaching of Pu-239,240 was observed even at the high concentration of 1M-HNO_3. These two data are somewhat different from the result for the Pu-239,240 in the Nishiyama soil. The situation for Pu may be more complicated as compared with that for Am because Pu is liable to hydrolyze and polymerize. Comparison of these results with the data for the Nishiyama soil suggests that the Pu in the Nishiyama soil is likely to be present not as the refractory plutonium oxide ignited at high temparature, but rather as the hydrolyzed forms of Pu which are difficult to be desorbed with weak acidic solution.

From these experimental results, it seems most reasonable to conclude that the Am and Pu in the Nishiyama soil are present in different forms, and the Am is more leachable than the Pu. Although the Am and Pu in the Nishiyama soil were unlikely to be dissolved under conditions studied at the low concentration of nitric acid solution, plausible environmental leaching conditions might be expected to mobilize the Am much more easily than the Pu present in the soil. These findings may be useful for the radioactive waste disposal problems.

ACKNOWLEDGMENTS. A part of this work was supported by a Grant-in-Aid for Scientific Research from the Ministry of Education, Science and Culture,Japan. We would like to express our thanks to Prof.K.Kodaira, Ashikaga Institute of Technology, Prof.S.Okajima, Atomic Disease Institute,Nagasaki University, and Dr.T.Nakanishi,Faculty of Science, Kanazawa University, for supplying the valuable samples of the Japanese rice-field surface soils and the Nagasaki soils.

REFERENCES

[1] PERKINS,P.W., THOMAS,C.W., 'Worldwide fallout', in Transuranic Elements in the Environment (W.C.Hanson,Ed.), pp.53-81. USDOE Technical Information Center, Oak Ridge,Tenn.,1980.
[2] THEIN,M.,BALLESTRA,S.,YAMATO,M.,FUKAI,R.,'Delivery of transuranic elements by rain to the Mediterranean Sea', Geochim. Cosmochim. Acta **44**(8), 1091(1980).
[3] YAMATO,A., ' An anion exchange method for the determination of Am-241 and plutonium in environmental and biological samples', J. Radioanal. Chem. 75,265(1982).
[4] KOMURA,K.,YAMAMOTO,M.,SAKANOUE,M., 'Determination of Pu-240/Pu-239 ratio in environmental samples based on the measurement of LX/ alpha-ray activity ratio', Health Phys. **46**(6),1213(1984).
[5] YAMAMOTO,M.,KOMURA,K.,SAKANOUE,M., 'Am-241 and plutonium in Japanese rice-field surface soil', J. Radiat. Res. **24**,237(1983).

[6] YAMAMOTO,M.,KOMURA,K.,SAKANOUE,M., 'Distribution and characteristics of Pu and Am in soil', in Environmental Migration of Long-lived Radionuclides, pp.481-489,STI/PUB/59(Vienna,IAEA),1982.

[7] YAMAMOTO,M.,IGARASHI,S.,YOSHIOKA,M.,KITAGAWA,T.,SAKANOUE,M., 'Transuranic elements in Niu Bay sediments', Health Phys. **46**(2), 455(1984).

[8] BENNETT,B.G., 'Environmental aspects of americium', Thesis New York Univ.,1979.

[9] YAMAMOTO,M.,KOMURA,K.,HOSHI,M.,SAWADA,S.,OKAJIMA,S.,SAKANOUE,M., 'Pu isotopes, Am-241 and Cs-137 in soils from the atomic bombed areas in Nagasaki and Hiroshima', J. Radiat. Res. in press.

[10] YAMAMOTO,M.,KOMURA,K.,SAKANOUE,M., 'Discrimination of the plutonium due to atomic explosion in 1945 from global fallout plutonium in Nagasaki soil', J. Radiat. Res. **24**,250(1983).

[11] SAKANOUE,M.,KOMURA,K., 'Application of in-situ gamma-spectrometry to the estimation of radiation dose and environmental radioactive nuclides', Report of Kyoto Univ. Reactor Lab.,KURRI-TR-**155**, 20(1977).

[12] SAKANOUE,M.,MARUO,Y.,KOMURA,K., 'In-Situ low-level gamma-ray spectrometry and X-ray fluoresence analysis', in Methods of Low Level Counting and Spectrometry, IAEA,Vienna STI/PUB/592, pp.105-124(1981).

[13] ESSINGTON,E.H.,FOWLER,E.B.,GILBERT,R.O.,EBERHARDT,L.L.,'Plutonium, americium and uranium concentrations in Nevada Test Site soil profiles', in Transuranium Nuclides in the Environment, IAEA, Vienna,pp.157-173(1976).

PRODUCTION AND RECOVERY OF AMERICIUM-241

James D. Navratil
Rockwell International
Rocky Flats Plant
P. O. Box 464
Golden, Colorado 80402-0464

ABSTRACT. Nuclear reactors produce mixtures of ^{241}Am and ^{242}Am. However, beta decay of ^{241}Pu produces radioactively-pure ^{241}Am. Separation and purification processes for ^{241}Am include ion exchange, precipitation, and both liquid-liquid and molten-salt extraction. This paper reviews the principal production and recovery processes.

1. INTRODUCTION

Americium has isotopes with mass numbers between 232 and 247.[1] Only three isotopes have half-lives greater than a few hours: ^{243}Am (7400 yr), ^{241}Am (432 yr) and ^{242}Am (152 yr). ^{241}Am decays to ^{237}Np, emitting alpha and gamma radiation. Because of its essentially monoenergetic alpha and gamma radiations, ^{241}Am is particularly suited for use as an excitation source and in a multitude of industrial and scientific thickness and density measurements. The list of applications for ^{241}Am is one of the largest for any actinide isotope.

Nuclear reactors produce mixtures of ^{241}Am and ^{242}Am. However, ^{241}Am is a by-product of plutonium scrap recovery operations as it is produced from the beta decay of the ^{241}Pu present in small quantities in the ^{239}Pu stream. High purity americium oxide has been produced at the U.S. Department of Energy's Rocky Flats Plant (RFP) since 1962.

The diverse operations at RFP for separating and purifying americium have included aqueous precipitation, ion exchange and both conventional liquid-liquid as well as molten salt extraction processes. An overview of these production and recovery processes is presented.

2. PRODUCTION

Americium, atomic number 95, was the fourth transuranium element to be discovered.[1] ^{241}Am was identified by Seaborg, James, Morgan and Ghiorso late in 1944 at the wartime Metallurgical Laboratory (now Argonne National Laboratory) of the University of Chicago as a result

N. M. Edelstein et al. (eds.), Americium and Curium Chemistry and Technology, 293–299.

of successive neutron capture reactions by plutonium isotopes in a nuclear reactor.

Present-day nuclear reactors yield mixtures of americium isotopes, whereas isotopically pure ^{241}Am is produced by the beta decay of the ^{241}Pu present in small quantities in the ^{239}Pu stream. Kilogram amounts of ^{241}Am are separated and purified by appropriate chemical processes from plutonium and other impurity elements.

^{241}Am was first offered for sale in March 1962 at $1500 per gram by the U.S. Atomic Energy Commission (AEC). The first allotment of 200 grams, offered through the US AEC Isotope Pool at Oak Ridge National Laboratory (ORNL), came from RFP. In the past, RFP as the sole source of ^{241}Am has been supplemented from nuclear fuel reprocessing facilities at US DOE Hanford and Savannah River facilities, and more recently from the Los Alamos National Laboratories plutonium scrap recovery facility.

3. EARLY AQUEOUS PROCESSES

Production scale plutonium processing commenced at RFP in 1952 and was based on processes developed at Los Alamos.[2] Originally the only means of separating the americium from the main plutonium stream was by precipitation of plutonium with hydrogen peroxide. Americium contained in the plutonium peroxide filtrate was precipitated with ammonia or sodium hydroxide and stored until 1959 when it was recovered by a thiocyanate ion exchange process.[3] The thiocyanate process produced high purity AmO_2 low in chromium, iron, lead, nickel, plutonium and rare earths. This very inefficient process caused excessive corrosion to process equipment and glove-boxes and generated considerable aqueous waste. To eliminate the necessity of recovering americium from plutonium peroxide filtrates, a molten salt extraction (MSE) process was developed in 1967 to extract americium directly from aged plutonium metal.[4,5] The MSE residue salts (NaCl, KCl and $MgCl_2$) were originally processed through six operations: dissolution in dilute acid or water, precipitation of the americium and plutonium with sodium hydroxide, separation of the americium from the plutonium by anion exchange, purification of the americium through the thiocyanate process, oxalate precipitation of the americium and calcination of the americium oxalate to americium oxide.

Through optimization of the MSE process, a sixfold increase in the $MgCl_2$ content of the residue salt resulted. As a consequence, filtration times for the hydroxide precipitate increased by a factor of 25. However, by 1975, sodium hydroxide had been replaced by a cation exchange process and a double oxalate precipitation had replaced the thiocyanate process.[6]

The MSE residues did not contain aluminum, calcium, lead, rare earths, zinc or other interfering impurities. But the process changes resulted in AmO_2 with a purity comparable to that obtained using the thiocyanate process. This purity met the specifications (above 95 wt.% AmO_2 and below 1 wt.% individual contaminant elements) established by the US DOE Isotope Pool at ORNL.

4. MOLTEN SALT EXTRACTION PROCESS

MSE has been used successfully at RFP to remove americium from plutonium metal since 1967.[4,5] The current process places 2 kg batches of molten plutonium metal in contact with a molten salt mixture containing 35 mol % NaCl, 35 mol % KCl and 30 mol % $MgCl_2$ under an argon atmosphere. The alkali metal salts act as a diluent and $MgCl_2$ reacts with americium according to the reaction

$$Am + 3/2MgCl_2 \xrightarrow{750\ °C} AmCl_3 + 3/2Mg \qquad (1)$$

A small amount of plutonium is also lost to the salt as $PuCl_3$. To minimize both the plutonium losses and the quantity of salt used, and to achieve the proper level of americium removal, a two-step counter-current extraction scheme is used (Fig. 1). The residue salts, processed by a variety of aqueous methods, are currently the only recoverable source of americium at RFP.

5. ROCKY FLATS PLANT AMERICIUM RECOVERY PROCESS

The present method of processing MSE residue salt at RFP is outlined in Figure 2. MSE residues are first dissolved in dilute hydrochloric acid, and the resulting solution is filtered. After allowing for acidity adjustments, the filtrate is passed through Dowex 50-x8 cation exchange resin to convert from a chloride to nitrate system and remove gross amounts of monovalent impurities.[6] The trivalent actinides are sorbed along with trivalent and divalent impurities. The loaded resin bed is washed with dilute nitric acid; subsequently the actinides are eluted with 7M HNO_3. This eluate is passed through an Amberlite IRA-938 resin column where plutonium loads on the resin and americium follows the effluent. Americium in the effluent is precipitated as the oxalate and calcined to AmO_2 at 600 °C.

RFP has found the present cation exchange process very efficient for removing chlorides and relatively insensitive to fluctuations in process parameters. However, large volumes of acid are required to elute the plutonium. Use of a macroreticular resin (Dowex MSC-1) would decrease the magnitude of this latter problem.[7] Ion selectivity is another problem which results in residual cations such as calcium, lead, magnesium and potassium being separated from the plutonium in the anion exchange step but are carried to a limited extent into the americium oxalate precipitation step.

To replace cation exchange, a carbonate precipitation process is presently being tested at RFP.[8] The actinides are first precipitated as carbonates[9] and then redissolved in 7M HNO_3 before being sent to an anion exchange step. However, the narrow pH range (5 - 6) required is difficult to control, and calcium and lead contamination in the americium product create a further problem.

6. NEW SOLVENT EXTRACTION PROCESS

A variety of experimental MSE residues and alloys contain aluminum,

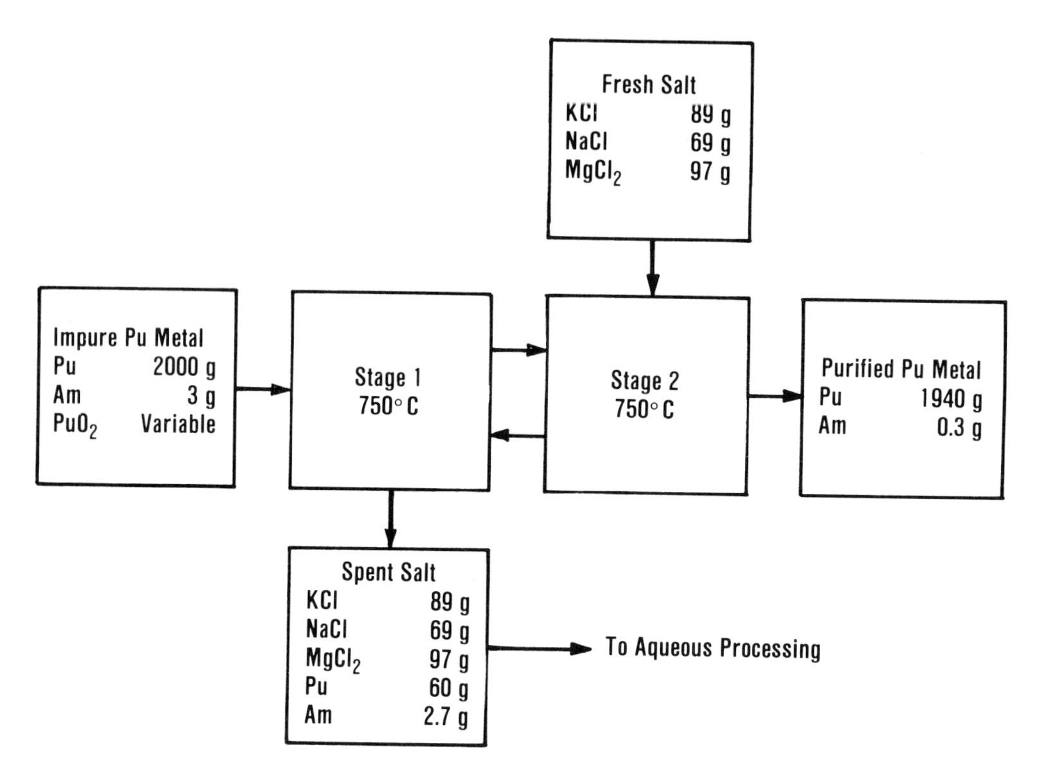

Figure 1. RFP MSE process (the asterisk indicates that the salt-to-metal ratio is variable, depending on the americum content of plutonium).

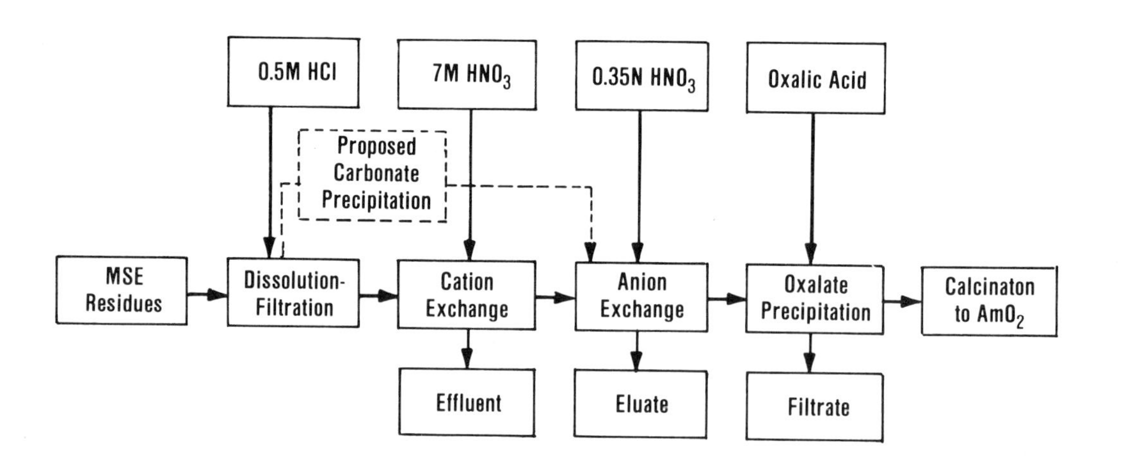

Figure 2. RFP americium recovery process.

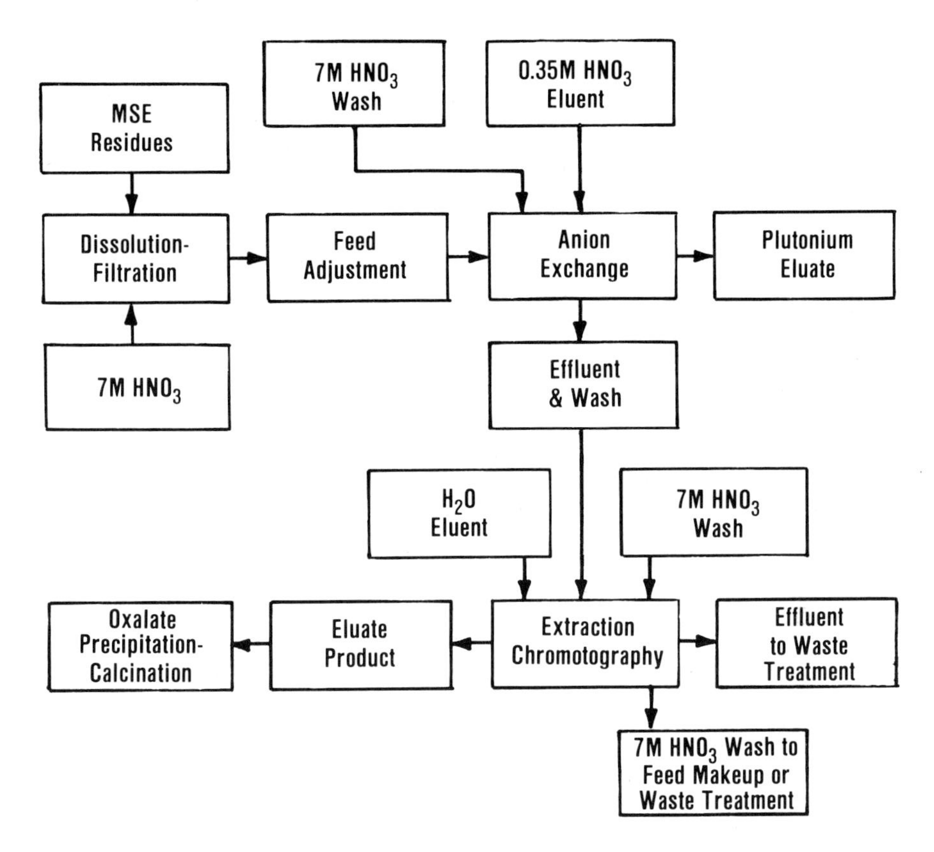

Figure 3. CMP process.

calcium, zinc and other minor impurities such as lead. Because the present RFP americium recovery process does not effectively separate these elements, it was necessary to develop a new carbamoylmethyl-phosphonate (CMP) solvent extraction process.[10,11] In this process, anion exchange and extraction chromatography techniques are combined (Fig. 3) in the first large-scale application of a bifunctional solvent extractant in an extraction chromatography mode. The key to the process is the selective extraction of americium by dihexyl-N, N-diethylcarbamolymethylphosphonate which is sorbed on an inert solid support of Amberlite XAD-4 (a non-ionic polystyrene divinylbenzene resin), manufactured by the Rohm and Haas Co.

Plutonium is removed from the 7M HNO_3 feed solution containing actinide and impurity elements by sorption on an anion exchange process. To selectively sorb the americium, the anion column effluent and wash solutions pass through the supported CMP extractant. The extractant is then washed with 7M HNO_3 to remove residual impurities. Following its elution with water, americium is finally precipitated from solution as the oxalate and calcined to AmO_2.

That americium can be effectively separated and purified from impurities such as aluminum, copper, chlorine, fluorine, iron, lead, magnesium, plutonium, potassium, sodium and zinc was demonstrated by CMP pilot plant operations at RFP.[12] The oxide produced (100 g) contained 96.5 wt % AmO_2 with 0.085 wt % Pu and less than 0.15 wt % of any other individual impurity element. Throughout the trial period, no significant hydrolysis or radiolysis effects on the separation materials were observed. Because of the excellent americium recovery, low waste volumes and the ability to use existing ion exchange equipment, RFP found this process to be very attractive. Moreover, by sorbing the relatively expensive solvent extractant on a support, significantly less solvent inventory is required compared to liquid-liquid extraction. Furthermore, a small back-up column of Amberlite XAD-4 support prevents loss of extractant to the aqueous stream.[13]

Production testing of the CMP process is presently underway at RFP. The results will determine the length of time the extractant support will maintain its capacity for americium, the necessary frequency of hydrolysis product removal by washing with NaOH and the overall deterioration and degradation of the support.

7. CONCLUSIONS

RFP has played an important role in developing new ^{241}Am separation and purification processes. These processes, which include both molten salt and organic solvent extraction, precipitation and ion exchange, have provided high purity americium oxide to the US DOE Isotope Pool. The production of americium as a by-product of plutonium scrap recovery operations will continue to require improved recovery methods, however.

8. ACKNOWLEDGMENTS

This paper was prepared under contract with the US Department of Energy.

9. REFERENCES

1. W. W. Schulz, The Chemistry of Americium, Rep. TID-26971, Energy Research and Development Administration, Washington, DC, 1976.

2. E. L. Christensen, L. W. Gray, J. D. Navratil and W. W. Schulz, 'Present Status and Future Directions of Plutonium Process Chemistry'. In W. T. Carnall and G. R. Choppin (eds.), Plutonium Chemistry, American Chemical Society, Washington, DC, 1983, p. 349.

3. V. A. Ryan and J. W. Pringle, 'Preparation of Pure Americium', USAEC Rep. RFP-130, 1960 (Dow Chemical Company, Golden, CO).

4. J. B. Knighton, R. C. Auge, J. W. Berry and R. C. Franchini, 'Molten Salt Extraction of Americium from Molten Plutonium Metal', U.S. ERDA Rep. RFP-2365, March 1976 (Dow Chemical Company, Golden, CO).

5. J. B. Knighton, P. G. Hagan, J. D. Navratil and G. H. Thompson, 'Status of Americium-241 Recovery at Rocky Flats Plant'. In J.D. Navratil and W. W. Schulz (eds.), Transplutonium Elements - Production and Recovery, American Chemical Society, Washington, DC, 1981, p. 53.

6. S. G. Proctor, 'Cation Exchange Process for Molten Salt Extraction Residues', U.S. ERDA Rep. RFP-2347, March 1975 (Dow Chemical Company, Golden, CO).

7. R. A. Silva and J. D. Navratil, Solvent Extr. Ion Exch., **1** (1983) 824.

8. P. G. Hagan and F. J. Miner, U.S. Patent 3,996,331, 1976.

9. P. G. Hagan, J. D. Navratil and R. S. Chichorz, J. Inorg. Nucl. Chem., **43** (1981) 1054.

10. J. D. Navratil, L. L. Martella and G. H. Thompson, 'Americium Recovery and Purification at Rocky Flats'. In J. D. Navratil and W. W. Schulz (eds.), Actinide Separations, American Chemical Society, Washington, DC, 1980, p. 455.

11. J. D. Navratil, L. L. Martella and G. H. Thompson, Proc. Int. Solvent Extraction Conf., Vol. 3, Univ. of Liege, Liege, 1980, Paper 106.

12. W. I. Yamada, L. L. Martella and J. D. Navratil, J. Less-Common Met., **86** (1982) 211.

13. J. D. Navratil, J. Nucl. Sci. Technol., **18** (1981) 561.

PRODUCTION OF AMERICIUM ISOTOPES IN FRANCE

G. KOEHLY, J. BOURGES, C. MADIC, T.H. NGUYEN, M. LECOMTE
Département de Génie Radioactif, Service des Etudes de Procédés, Section des Transuraniens, Centre d'Etudes Nucléaires de Fontenay-aux- Roses, B.P. n° 6 92260 Fontenay-aux-Roses, France

ABSTRACT

The program of productions of americium 241 and 243 isotopes is based respectively on the retreatment of aged plutonium alloys or plutonium dioxide and on the treatment of plutonium targets irradiated either in CELESTIN reactors for Pu-Al alloys or OSIRIS reactor for plutonium 242 dioxide. All the operations, including americium final purifications, are carried out in hot cells equipped with remote manipulators. The chemical processes are based on the use of extraction chromatography with hydrophobic SiO_2 impregnated with extracting agents. Plutonium targets and aged plutonium alloys are dissolved in nitric acid using conventional techniques while Plutonium dioxide dissolutions are performed routine at 300 grams scale with electrogenerated silver II in 4 M HNO_3 at room temperature. The separation between plutonium and americium is performed by extraction of Pu(IV) either on TBP/SiO_2 or $TOAHNO_3/SiO_2$ column. Americium recovery from waste streams rid of plutonium is realized by chromatographic extraction of Am(III) using mainly TBP and episodically DHDECMP as extractant. The final purification of both americium isotopes uses the selective extraction of Am(VI) on $HDDiBMP/SiO_2$ column at 60 grams scale. Using the overall process a total amount of 1000 grams of americium 241 and 100 grams of americium 243 has been produced nowadays and the AmO_2 final product indicates a purity better than 98.5 %.

Introduction

The French Atomic Energy Commission is engaged in two programs of producing americium 241 and 243 mainly to satisfy domestic and foreign requirements of these isotopes and to contribute to the solution of the elimination americium 241 from industrial nuclear wastes.

Americium 243 - Curium 244 mixtures were originally produced by irradiation of plutonium 239/Al targets in the Celestin reactors at the Marcoule Plant /1/ and are currently being produced by irradiation of $^{242}PuO_2$ targets in the Osiris reactor at Saclay.

N. M. Edelstein et al. (eds.), Americium and Curium Chemistry and Technology, 301–320.

Americium 241 is typically recovered from two sources of supply :

- recycling of aged plutonium alloys or dioxide,
- processing of large volumes of alpha active wastes. /2/

All the chemical processes for the production and the purification of these isotopes are carried out in hot cells using nitric acid media and are performed by means of the liquid-liquid chromatography (L.L.C.) technique /3/ /4/.

The total amounts of americium and curium isotopes produced since the beginning of the program are respectively :

Americium 243 100 grams
Americium 241 one kilogram
Curium 244 100 grams

This paper describes the experience gained over ten years and focuses mainly on some original processes developed at the pilot plant scale.

EXPERIMENTAL

Americium 243 and Americium 241 sources

The main characteristics of the targets irradiated to produce americium 243 are given Table I.

TABLE I
Characteristics of irradiated targets

	$^{239}Pu/Al$ targets	$^{242}PuO_2$ targets
Fuel element		
Total mass	5445 g	375 g with 360 g Ni
Mass of plutonium	400 g	12.6 g
Dimensions	1067 x 79.7 x 67.1 mm	Ø 22 mm, L 90 mm
No. of elements	11 plates	3 orthocylindrical containers
Irradiation, Cooling		
Reactor	Celestin (Marcoule)	Osiris (Saclay)
Integrated flux	11.28 n·kb-1	1.9 n·kb-1
Cooling time	3 years	2 months
Composition after cooling		
Actinide mass :		
Plutonium 242	44 to 50 g	10.280 g
Americium 243	8 to 10 g	1.600 g
Curium 244	7.5 to 9 g	0.780 g
Fission products :		
Total mass	340 g	0.560 g
Rare earths	240 g	-
β , γ Activity	3.7×10^4 Ci	150 Ci

A $^{242}PuO_2$ target element consists of an annular nickel container, the annular region of which has been filled with 5 g of plutonium dioxide of isotopic composition exceeding 92 % ^{242}Pu. Irradiated Pu/Al targets are generally processed after three years cooling while PuO_2 targets are processed after one or two months cooling.

Two main sources provided the americium 241 :

- Processing of aged stocks of plutonium metal alloys or plutonium dioxide rich in plutonium 241 on the scale of 300 g of fissile material.
- Recovery of americium from alpha aqueous liquid wastes resulting from the reprocessing of scrap from fabrication of (U, Pu)O_2 fuels.

The annual volume of these liquid wastes are estimated to be several tens of m^3 of highly varied chemical composition. The compositions of two americium 241 liquid wastes are given Table II.

TABLE II
Characteristics of alpha wastes solutions

Component		VALDUC criticality station	MASURCA
HNO_3 (N)		7.85	1.1
Uranium	g/l	2	12
Neptunium	g/l	-	0.18
Plutonium	g/l	0.50	0.07
Americium 241	g/l	2.36	0.108
Iron	g/l	9.2	11.1
Cadmium	g/l	-	35.4
Nickel	g/l	4.5	1.01
Chromium	g/l	4	1.5
Gadolinium	g/l	0.07	-
Cerium 144	mCi/l	*	0.28
Ruthenium 106	mCi/l	*	1.34
Cesium 137	mCi/l	*	9.1

* At very low concentration.

Hot Cells.

The equipment and operation of the Hot Cells have been described in detail in a previous paper /3/, so the working principles will only be summarized here :

- Nitric acid is used to minimize corrosion of stainless steel equipment.
- The dissolution of irradiated targets are performed and the high beta-gamma activity cycle in the hot cell named Petrus, surrounded by 80 cm of barytes concrete. The process for the medium beta-gamma activity cycle is carried out in Candide, a hot cell shielded with 15 cm of lead.

. The recovery and purification of americium 241 from waste solutions is carried out in hot cells shielded with 10 cm of steel : Castor, Petronille and Irene.

Dissolvers

^{243}Am production

The irradiated targets are dissolved as indicated :

. For Pu/Al : in Uranus 55 stainless steel dissolver of 50-liter capacity using nitric acid with traces of mercuric ions.

. For PuO_2 : in a tantalum dissolver of 4-liters capacity using nitric acid and traces of HF.

To handle safely the various gaseous products that escape during the dissolution, the Uranus dissolver is equipped with soda lime, charcoal and iron filings absorbers, and the tantalum vessel with charcoal and silver zeolite columns for iodine trapping.

^{241}Am production

Plutonium alloys were generally dissolved in a tantalum reactor under reflux conditions with 14 M HNO_3, 0.05 M HF, while pure plutonium dioxide and plutonium metal scrap, previously converted to oxide by calcination in air at 500-600°C, were dissolved at room temperature in an electrolytic dissolver by electrogenerated silver(II) as described in the next paragraph.

Plutonium Oxide Dissolution with Electrogenerated Silver(II).

Recently, Bray and Ryan /5/ elucidated the conditions needed to perform a PuO_2 dissolution with electrogenerated strong oxidizing agents such as Ce^{4+}, Co^{3+}, Ag^{2+}, in nitric acid solutions free from fluoride ions. For our own purpose, the principle of a PuO_2 dissolution with Ag^{2+} was found sufficiently attractive that an extensive parametric study of this technique was undertaken. That study demonstrated that it was possible to define the experimental conditions to perform a quantitative plutonium dioxide dissolution with a current efficiency, equal to unity /6/. An experimental reactor that allows the dissolution of 300 g. of pure aged plutonium dioxide per day has been tested and determined to be useful to the americium production program.

The plutonium dioxide dissolution with Ag(II) at the 300-gram scale is typically performed under in the following chemical conditions :

Nitric acid : 4 M
Temperature : 30 ± 10°C
Silver concentration : 0.05 M
Current density : 35 $mA \cdot cm^{-2}$

The dissolver used for this technique is a 3.5-liter-capacity glass electrolyzer composed of two separated compartments :

- An anodic compartment equipped with a platinum grid having a surface area of 850 cm^2.

- A cathodic compartment consisting of a 0.2-liter-capacity aluminum silicate diaphragm in which is immersed a tantalum rod 6 mm in diameter.

The heat generated by the electrolysis is removed by a coolant flowing through a glass coil. The plutonium dioxide powder and the anolyte solution are vigorously stirred by a teflon propeller driven by an electric motor. The electrolytic dissolver is also equipped with a temperature sensor to control the cooling unit dissolver and with a liquid draw-off line.

Chromatographic equipment

The chromatographic equipment includes : a plexiglass column packed with a stationary phase impregnated with a suitable extractant, positive-displacement proportioning pumps of Prominent Label, storage tanks for various solutions and a detector for in-line analysis at the column outlet.

As a rule, the L.L.C. columns are discarded after use. The dry packed columns are placed in the hot cell where they are pre-equilibrated with solutions that displace the air and produce the chemical conditions for extraction. When a column has been used to the point of a 50 % decrease in exchange capacity, it is dried with compressed air, cut into sections and removed as solid waste.

Our studies have been directed toward the application to L.L.C. of pure compounds for which extraction properties are well known as a result of prior use in liquid-liquid extraction processes.

The preparation and characteristics of the stationary phase was reported in references /3/ and /4/, so, the characteristics of typical columns used for actinide extraction will only be summarized in Table III

TABLE III

Characteristics of liquid-liquid chromatographic (L.L.C.) columns used in extraction of uranium, plutonium, and americium.

Stationary phase (%) by weight	Extracted compound	Exchange capacity $mole \cdot kg^{-1}$	Exchange capacity (g) for actinides (2.8 kg column)
TBP (27 %)	$UO_2(NO_3)_2(TBP)_2$	0.507	337
	$Pu(NO_3)_4(TBP)_2$	0.507	339
	$Am(NO_3)_3(TBP)_3$	0.338	222
TOA (25 %)	$Pu(NO_3)_6(TOAH)_2(TOAHNO_3)_2$	0.177	119
POX 11 (30 %)	$Pu(NO_3)_4(POX\ 11)_2$	0.433	290
	$Am(NO_3)_3(POX\ 11)_4$	0.216	145
DHDECMP (30 %)	$Am(NO_3)_3$ $(DHDECMP)_3$	0.275	185
HD(DiBM)P (30 %)	$AmO_2(D(DiBM)P_2(HD(DiBM)P)_2$	0.214	144

TBP = Tributylphosphate
TOA = Trioctylamine
POX 11 = Di-n-hexyloctoxymethylphosphine oxide
DHDECMP = Di-hexyl di-ethyl carbamoyl methylene phosphonate
DH(DiBM)P = Bis-2,6-dimethyl-4-heptyl phosphoric acid

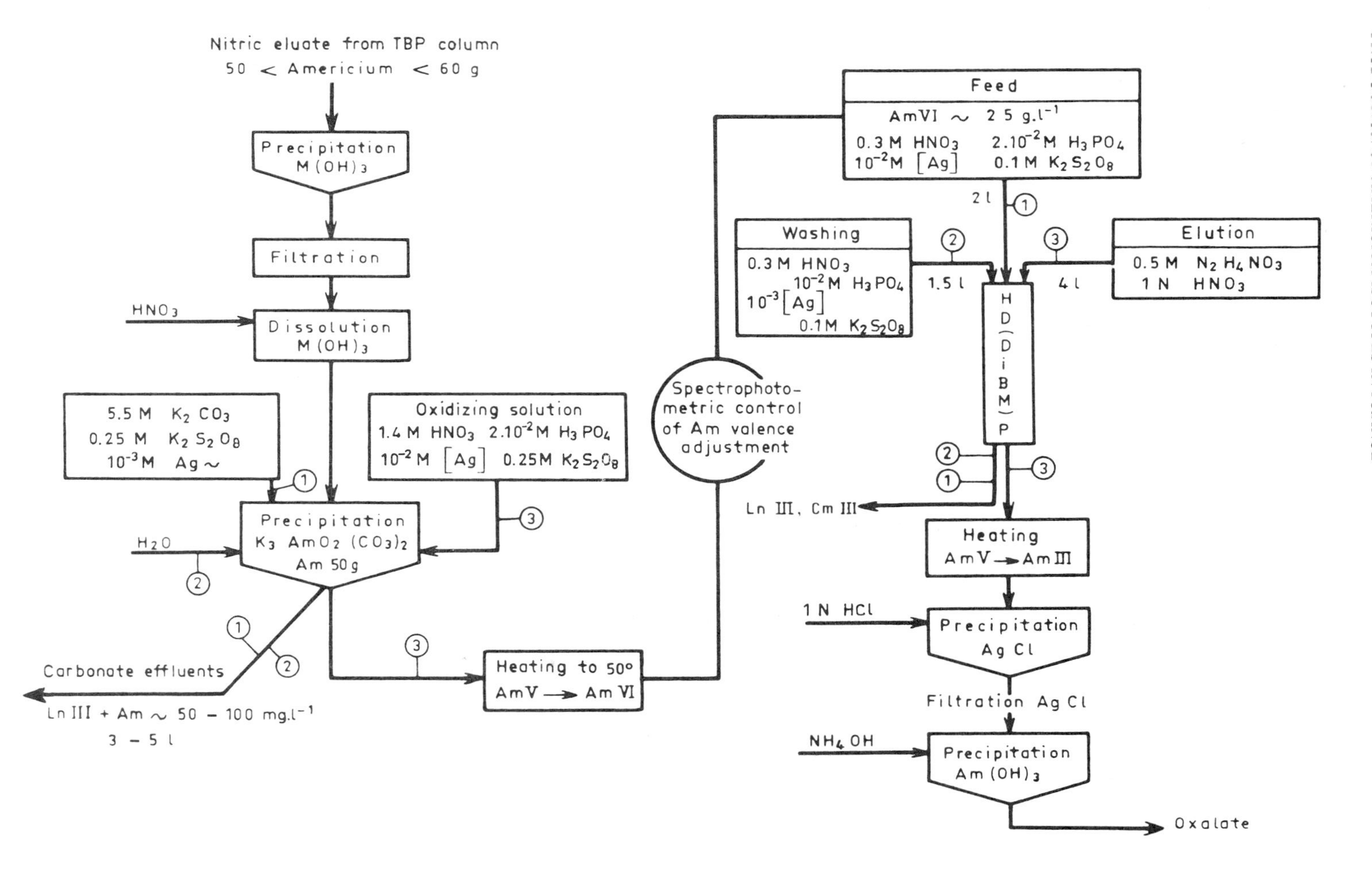

Figure 1 - Americium purification process combining precipitation of potassium americyl carbonate and Am(VI) L.L.C. extraction

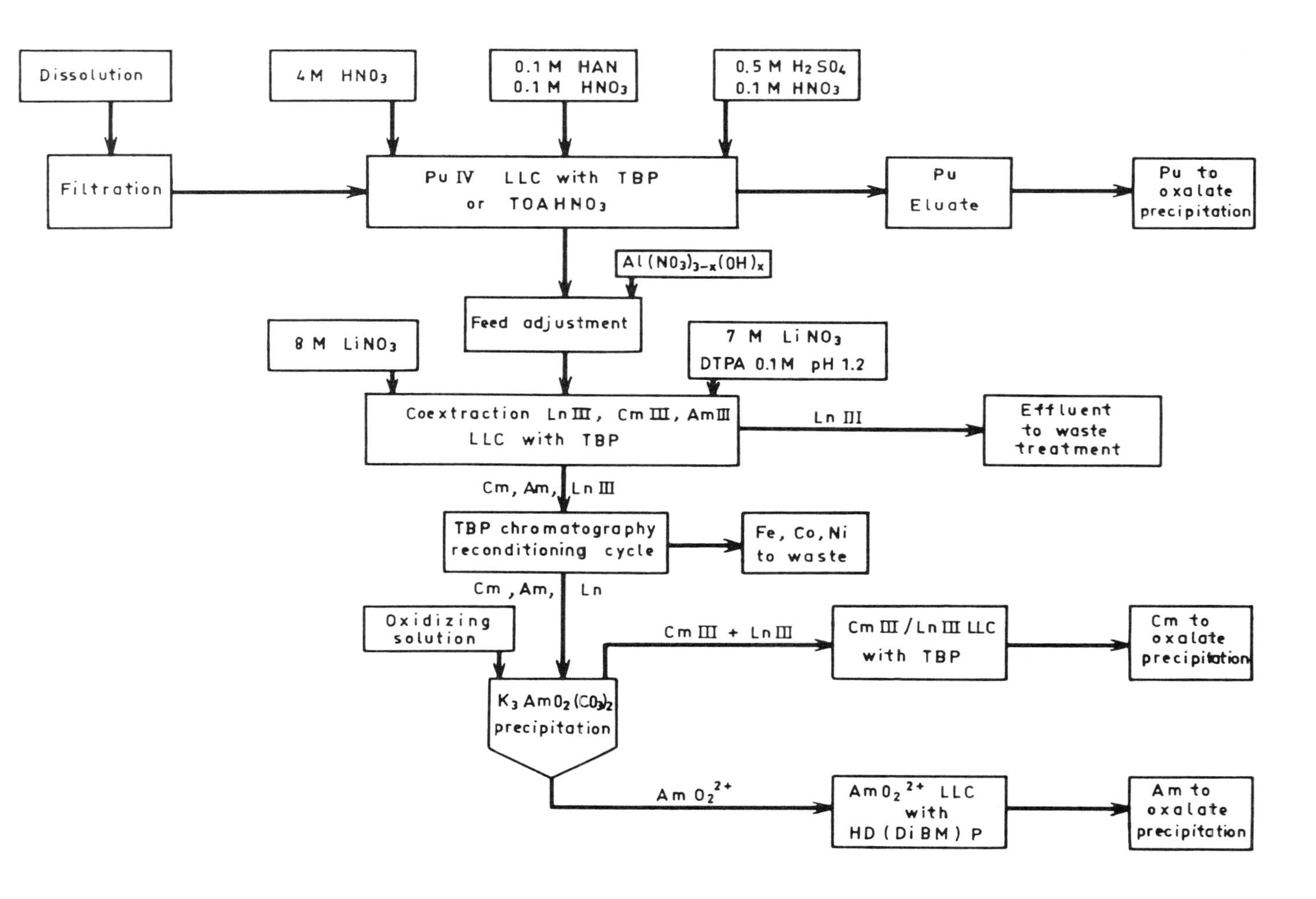

Figure 2 - Conceptual FLOWSHEET for chemical treatment of Pu-Al and PuO_2 targets

Am(VI) Extraction Chromatography Process

This process is based on a combination of two processes : the precipitation of potassium americyl carbonate, first described at the multigram scale by Burney /7/, and the Am(VI) extraction chromatography by HD(DiBM)P solvent discovered by Peppard /8/ and applied first in L.L.C. to americium 243 at the multigram scale /1/.

The steps of the process, which is illustrated schematically in **Figure 1** and routinely carried out at the 50-gram scale are the followings :

- Precipitation of potassium americyl carbonate in 3.5 M K_2CO_3 solutions at 80 ± 10°C by addition of $K_2S_2O_8$ /5/.
- Dissolution of the water-washed precipitate with an oxidizing acid solution containing $K_2S_2O_8$ and Ag(II) ions acting as a catalyst in the oxidation of Am(V) to Am(VI) by $S_2O_8^{2-}$ ions.
- Total conversion of Am(V) to Am(VI) by warming the resultant solution to 50°C and checking the oxidation quantitativity by spectrophotometric measurement.
- Rapid Am(VI) extraction on a HD(DiBM)P column previously treated by an oxidizing solution containing $S_2O_8^{2-}$ and Ag(II) ions until a persistent dark color of Ag(II) ions is observed in the effluent.
- Slow elution of Am(VI) from the column by a reducing solution 0.5 M, $N_2H_4NO_3$, 1 M, HNO_3 (after oxidizing washes eliminate the impurities).

Under ideal conditions, the americium is recovered in 2.5 to 3 liters of eluate.

Analytical controls

In addition to the routine controls, α and γ counting and spectrometry are used as well as in-line gamma detectors /3/ monitoring the operation of the chromatographic columns. The quality of the final products (AmO_2 or CmO_2) is determined by mass spectrography and calorimetry.

The oxidation of americium to Am(VI) is controlled by a Hewlett Packard 8450 A spectrophotometer.

RESULTS AND DISCUSSION

Processing of irradiated targets

The production of americium 243 and curium 244 was initiated around 1970 by the irradiation of Pu/Al targets. The chemical processing of the industrial targets /3/ has generated a stock of 400 grams of plutonium 242 of average isotopic purity around 90 % which now supplies a high isotopic purity americium 243 by irradiation of $^{242}PuO_2$ targets.

The conceptual flowsheet for the chemical processing of both types of targets **(Figure 2)** is the same.

After nitric acid dissolution of the targets (see Tables I and II), the first step of the process is the recovery and purification of plutonium 242 which is accomplished by extraction chromatography using $TOAHNO_3$ or TBP columns. To prevent clogging of the columns by solids (essentially fission product precipitates) the feed solution is filtered through a small pre-column containing a silica bed.

After the column is washed with nitric acid solution, the elution of plutonium 242 is performed with sulfonitric acid for TOAHNO3 or hydroxylammonium nitrate (HAN) reducing solution for a TBP column. The plutonium eluate undergoes the classical procedure of Pu(IV) oxalate precipitation to obtain plutonium dioxide after calcination of the oxalate. The second step of the process concerns the separation of americium-curium from lanthanides, which is performed on a TBP column in high nitrate medium using DTPA. The L.L.C. TBP process /1/ /3/ /4/ implies three successive stages :

- Coextraction of lanthanides and americium-curium trivalent elements on a TBP column from a feed solution previously adjusted to 0 ± 0.05 free acid and to 1.8 M in $Al(NO_3)_3$ by use of aluminum nitrate solution deficient in nitrate ion ($Al(NO_3)_{3-x}(OH)_x$.)
- Elimination of certain fission products : (e.g. Ru, Cs) by scrubbing the column with 8 M $LiNO_3$ salting-out solution.
- Sequential elution of curium and americium by 7 M $LiNO_3$, 0.1 M DTPA, pH 1.2, permitting the collection of curium, americium, europium and cerium. (in order of appearance).

As the intergroup separation factor is not sufficiently high /1/, complete decontamination of the mixture of americium and curium from lanthanides necessitates several cycles using 7 M $LiNO_3$, 0.1 M DTPA, pH = 1.2 elution solution. A new TBP cycle can be carried out from a previous eluate by neutralizing the complexing power of DTPA by addition of 0.5 M $Al(NO_3)_3$.

The separation of americium 243 from curium 244 constitutes the third step of the process. This separation can be accomplished either by the L.L.C. TBP process or by precipitation of potassium americyl carbonate. However, as the final americium purification is performed by Am(VI) extraction, the americium-curium separation is preferably done by precipitation of $K_3AmO_2(CO_3)_2$ which supplies a suitable product for the Am(VI) L.L.C. process. The L.L.C. TBP process is mainly used to purify curium. The purity of the americium 243 ranges from 97 to 98 %.

The processing of Pu/Al target by means of the L.L.C. technique does not involve major problems and is typically accomplished in a three month campaign. The overall process starting from 88 liters of dissolution solution generates around 250 liters of highly salted liquid wastes. The overall actinide recovery is 96 %. The amount of insolubles during dissolution is estimated to be less than 2 %. In comparison, the PuO_2 target processing, which was expected to be easier to perform because of lower initial concentrations of lanthanides, encountered two major difficulties :

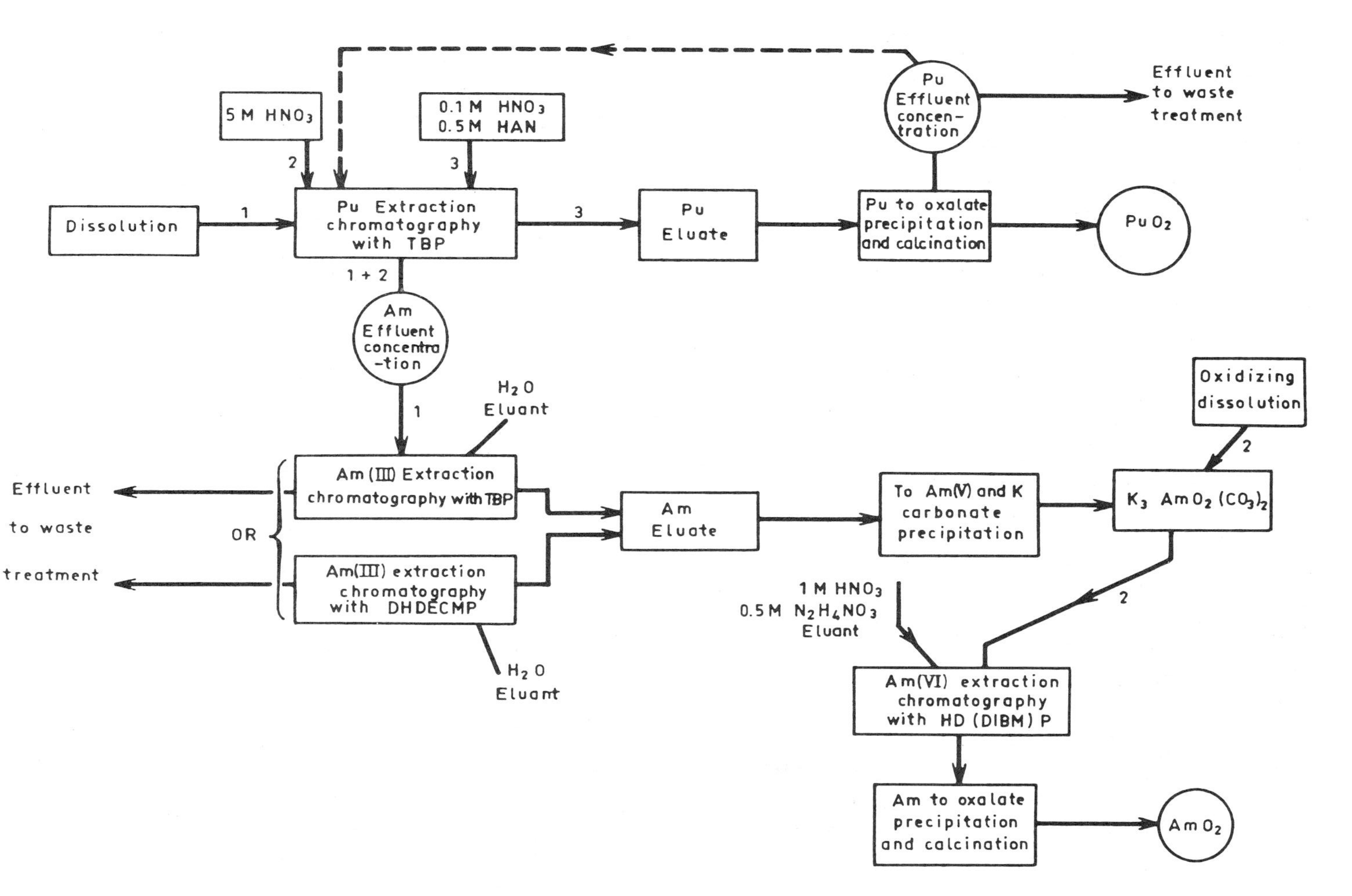

Figure 3 - Conceptual flow diagram for recycling of aged plutonium alloys on plutonium dioxide

- Potential releases of radioactive iodine outside the hot cell during the dissolution step must be monitored.
- The presence of high nickel concentration in solution (1.7 M) ahead of the first cycle of americium and curium extraction complicates the adjustment of the feed solution to 0.0 ± 0.2 H^+ and 1.7 M Al^{3+}.

The main advantage of conducting the final americum 243 purification by the AmO_2^{2+} L.L.C. process is to obtain nearly complete decontamination of americium from curium and trivalent lanthanides. However, poor decontamination factors of americium from plutonium and cerium (^{144}Ce) are observed in this process, since both elements are extracted on the HD(DiBM)P column (PuO_2^{2+} for Pu, Ce(IV) for Ce) during the fixation step and are recovered with the americium during the reducing elution (partially in the case of plutonium). Therefore, traces of plutonium and cerium should be eliminated in prior steps before using this process.

Recycling of aged plutonium alloys or plutonium dioxide

A total amount of 5 kilograms of plutonium alloy or metal scrap or plutonium dioxide containing 5 % of americium by weight has been recycled according to the flow diagram shown in **Figure 3.**

An example of the kinetics of the dissolution of 300 g of PuO_2 is presented in Figure 4. Quantitative dissolution of this PuO_2 is obtained in about 210 minutes; this corresponds to an average current efficiency of 56 %.

After dissolution : the resultant solution undergoes a filtration and a chemical reduction by H_2O_2 to adjust the plutonium to the tetravalent state. Aluminium nitrate is added to complex excess fluoride when the dissolution is performed with HF. Pu(IV) is extracted at the 300 g-scale on two TBP columns in series (2.7 kilograms each). Americium passes through the columns during this step. After washing the columns, the plutonium is eluted with a reducing solution of 0.05 M HNO_3, 0.5 M HAN, and 0.1 M $N_2H_5NO_3$ at 50°C. The plutonium eluate is acidified then heated slowly to 60°C in order to destroy excess reducing agents and to oxidize plutonium(III) to plutonium(IV). Plutonium is converted to the oxalate form by precipitation at 60°C. The oxalate is converted to oxide by calcination at 500-600°C for 4 hours. The plutonium losses are minimized by systematically concentrating and recycling the effluents.

The raffinate containing the americium is concentrated by a factor of ten by evaporation of the effluents to provide a stock of concentrated americium solution which is periodically treated by the L.L.C. technique using either TBP or DHDECMP as described in the next paragraph.

This recycling of aged plutonium has resulted in the production of 250 grams of americium 241.

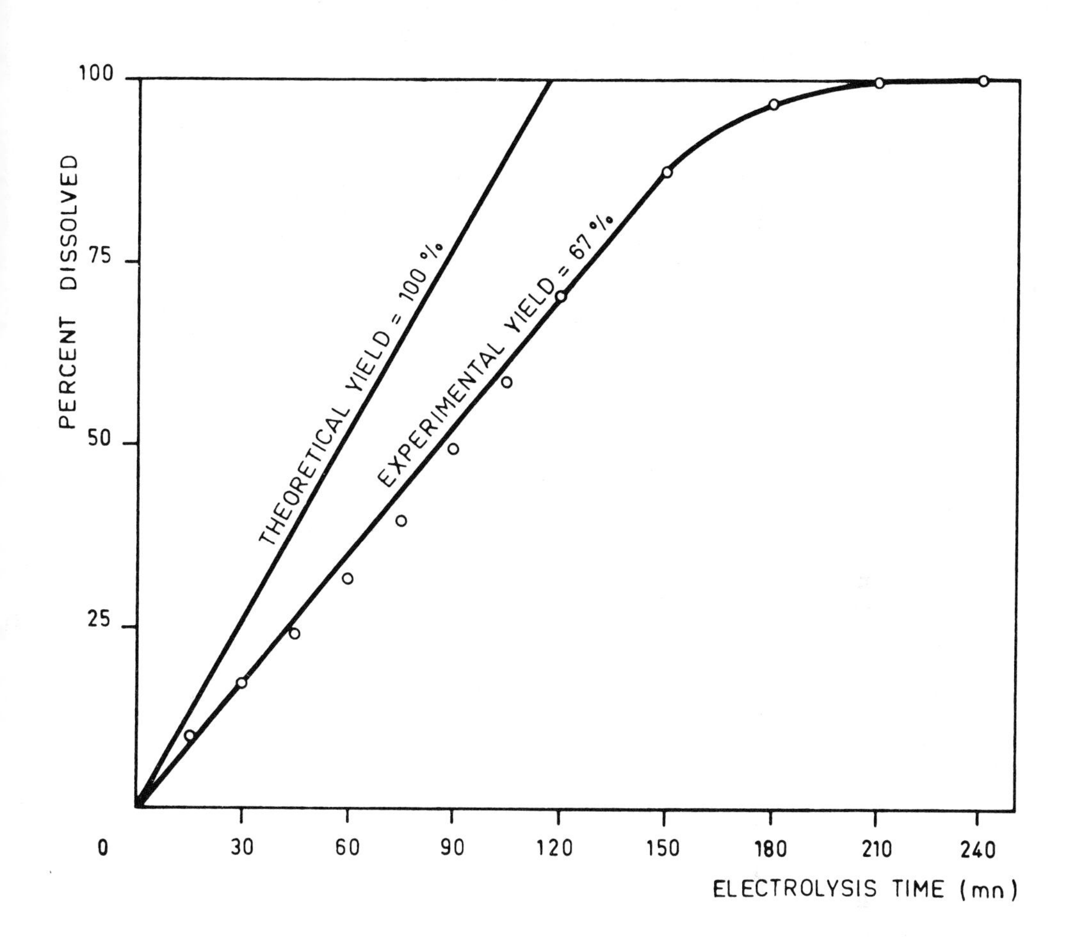

Figure 4 - PuO_2 dissolution with electrogenerated Silver II

Experimental conditions

PuO_2 295.5 g I = 30 ± 0.1 A V = 5 V T = 24 ± 0.5°C

Anolyte solution

4 M HNO_3 0.1 M Ag Volume 3.5 liters

Catholyte solution

8 M HNO_3 Volume 0.2 liter

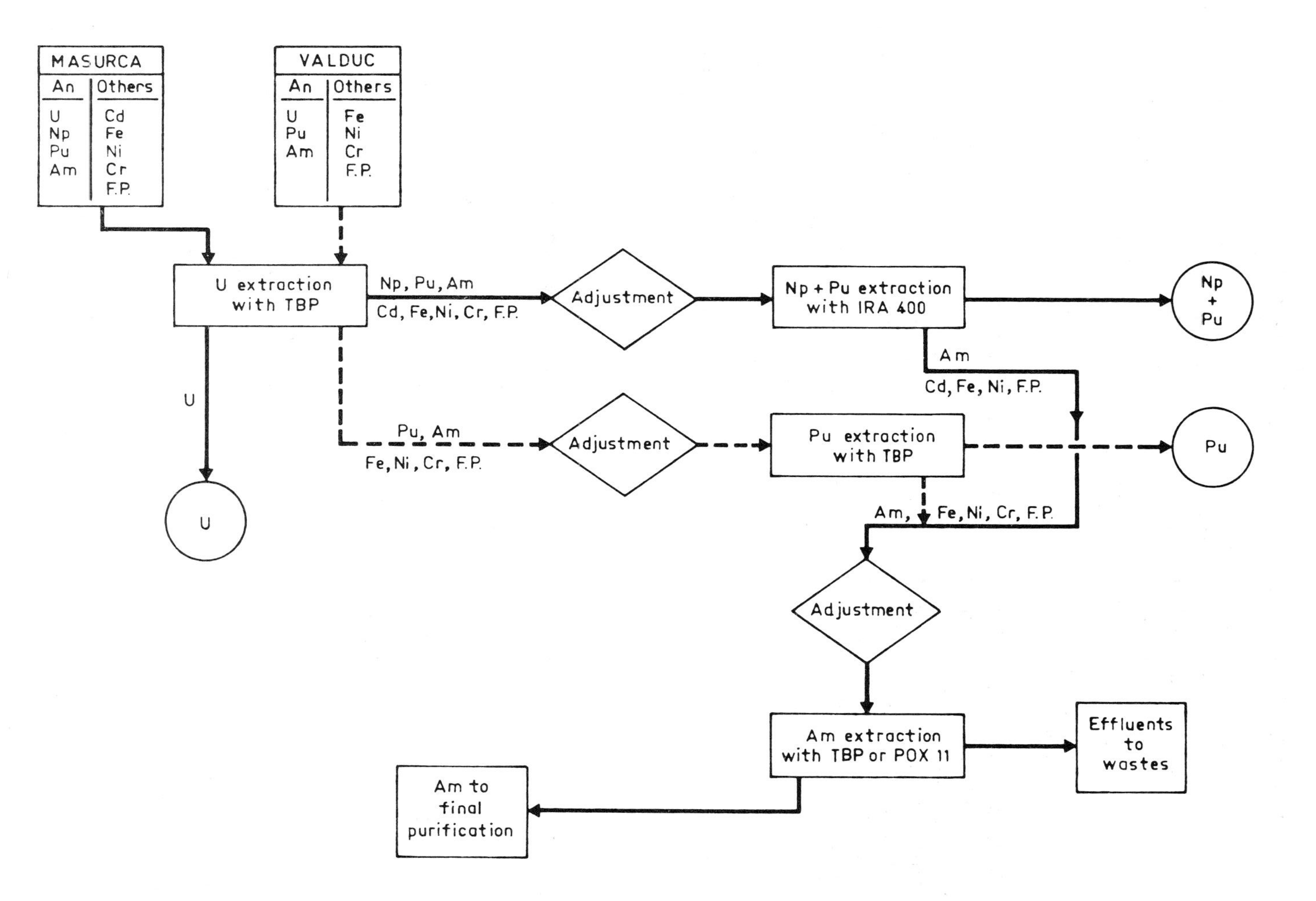

Figure 5 - Conceptual flow diagram for decontamination of alpha liquid wastes

Recovery of americium 241 from alpha aqueous wastes

The objective of the waste treatment is first, the elimination of alpha emitters from the waste and second, the recovery of americium 241. The conceptual flow diagram for both types of liquid waste (Table III) is presented in **Figure 5.** As the chemical process flowsheet for Masurca wastes has been described earlier /3/, we shall only discuss the major principles here in dwelling on the three extractants tested in L.L.C. technique for Am(III) extraction : TBP, POX 11 and DHDECMP, an extracting molecule developed in the U.S.A. /9/ /10/.

Because the uranium and plutonium present in such effluents generally exhibit a higher affinity for these extractants than americium, their prior elimination is necessary. The U + Pu coextraction is performed with TBP either by liquid-liquid extraction in mixer-settlers when the U + Pu concentration is high (U = 12 $g \cdot l^{-1}$ for Masurca wastes) or by L.L.C. when the exchange capacity of the TBP column is sufficient (as for Valduc criticality station wastes).

This operation does not require prior feed adjustment because the acidity and nitrate ion concentration of the wastes are sufficiently high. After elimination of U + Np + Pu from the liquid wastes (neptunium present in Masurca wastes requires a special extraction cycle) the Am(III) extraction chromatography by the above three extractants requires that the feed solution to be adjusted to the chemical conditions given in **Table IV.**

TABLE IV

Usual chemical conditions for Am(III) extraction chromatography

Extractant	Feed solution adjustment			k_D Am(III) ($ml \cdot g^{-1}$)
TBP	1.8 $\underline{M}$ $Al(NO_3)_3$	0.00 ± 0.02 $\underline{M}$	H^+	200
POX 11	3.6 $\underline{M}$ $LiNO_3$	0.1 $\underline{M}$	HNO_3	5×10^3
DHDECMP	5 $\underline{M}$ HNO_3			25

However the above chemical conditions for efficient americium recovery were found inadequate for the treatment of Masurca and Valduc liquid wastes containing high concentrations of hydrolyzable cations such as Fe(III), Ni(II), Co(II), and Cd(II). Nevertheless, complementary studies demonstrated :

- L.L.C. with TBP could be performed if the neutralization of the feed solution to a free acidity around zero is done in the presence of DTPA or ammonium oxalate in equal concentration to the hydrolyzable cations (Fe, Ni, Co). The hydrolysis of cations is indeed suppressed and the k_D Am(III) value of 100 is preserved in presence of high aluminum nitrate concentration.

- L.L.C. with POX 11 is efficient in the presence of large amounts of iron(III) if EDTA has been added in equal concentration to the Fe(III). This avoids the slow extraction of Fe(III) by POX 11 and restores all the extractive properties of POX 11 for Am(III) /3/.
- L.L.C. with DHDECMP can be used to extract trivalent americium from moderate volumes of 5 to 7 M HNO_3 in spite of the relatively low maximum k_D value of Am(III). If americium recovery from larger volumes is required, it is necessary to increase the k_D Am(III) value by adding a salting-out agent such as $LiNO_3$ and by simultaneously lowering the acidity of the feed solution /11/. In related studies, investigations have shown that the exchange kinetics is faster in L.L.E. than in L.L.C. For instance, equilibrium is reached in two minutes in liquid-liquid extraction in HNO_3 or $LiNO_3$ salted medium but may require ten minutes in L.L.C. (DHDECMP 0.82 M/SiO_2) with the same media.

In spite of complicated feed solution adjustments, POX 11 and TBP extractants used in the L.L.C. technique have worked out satisfactorily for americium recovery from 4 m^3 of Masurca wastes and 0.4 m^3 of Valduc wastes. In a period of two years working at a normal pace, six chemist-technicians have recovered 1.020 kg of americium 241. Both processes have produced a total amount of 8 m^3 of highly salted waste solutions containing less than 1 mg/l of americium : the acceptable release limit for a subsequent waste treatment. The L.L.C. with DHDECMP absorbed on SiO_2, tested episodically on the first americium recovery cycle from Valduc wastes, was found inadequate because of its relatively low extraction power and its slow exchange kinetics. But DHDECMP in L.L.C. was much more effective in americium recovery from reacidified carbonate effluents resulting from the potassium americyl carbonate precipitation process. Generally, a second cycle of Am(III) chroma-tographic extraction with TBP was necessary to improve the americium decontamination from hydrolyzable cations such as Fe, Co, Ni, Cd, etc... before proceeding to the final americium purification by AmO_2^{2+}extraction.

The final americium purification, beginning with the process combining $K_3AmO_2(CO_3)_2$ precipitation and Am(VI) L.L.C. extraction, routinely performed at the 50-g scale as illustrated in Table V, has required the attainment of the following experimental conditions :

- The Am(III) chromatographic extraction preceeding the potassium americyl carbonate precipitation must be efficient enough to leave in the Am(III) eluate low concentrations of transition elements able to form insoluble hydroxides or carbonates which could hinder Am(III) solubilization in potassium carbonate medium.
- The correct AmO_2^{2+} chromatographic extraction procedure requires that the Am(V) present in $K_3AmO_2(CO_3)_2$ be quantitatively oxidized to Am(VI) by an oxidizing acid dissolution and the subsequent Am(VI) extraction be carried out rapidly to limit the radiolytic reduction of Am(VI) to Am(V) or to Am(III) in americium 241 concentrations such as 25 $g \cdot l^{-1}$.

Experiment No	^{241}Am feed (g)	VOLUMES OF SOLUTIONS: Feed (l)	Wash (l)	Elution (l)	Reconditioning (l)	MASS OF 241 AMERICIUM: Fixation loss (g)	Wash loss (g)	Reconditioning loss (g)	Purified weight (g)	Purication yield (%)
1	39.47	2.0	1	3.2	-	0.690	0.66	.	38.29	97
2	46.5	1.75	1.3	5.68	4	1.39	3.22	0.230	39.78	85
3	55	2.18	1.70	3.39	5	1.36	4.87	0.160	49,59	90
4	36.35	1.50	0.80	3.86	.	1.37	2.71	.	31.32	86
5	50	2.0	1.35	5.13	4.3	0.262	1.02	0.242	49.50	99
6	33.3	2.25	1.25	4.48	3.1	1.016	3.26	0.564	28.46	85
7	59.1	1.8	1.50	3.50	3	2.28	5.68	0.702	48.10	81
8	51.35	1.65	0.40	4.20	3	0.850	1.46	1.260	51.80	100
9	58.88	1.90	1.20	3.20	4.5	3.09	6.08	1.340	49.47	84
10	76.15	3.5	1.55	5.08	-	5.37	8.28	-	62.29	81

TABLE V

AmO_2^{2+} extraction chromatography on HD(DiBM)P column
Characteristics of columns used :
- in runs 1 to 4 = 0,8 kg column
- in runs 5 to 9 = 2,9 kg column

Flow rates : feed and wash 5 to 10 l,h^{-1} Elution : 0,5 l,h^{-1}

The average yield of the potassium americyl carbonate precipitation process was found 99.5 % while the AmO_2^{2+} L.L.C. yield has fluctuated from 80 to 100 % in spite of the above experimental precautions. The americium losses which occur mainly in the fixation and wash steps, were linked to incomplete oxidation and to radiolytic reduction of AmO_2^{2+} to AmO_2^{+} and Am^{3+} as detected by in-line spectrophotometric measurements.

In other studies, HD(DiBM)P adsorbed on SiO_2 at 30 % by weight was found to retain a small fraction of the AmO_2^{2+} randomly fluctuating from 0.5 to 2 % as observed in a number of chromatographic runs (Table V).

The subsequent steps of the americium 241 final purification include :

- Americium hydroxide precipitation by NH_4OH after elimination of Ag^+ by heating the eluate reducing solution at 60°C (see Figure 1).
- The conventional Am(III) oxalate precipitation /12/.
- The conversion of Am(III) oxalate to AmO_2 by calcination of the oxalate in a furnace at 800°C for 4 hours.

The above final steps involve a minimal loss of americium of 0.1 %, attributed to a slight americium carrying during AgCl precipitation and a slight solubility of americium in ammoniacal effluents (2 to 10 mg/liter).

The americium 241 purity, as determined by calorimetric measurements on weighed samples of 70 to 100 mg, was found to average better than 99 %.

CONCLUSIONS

Initially recovered from irradiated Pu-Al targets, americium 243 is now produced by processing irradiated $^{242}PuO_2$ targets. The chemical processing of both types of targets, performed mainly by extraction chromatography, remains the same. However, PuO_2 target dissolution provides moderate solution volumes with small quantities of fission products and permits a reduction in the processing time and a decrease in the size of the equipment. The typical isotopic purity of americium 243 products has increased from 96 to 99.2 % by weight.

The principal sources of supply of americium 241 have been and remain alpha active and aged plutonium scrap. Most of the chemical processes involve liquid-liquid chromatography operated on the scale of 100 grams of actinides. The L.L.C. technique was found quite appropriate for our requirements : most of our personnel work a normal schedule and the equipment is simple and light and easy to manage (easy to discard or to maintain). During the previous decade, several extractants, such as TBP, POX 11, DHDECMP, and HD(DiBM)P, have been tested for americium 241 recovery and purification at the 50 gram scale. Conventional TBP and POX.11 extractants were found effective for Am(III) recovery (yield 99.3 %) but generate problems for liquid waste management. The DHDECMP molecule, which theoretically allows direct Am(III) extraction from high nitric acid effluents, did not work out properly in the first cycle of americium recovery from highly salted effluents. This extractant adsorbed on SiO_2 exhibits moderate extraction power for Am(III) and slow adsorption/desorption kinetics.

The HD(DiBM)P has proved to be a reliable extractant in the final americium purification when the quantitative oxidation of americium to the hexavalent state was ensured.

The main progress in americium technology has been the development of two original processes :

- The Am(VI) chromatographic extraction with HD(DiBM)P at the 50-gram scale assures the purity of the americium product.
- The rapid and quantitative plutonium dioxide dissolution using electrogenerated silver(II), now performed at the 300-gram scale in Fontenay-aux-Roses, appears to be a promising technique for industrial applications.

LITERATURE CITED

1 BOURGES, J.; MADIC, C.; KOEHLY, G.; A.C.S. Symposium Series 117, (1980) 83.

2 MADIC, C.; KERTESZ, C.; SONTAG, R.; KOEHLY, G.; Separation Science and Technology 154 (1980) 745.

3 KOEHLY, G.; BOURGES, J.; MADIC, C.; SONTAG, R.; KERTESZ, C.; A.C.S. Symposium Series 161, (1981) 19.

4 MADIC, C.; BOURGES, J.; KOEHLY, G.; IAEA TC 518 (1984) 14.

5 BRAY, L.A. ; RYAN, J.L. ; Actinide Recovery from Waste and Low grade sources. Navratil, J.D. ; SCHULTZ, W.W. Eds. Harwood Academic Publishers - New-York (1982), 129

6 Commissariat à l'Energie Atomique - Brevatome N° 84.4766 March, 27th, (1984) Patent.

7 BURNEY, G.A.; Nucl. Appl. (1968) 217.

8 MASON, G.W.; BOLLMEIER, A.F.; PEPPARD, D.F.; J. Inorg. Chem. 32 (1970) 1011.

9 Mc ISAAC, L.D. ; BAKER, J.D. ; KRUPA, J.F. ; MEIKRANTZ, D.H. ; SCHROEDER, N.C. "A.C.S. Symposium Series 117, (1980), 395.

10 KNIGHTON, J.B. ; HAGAN, P.G. ; : NAVRATIL, J.D. ; THOMPSON, G.H. ; ACS Symposium Series 161, (1981), 53.

11 HORWITZ, E.P. ; KALINA, D.G. ; MUSCATELLO, A.C. ; Separation Science and Technology 16, (1981), 403.

12 CLEVELAND, J.M. ; Plutonium handbook Edited by Wick, O.J., Vol. 1 (1967) Chapter 12.

AMERICIUM METAL PREPARATION ON THE MULTIGRAM SCALE*

J. G. Reavis, S. A. Apgar III, and L. J. Mullins
Materials Science and Technology Division,
Los Alamos National Laboratory,
Los Alamos, NM 87545

ABSTRACT. The Materials Science and Technology Division of Los Alamos National Laboratory has responded to periodic requests for tens of grams of americium metal. The apparatus used for americium preparation has grown progressively simpler as experience has been gained with the preparation. The metal was prepared by heating a mixture of americium oxide and excess lanthanum metal under an atmosphere of argon in a tantalum crucible. A simple entrainment baffle rested on the mixture and the crucible was covered by a loose-fitting lid which precluded line-of-sight exit of a gas but allowed evacuation. The americium metal produced by reduction of the oxide was distilled to the cooler upper portion of the crucible wall _in vacuo_, the crucible was cut into two pieces, and the top section was reheated to melt and cast the americium into a ceramic crucible. A coating of yttrium oxide applied to the tantalum crucible before use improved the yield of cast americium.

1. INTRODUCTION

Americium metal was apparently first prepared on the microgram scale by Westrum and Eyring (1) in 1946, although this work was not reported in the open literature until 1951. There were several reports (2, 3, 4, 5) of preparation and study of the metal on the microgram and milligram scales during the ensuing years up to 1962. Reduction of AmF_3 at elevated temperatures by barium or lithium vapor in a vacuum system was the method most often used, but reduction of the dioxide by lanthanum was also used (4) in these studies.

Preparation of americium metal on the multigram scale was first reported by Johnson and Leary (6) at Los Alamos in 1964. Since that time, multigram batches of americium metal have been prepared at Los Alamos, Livermore (7), Karlsruhe (8), and Rocky Flats (9, 10). There have also been reports of metal preparation on the 1-g scale (11, 12).

*Work done in the Materials Science and Technology Division of the Los Alamos National Laboratory under the auspices of the United States Department of Energy.

N. M. Edelstein et al. (eds.), Americium and Curium Chemistry and Technology, 321–330.

This report describes the evolution of the preparative techniques used at Los Alamos.

2. CHOICE OF THE METAL PREPARATION PROCESS

The commonly used quantities for prediction of the course of reactions at high temperatures between halides or oxides and active metals are the free energies of formation of the reactants and products of reactions. Free energies of formation of several oxides, chlorides, and fluorides that may be considered in the preparation of americium metal are listed in Table I. The values listed in Table I are in general based on the compilation by Glassner (13), although extrapolations were made for AmO_2 and AmF_4 by comparison with other actinides. One would predict that the americium compounds listed here could be reduced by heating them with an excess of metals that would form compounds having more negative free energies of formation (those found

TABLE I. FREE ENERGIES OF FORMATION OF SELECTED OXIDES, CHLORIDES, AND FLUORIDES AT 1000 K
(Values stated in kcal per gram-atom of anion)

Oxide	-ΔF	Chloride	-ΔF	Fluoride	-ΔF
CaO	127	$BaCl_2$	84	CaF_2	125
ThO_2	123	LiCl	79	BaF_2	123
La_2O_3	122	$CaCl_2$	78	LiF	122
Am_2O_3	120	NaCl	76	LaF_3	121
BaO	111	$LaCl_3$	67	NaF	112
UO_2	109	$AmCl_3$	66	MgF_2	111
AmO_2	95	$PuCl_3$	59	AmF_3	110
H_2O	46	HCl	24	HF	65

higher in Table I) than the free energy of formation of the americium compound. Care must be used, however, in using this oversimplified method of predicting suitability of a reductant. This is illustrated by the experience with $AmCl_3$. At first glance, one would predict successful reduction of $AmCl_3$ by calcium, but this method failed (6). It should be pointed out, however, that this method might have been successful had it been known at that time that the melting point of americium is 1173°C instead of the much lower value (994°C) that had been reported (5). A metal regulus might have been successfully produced had higher temperatures been used. On the other hand, one would predict from Table I that hydrogen would not reduce americium compounds, but it has been demonstrated that under certain conditions of high gas purity, rapid flow rates, and high temperatures, AmO_2 can be reduced by hydrogen (8, 12).

In addition to the thermochemistry of the reactions, one must also consider certain engineering details when selecting a method for preparation of active metals such as americium. These details include availability of pure reductants and suitable container materials, separability of the products of the reaction, and radiation exposure of personnel performing the operation. After consideration of the many factors involved, those who wished to prepare americium metal on the multigram scale selected the methods listed in Table II. As was mentioned in the preceding paragraph, all of the reactions are energetically favored, with the exception of reaction No. 3. A number of factors combine to make this reaction proceed. The system is operated with ultrapure flowing hydrogen containing concentrations of oxygen and water in the parts-per-billion range (12). The free energy of formation of Pt_5Am helps to drive the reaction, but the free energies of formation of the actinide-noble metal intermetallic compounds

TABLE II. METHODS USED FOR PREPARATION OF AMERICIUM METAL AND THE ATTENDANT FREE ENERGY CHANGES

No.	Reaction	ΔF, kcal/ g-mole Am
1	$AmO_2 + 4/3\ La \rightarrow 2/3\ La_2O_3 + Am$	-40
2	$AmO_2 + Th \rightarrow ThO_2 + Am$	-55
3	$AmO_2(Pt) + 2H_2 \rightarrow 2H_2O + Pt_5Am$	+100[a]
4	$AmF_4 + 2Ca \rightarrow 2CaF_2 + Am$	-140
5	Pu + Am distillation	--

[a]This value of ΔF was calculated for the reaction in the absence of platinum.

are probably only -10 kcal/mole (14). Another significant effect may be the introduction of a high activity coefficient of hydrogen in the presence of platinum. Americium is distilled away from the mixed products of reactions 1-3. The feasibility of this procedure can be seen by comparing vapor pressure curves shown in Fig. 1. The vapor pressure of americium is greater than that of lanthanum by a factor of approximately 3×10^4, and vapor pressures of platinum and thorium are much lower than that of lanthanum. Figure 1 also gives an indication of the feasibility of separation of americium from certain nonvolatile impurities as well as an indication of the difficulty of separation from certain more-volatile impurities such as calcium, samarium, and magnesium.

Reaction No. 4 of Table II is energetic and is used only with the sealed pressure vessel or "bomb" technique (9). The products of reaction are mechanically separable and the americium appears in the attractive form of a regulus having almost the minimum possible surface area per gram of product. The use of the fluoride, however, has a serious drawback. The alpha particles emitted by americium interact with fluorine atoms and generate an intense field of neutrons. It is much more difficult to shield personnel from these neutrons than from the alpha activity or the 60-keV gamma radiation of ^{241}Am. A few millimeters of lead will almost totally attenuate the gamma radiation, whereas several centimeters of hydrogenous material are required to significantly attenuate neutrons.

The fifth process listed in Table II is an attractive method of preparing americium metal, but it requires a special feed material that has a limited availability. Hundreds of kilograms of ^{241}Am are formed

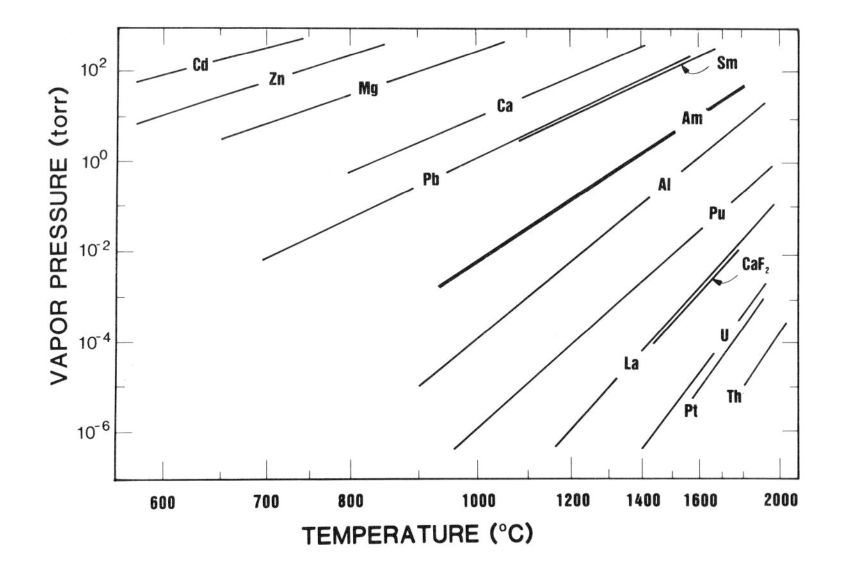

Figure 1. Vapor pressures of selected metals (and CaF_2) encountered in americium preparation as reactants, products, or impurities.

worldwide each year (15) by decay of ^{241}Pu, but the concentration is very low, and direct separation by distillation of the americium is not feasible. One product of a process called "molten salt extraction" consists of a plutonium/americium alloy containing about 5% americium (10, 16). As can be seen from Fig. 1, the vapor pressure of americium is more than three orders of magnitude greater than that of plutonium, allowing separation of the two by a double-distillation process.

Consideration of the various factors discussed above led the authors to prefer lanthanum reduction of AmO_2 over the other methods of metal production that are discussed. Factors that influenced this decision were availability of equipment, facilities, and feedstocks. Other investigators may select other methods because of their particular situations.

3. EXPERIMENTAL

3.1. Equipment

Americium metal has been prepared in our laboratories in batches of 10-25 g in stainless steel gloveboxes routinely used for plutonium processing. The design of these gloveboxes includes walls consisting of a sandwich of 0.25 in. of lead between an inner wall of 0.19 in. stainless steel and an outer skin of 0.0625 in. stainless steel.

Leaded glass, 0.25-in. thick, was added to the standard 0.25-in.-thick safety glass windows. The gloves were 0.030-in.-thick leaded neoprene with Hypalon coatings (Hypalon is a radiation-resistant modified polyethylene manufactured by the Du Pont Company.) both inside and outside. The atmosphere inside the glovebox was argon containing less than 500 ppm oxygen and water. The pumping system consisted of a two-stage mechanical pump and an oil diffusion pump equipped with a water-cooled baffle. The vacuum envelope of the furnace was a 55-mm-o.d. quartz tube sealed to metal flanges at both ends by means of Viton O-rings. Heating of the reactants and products was by coupling of a 10-kHz induction field directly to the walls of the tantalum furnace.

3.2. Evolution of Furnace Design

The evolution of the design of the furnace for americium metal preparation at Los Alamos has been toward simplification. This evolution is illustrated in Fig. 2. All parts of the furnace were fabricated

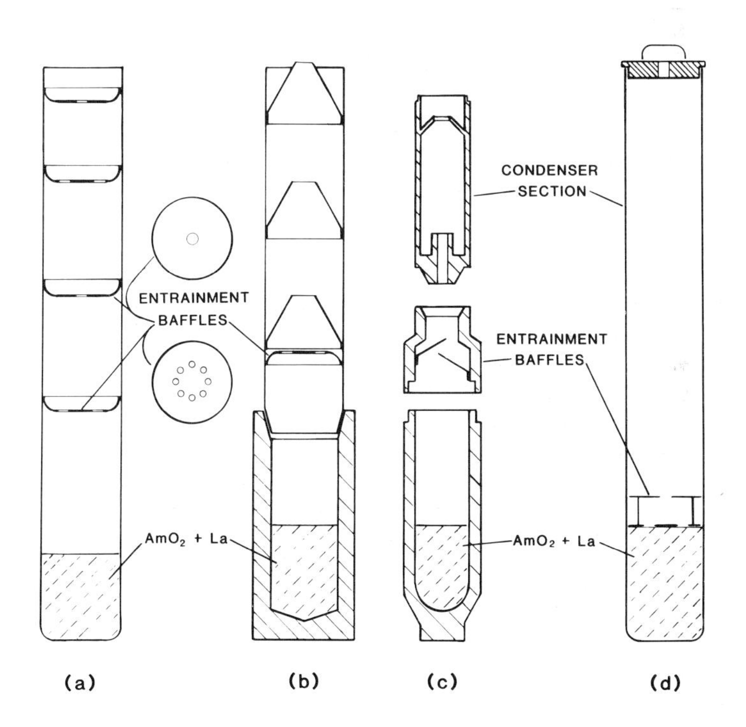

Figure 2. Evolutionary stages of the tantalum furnace used for americium metal preparation.

from tantalum. The techniques used were deep drawing, machining and welding. All furnaces contained the reactants in a strongly-heated reaction zone at the bottom, and all had top condenser sections which were heated much less strongly by using nonuniform spacing of the induction field coils (Fig. 3) whose field coupled directly to the tantalum furance walls. All furnaces had entrainment baffles of varying degrees of simplicity. Designs (a) and (b) were used in the earliest multigram preparations. Design (c), which was originated at Livermore (7), was used at a later date. There were two major problems with these designs. One problem was high expense of fabrication of a device that had a low rate of reuse (usually zero to three times). The second problem was retention of 25-35% of the product metal because of formation of a surface film that would not drain from the condenser into the casting crucible. Later preparations were performed in design (d). In this design, the reactor/di·tillation pot and the distillate condenser are combined in the form of a single flat-bottom, welded-construction tantalum crucible that is 25 mm in diameter by 175 mm deep with 0.5-mm-thick walls (Fig. 3). A loose-fitting entrainment baffle consists of two thin tantalum discs penetrated by small random holes. The discs are held approximately 6 mm apart by tantalum wire,and the assembly rests on the surface of the reacting mixture. The crucible is covered by a loose-fitting tantalum lid and the top (condenser) section is cooled by a water-cooled collar. The furnace of design (d) was the only furnace that used water cooling. The condenser sections of the other furnaces cooled by radiation of heat to the surroundings. In the furnace of design (d), a thin (0.25-mm-thick) split cylinder of tantalum in direct contact with the quartz vacuum envelope served as a heat reflector. The temperature of the tantalum crucible was determined by sighting an optical pyrometer through the 0.1-in.-wide vertical slit in the heat reflector. Pyrometry corrections were made for light absorption by the quartz tube and the glovebox window. Metallic mirror formation on the inside of the quartz tube during distillation sometimes gave rise to uncertainties in temperature measurement. Whenever metallic mirror formation became detectable, distillation was discontinued, the system was disassembled, the quartz tube and tantalum reflector were repositioned to allow observation through a clean quartz surface, and distillation continued.

3.3. Operation of the System

The empty apparatus was degassed before use by heating to approximately 1500°C *in vacuo* until the pressure had fallen to the background of about 10^{-5} torr. An excess (about 75%) of finely divided lanthanum chips was mixed with AmO_2 powder by combining and tumbling in a lead-shielded steel container. The mixture was transferred to the tantalum crucible by use of a long-stem stainless steel funnel (to avoid dust transfer to the walls) and was tamped for intimate contact. These operations were performed in the argon atmosphere containing less than 500 ppm oxygen and water. The loaded crucible was placed in the vacuum system and alternately evacuated slowly to the background pressure and backfilled with high-purity argon three times. The last filling of argon was left in the furnace and the system was slowly heated. A temperature spike of about 350°C was usually observed when the reaction started at about 800°C. Heating was continued approximately 1 min, then discontinued. The system was then evacuated and heated at a rate slow enough to maintain observed pressures only slightly above the background pressure. The temperature was held at about 1400°C for times sufficient for distillation of all the americium, assuming a distillation rate of 3 g/h. This rate is an overall batch rate that was determined empirically. The actual rate probably varied widely during the course of distillation. After distillation, the system was cooled, disassembled, and inspected by gamma scanning to verify that the americium had indeed been transferred from the reaction zone to the condenser zone. Figure 3 shows the spatial relationship between the induction coil and the furnace, the temperature profile during distillation, and the gamma activity profile at the end of distillation.

After verification that the americium had been transported to the condenser section, the reactor and condenser sections of the apparatus

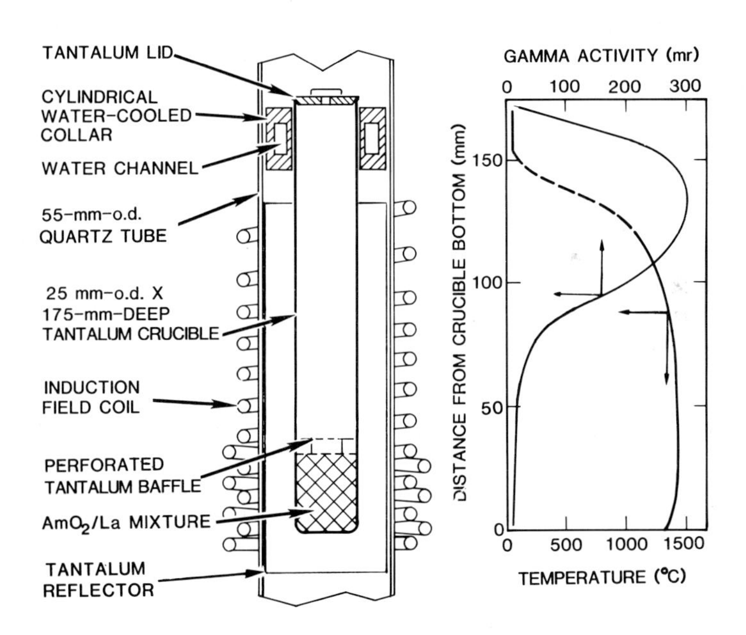

Figure 3. Americium preparation furnace, the crucible temperature profile during the distillation phase of the preparation, and the gamma activity profile of the americium at the end of distillation, all shown with the same scale and juxtaposition.

were separated and the condenser section was heated in an argon atmosphere to melt the americium and cast it into a Y_2O_3 crucible. To lessen thermal shock of the crucible by the molten metal, a surrounding tantalum susceptor was used to heat the crucible simultaneously. In the use of furnace designs (b) and (c) (Fig. 2), the condenser sections were inverted for the casting operation. In the use of design (d), the bottom 50 mm of the crucible containing the La_2O_3, excess lanthanum, and the entrainment baffle was cut off by use of tubing cutters. This bottom section was discarded and the condenser section was inverted and heated to 1200°C for 1.5 min under an atmosphere of argon to cast the americium product.

The yield of the casting operation was increased by using coatings to reduce the degree of wetting of the tantalum. Best results were obtained by use of Y_2O_3 applied as a slurry to the inside and outside of the upper section of the crucible after it has been heated briefly to dull red heat in air. This heating seems to produce a surface that can be more evenly coated with the Y_2O_3 slurry. The slurry was applied at a crucible temperature slightly less than 100°C. The coated crucible was first baked at a low temperature in air, then outgassed _in vacuo_ at 1500°C just before use.

4. PRODUCT YIELD AND PURITY

The overall yields of americium in the early work (6) using designs (a) and (b) (Fig. 2) averaged about 75% per batch. The average yield using design (d) on the 25-g scale was 89%. It was feared that the simplified design (d) would allow greater contamination of the product by lanthanum, but the change was hardly significant. Americium product from design (b) contained 2500 ppm lanthanum, whereas that from design (d) contained 4000 ppm. The only other impurities in the latter product that could be detected by spectroscopic analysis were 20 ppm magnesium and 20 ppm calcium. The yttrium concentration was reported as less than 200 ppm. No analyses for oxygen or other anionic impurities were made.

5. SUMMARY

It was demonstrated that americium metal of approximately 99.5% purity can be produced on the 20- to 25-g scale by lanthanum reduction of the oxide in a very simple apparatus. As was predicted from relative vapor pressures, volatile metal impurities such as calcium and magnesium accompany the americium during distillation. The major impurity detected was lanthanum at a concentration that would be predicted for the equilibrium vapor composition above lanthanum metal containing 1% americium. It is speculated that higher-purity americium might be produced if distillation were to be discontinued earlier, with no attempt at quantitative recovery of the metal.

LITERATURE CITED

1. Westrum, E. F.; Eyring, LeRoy, J. Am. Chem. Soc., 1951, 73, 3396-3398.

2. Carniglia, S. C.; Cunningham, B. B., J. Am. Chem. Soc., 1955, 77, 1502.

3. Graf, P.; Cunningham, B. B.; Dauben, C. H.; Wallmann, J. C.; Templeton, D. H.; Ruben, H., J. Am. Chem. Soc., 1956, 78, 2340.

4. McWhan, D. B.; Wallmann, J. C.; Cunningham, B. B.; Asprey, L. B.; Ellinger, F. H.; Zachariasen, W. H., J. Inorg. Nucl. Chem., 1960, 15, 185-187.

5. McWhan, D. B.; Cunningham, B. B.; Wallmann, J. C., J. Inorg. Nucl. Chem., 1962, 24, 1025-1038.

6. Johnson, K. W. R.; Leary, J. A., 'Preparation of Americium Metal', Los Alamos Scientific Laboratory report LA-2992, January 1964.

7. Wade, W. Z.; Wolf, T., J. Inorg. Nucl. Chem., 1967, 29, 2577-2587.

8. Spirlet, J. C.; Muller, W., J. of the Less-Common Met. 1973, 31, 35-46.

9. Conner, W. V., J. of the Less-Common Met. 1974, 34, 301-308.

10. Berry, J. W.; Knighton, J. B.; Nannie, C. A., 'Vacuum Distillation of Americium Metal', Rocky Flats report RFP-3211, January 1982.

11. Adair, H. L., J. Inorg. Nucl. Chem., 1970, 32, 1173-1181.

12. Erdmann, B.; Keller, C., Inorg. Nucl. Chem. Lett., 1971, 7, 675-683.

13. Glassner, A., 'The Thermochemical Properties of the Oxides, Fluorides and Chlorides to 2500 K', Argonne National Laboratory report ANL-5750 (1957).

14. Peterson, D. E. in Plutonium Chemistry (W. T. Carnall and G. R. Choppin, eds.), pp. 99-108, ACS Symposium Series 216, Am. Chem. Soc., Washington, D. C. (1983).

15. Schulz, W. W., The Chemistry of Americium, p. 28, ERDA (now DOE) Technical Information Publication TID-26971 (1976).

16. Knighton, J. B.; Hagan, P. G.; Navratil, J. D.; Thompson, G. H., in Transplutonium Elements -- Production and Recovery (J. D. Navratil and W. W. Schulz, eds.), pp. 53-64, ACS Symposium Series 161, Am. Chem. Soc., Washington, D. C. (1981).

INDEX